Halophiles

Genetics and Genomes

Edited by

R. Thane Papke

Department of Molecular and Cell Biology
University of Connecticut
Storrs, CT
USA

and

Aharon Oren

Department of Plant and Environmental Sciences
The Hebrew University of Jerusalem
Jerusalem
Israel

Caister Academic Press

Caister Academic Press
Norfolk, UK

www.caister.com

British Library Cataloguing-in-Publication Data
A catalogue record for this book is available from the British Library

ISBN: 978-1-908230-42-3 (hardback)
ISBN: 978-1-908230-65-2 (ebook)

Cover design adapted from photos taken by Scott Chimileski, Department of Molecular and Cell Biology, University of Connecticut, Storrs, CT, USA.

Contents

Contributors

Josefa Antón
Department of Physiology, Genetics and Microbiology
University of Alicante
Alicante
Spain

anton@ua.es

Adi Arbiv
Department of Life Sciences
Ben Gurion University
Beersheva
Israel

arbiva@post.bgu.ac.il

Anke Baumann
Institute for Molecular Biosciences
Goethe University Frankfurt am Main
Frankfurt am Main
Germany

anke.baum@gmx.de

Chen Cohen-Rosenzweig
Department of Life Sciences
Ben Gurion University
Beersheva
Israel

chencoh@post.bgu.ac.il

Jerry Eichler
Department of Life Sciences
Ben Gurion University
Beersheva
Israel

jeichler@bgu.ac.il

Matthew S. Fullmer
Department of Molecular and Cell Biology
University of Connecticut
Storrs, CT
USA

matthew.fullmer@uconn.edu

Toni Gabaldón
Pompeu Fabra University (UPF); and
Bioinformatics and Genomics Programme
Centre for Genomic Regulation (CRG)
Barcelona
Spain

toni.gabaldon@crg.es

J. Peter Gogarten
Department of Molecular and Cell Biology
University of Connecticut
Storrs, CT
USA

gogarten@uconn.edu

María Gomariz
Evolutionary Genomics Group
Department of Physiology, Genetics and Microbiology; and
Department of Materials
Optics and Electronics
University Miguel Hernández
Elche
Alicante
Spain

maria.gomariz@umh.es

Pedro González-Torres
Department of Physiology, Genetics and Microbiology
University of Alicante
Alicante
Spain

pedro.gonzalez@ua.es

Inga Hänelt
Department of Molecular Microbiology and Bioenergetics
Goethe University Frankfurt am Main
Frankfurt am Main
Germany

haenelt@bio.uni-frankfurt.de

Jaime Huertas-Cepa
Pompeu Fabra University (UPF); and
Bioinformatics and Genomics Programme
Centre for Genomic Regulation (CRG)
Barcelona
Spain

jhuerta@crg.es

Lina Kaminski
Department of Life Sciences
Ben Gurion University
Beersheva
Israel

linaka@post.bgu.ac.il

Lina Kandiba
Department of Life Sciences
Ben Gurion University
Beersheva
Israel

lina.kandiba@gmail.com

Zvia Konrad
Department of Life Sciences
Ben Gurion University
Beersheva
Israel

conrad@bgu.ac.il

Marianna Lucio
Research Unit Analytical BioGeoChemistry
Helmholtz Zentrum Munich, German Research Center for Environmental Health
Neuherberg
Germany

marianna.lucio@helmholtz-muenchen.de

Manuel Martínez-García
Department of Physiology, Genetics and Microbiology
University of Alicante
Alicante
Spain

m.martinez@ua.es

Julie Maupin-Furlow
Department of Microbiology and Cell Science
University of Florida
Gainesville, FL
USA

jmaupin@ufl.edu

Noha M. Mesbah
Department of Biochemistry
Faculty of Pharmacy
Suez Canal University
Ismailia
Egypt

noha_mesbah@pharm.suez.edu.eg

Volker Müller
Department of Molecular Microbiology and Bioenergetics
Goethe University Frankfurt am Main
Frankfurt am Main
Germany

vmueller@bio.uni-frankfurt.de

Shai Naparstek
Department of Life Sciences
Ben Gurion University
Beersheva
Israel

shainaparstek@gmail.com

Aharon Oren
Department of Plant and Environmental Sciences
The Institute of Life Sciences
The Hebrew University of Jerusalem
Jerusalem
Israel

aharon.oren@mail.huji ac.il

R. Thane Papke
Department of Molecular and Cell Biology
University of Connecticut
Storrs, CT
USA

thane@uconn.edu

Lejla Pašić
Department of Biology
Biotechnical Faculty
University of Ljubljana
Ljubljana
Slovenia

lejla.pasic@bf.uni-lj.si

Arantxa Peña
Department of Physiology, Genetics and Microbiology
University of Alicante
Alicante; and
Department of Biology-Microbiology
University of Balearic Islands
Palma de Mallorca
Balearic Islands
Spain

a.pena@uib.es

Francisco Rodríguez-Valera
Evolutionary Genomics Group
División de Microbiología
Universidad Miguel Hernández
San Juan de Alicante
Spain

frvalera@umh.edu

Elina Roine
Department of Biosciences
University of Helsinki
Helsinki
Finland

elina.roine@helsinki.fi

Ramon Rosselló-Móra
Marine Microbiology Group
Department of Ecology and Marine Resources
Institut Mediterrani d'Estudis Avançast IMEDEA (CSIC-UIB)
Esporles
Balearic Islands
Spain

rossello-mora@uib.es

Fernando Santos
Department of Physiology, Genetics and Microbiology
University of Alicante
Alicante
Spain

fernando.santos@ua.es

Phillippe Schmitt-Kopplin
Research Unit Analytical BioGeoChemistry
Helmholtz Zentrum Munich, German Research Center for Environmental Health
Neuherberg; and
Analytical Food Chemistry Department
Technische Universität München
Freising-Weihenstephan
Germany

schmitt-kopplin@helmholtz-muenchen.de\

Jörg Soppa
Institute for Molecular Biosciences
Goethe University Frankfurt am Main
Frankfurt am Main
Germany

soppa@bio.uni-frankfurt.de

Karolin Zerulla
Institute for Molecular Biosciences
Goethe University Frankfurt am Main
Frankfurt am Main
Germany

stehr@bio.uni-frankfurt.de

Baisuo Zhao
Graduate School Chinese
Academy of Agricultural Sciences
Beijing
China

bszhao@163.com

Preface

> Admittedly, it is trying to study microorganisms whose growth media fills the laboratory with steam, or the centrifuge heads with salt, or which grow so slowly that weeks, instead of hours, may be required for experiments and whose genetics are unknown or almost impossible to study.

These words were written by the late Donn Kushner (1927–2001) in his introduction to his book *Microbial Life in Extreme Environments* (Kushner, 1978a). Although much progress has been made in the 35 years that have passed since, the message is still perfectly true. Some of the most intriguing halophiles such as the flat square archaeon *Haloquadratum walsbyi* are still very difficult to grow and at best grow very slowly. Genetic systems have been developed only for a limited number of extremely halophilic Archaea, and a few moderately halophilic Bacteria. But thanks to the development of genetic systems for organisms such as *Halobacterium salinarum*, *Haloferax volcanii* (Euryarchaeota) and *Halobacillus halophilus* (Firmicutes) we now have a fair number of model organisms in which the molecular biology of diverse groups of halophilic microorganisms can be studied. Modern molecular techniques of genomics and culture-independent methods of metagenomics have also contributed much towards our understanding of the functioning of a wide range of halophilic microorganisms and their viruses.

This book contains eight essays covering different aspects of the genetics and the genomics of halophiles, with special emphasis on those topics in which significant progress has been made in recent years. Organisms featured are extreme as well as moderate halophiles, types that grow at neutral as well as alkaline pH, species that prefer moderate as well as high temperatures for growth.

The first chapters mainly deal with genomic and culture-independent metagenomic approaches to understanding the diversity and evolution of halophiles in their natural environment. Cultures of *Haloquadratum walsbyi* are now available, but they are difficult to handle. However, very much has been learned about their properties by culture-independent approaches, as shown in Chapter 1 by Lejla Pašić and Francisco Rodríguez-Valera. *Salinibacter ruber* is an extremely halophilic member of the Bacteria resembling in many properties the halophilic Archaea. It was discovered little over a decade ago, and also here genomic and metagenomic approaches have greatly increased our understanding of its biology (Chapter 2 by Aranxa Peña and co-workers). In Chapter 3, Matthew Fullmer and his colleagues document the frequency and wide-spread nature of horizontal gene transfer (HGT) among the haloarchaea in their natural environment, and provide conjecture on how that affects Halobacterial evolution and speciation. Our understanding of the viruses that infect halophilic prokaryotes, Archaea as well as Bacteria, has greatly increased in recent years, thanks to culture-dependent as well as on culture-independent techniques. Chapter 4 by Elina Roine gives an updated overview of this field.

> Though they are less exciting at first glance than the extreme halophiles, the moderately halophilic bacteria ... pose quite interesting questions, especially those implied by their ability to grow over wide ranges of solute concentrations. If the last decade has been that of the extreme halophiles, we can hope that the next one will see their more modest, moderate cousins (in the spiritual sense only) take their proper place in the scientific canon.

It is satisfying to note that the above sentences, another quote from Donn Kushner's book (Kushner, 1978b), were taken to heart by later generations of scientists. One of those 'modest, moderate cousins' is *Halobacillus halophilus,* whose genome sequence has been analysed and for which a system for genetic manipulation is now available. Chapter 5 by Noha Mesbah and her colleagues deals not only with the mechanisms of the adaptation of this neutrophilic mesophile to life in salt, but also with such mechanisms of *Natranaerobius thermophilus,* a 'polyextremophile' that thrives in the presence of high salt, high pH and high temperatures in the absence of molecular oxygen. Genetic manipulation studies with *Haloferax volcanii* as the model organism have greatly increased our understanding of the molecular biology of the biosynthesis of the glycoprotein S-layer cell wall of the haloarchaea. A review of our current knowledge of this topic is presented by Jerry Eichler and his co-workers in Chapter 6. In recent years we have learned much about the mechanisms of DNA replication and transcription in the haloarchaea. Is has become clear that many species contain multiple copies of the chromosome, and behave as polyploidic cells. The implications of this finding are discussed in-depth in Chapter 7 by Karolin Zerulla *et al.* Finally, Chapter 8 by Julie Maupin-Furlow presents our current understanding of cell regulation by proteolytic systems and protein conjugation in the halophilic Archaea.

We end with yet another quote from the Introduction chapter of Donn Kushner's 1978 book:

> Those who have persisted have found their rewards, both in the satisfaction and leisure for contemplation available to the student of an out-of-the-way field, and in the fascination afforded by the microorganisms themselves and the very clever ways they have found to adapt to such a wide range of environmental conditions.

It is this fascination that has been the basis for the many studies on which the chapters in this book were based. The halophiles, although being an 'out-of-the-way field', provide excellent opportunities to study some of the most basic questions in modern microbiology and molecular biology. We trust that the readers of this book will become fascinated as well, and we hope that the different chapters will serve as a source of inspiration for further in-depth studies on the genetics and genomics of halophilic microorganisms.

R. Thane Papke and Aharon Oren

References

Kushner, D.J. (1978a). Introduction: a brief overview. In Microbial Life in Extreme Environments, Kushner, D.J., ed. (Academic Press, London), pp. 1–7.

Kushner, D.J. (1978b). Life in high salt and solute concentrations. In Microbial Life in Extreme Environments, Kushner, D.J., ed. (Academic Press, London), pp. 317–368.

Ecology and Evolution of *Haloquadratum walsbyi* Through the Lens of Genomics and Metagenomics

1

Lejla Pašić and Francisco Rodríguez-Valera

Abstract

In this chapter we summarize the current knowledge on the ecology of natural populations of *Haloquadratum walsbyi* obtained through genomics and metagenomics. The cells of this enigmatic microbe populate crystallizer brines and deal with a high viral predation pressure. The natural populations of *Haloquadratum walsbyi* differ in genomic regions known as metagenomic islands that impart environmental adaptation: they are enriched in genes that are involved in transport of nutrients, but also in genes that code for cells surface components that can serve as viral recognition sites. Likewise, similar genomic variability is observed in natural populations of viruses that prey on this species. Natural populations of *Haloquadratum walsbyi* are not dominated by a single ecologically most successful lineage. Instead, they are composed of numerous clonal lineages that are preserved in space and time. The observed phenomena favour the 'Constant diversity' model of population dynamics which assumes that the expansion of metabolically superior clonal lineages will be selected against by predating viruses in a density-dependent fashion. This way, the microbial population would avoid catastrophic losses due to viral lysis, preserve intragenomic diversity and efficiently exploit niche resources.

Introduction

Owing to the pivotal role of microbes for ecosystem function, assessment of their diversity is often a central issue in ecological studies. Understanding microbial diversity, in particular at the genomic level, is needed to clarify many aspects of the nature of species evolution, including mechanisms of environmental adaptation and interaction with other biotic components. However, the evolutionary dynamics of a natural lineage, considered to be a basic unit of evolution (de Queiroz, 2005), remains largely unknown. This current gap in our knowledge is a consequence of a number of factors. In many ecosystems, most of the microbial activity is carried out by abundant organisms (Cottrell and Kirchman, 2003; Malmstrom *et al.*, 2005). These presumably major players in ecosystem functioning are often obtained in pure culture only after laborious and long-term studies (Pedrós-Alió, 2006). In some cases they remain uncultivated to date and are thus unavailable for laboratory experiments. Systematic investigations of natural populations of bacteria are further hampered by the very nature of microbes, as phenotypic characteristics and adaptations unique to individual lineages are often impossible to identify. Another obstacle comes from extrapolating the population genetics knowledge from multicellular eukaryotes to prokaryotic populations, as the extent of similarity in these processes remains controversial (e.g. Bapteste *et al.*, 2009).

Advances in sequencing technology provided means to study the extent of genomic heterogeneity in natural ecosystems without the need for prior cultivation. These molecular surveys revealed that the gene pool of a microbial species is much more complex than that found in individual clonal lineages (Palenik, 1994; Palys *et al.*, 1997; Schleper *et al.*, 1998; Whitaker *et al.*, 2003; Acinas *et al.*, 2004; Tyson *et al.*, 2004; Venter *et al.*, 2004; Thompson *et al.*, 2005). Accordingly, this led to the proposal of now widely accepted pan-genome concept (Tettelin *et al.*, 2005) in which the species genome is composed of the so-called 'core' genome which is shared by all individuals and the 'dispensable', 'flexible' or 'accessory' genome which consists of genes unique to individual lineages (Medini *et al.*, 2005, 2008). However, a number of questions remain unanswered. To date, it is unknown how many clonal lineages can be expected to co-exist in a given environment, what is the effective population size of these lineages or the extent of genetic diversity among them. Likewise, the mechanisms that generate and maintain diversity at this scale remain largely unknown. It is possible to tackle these issues by applying a metagenomic approach to the saturated brines of solar salterns (crystallizers). As we will see below, this low diversity biological system supports a dense microbial community composed of clonal lineages of *Haloquadratum walsbyi*. Such systems, naturally enriched in a single species, are considered good models for more complex communities. Over the years a wealth of sequence data were collected, both from metagenomes but also from the genomes of closely related lineages of *Haloquadratum walsbyi*. This chapter aims to summarize the findings on the ecology of this species.

The ecology of solar saltern crystallizer

Solar saltern crystallizers, one of the harshest environments with respect to NaCl concentration, form during commercial salt production through evaporation of seawater. This ancient salt production technology aims to gradually concentrate seawater by passing it through a series of interconnected ponds. This way, the salts that are undesired in NaCl production, such as calcium carbonate and calcium sulfate, are precipitated in the early stages of production, when the total salts in concentrating seawater reach about 10% (w/v) (Javor, 2002). The concentration of salts reaches their maximum in the crystallizer ponds and this is where pure crystals of NaCl are harvested. The crystallizer brine is saturated with NaCl but it is also rich in magnesium, chloride and potassium, providing a set of environmental conditions that are well suited only to highly specialized halophiles. At the end of salt production, potassium and magnesium salts can be retrieved in other special ponds called 'bitterns' that are devoid of life (Javor, 2002).

Although extreme in terms of NaCl concentration, the crystallizer brine is generally rich in nutrients that are brought in from lower salinities. The lower salinity ponds support extensive microbial communities, numerous cells of the algae *Dunaliella* that can produce massive amounts of glycerol and heavy populations of brine shrimp *Artemia*. An additional source of nutrients comes in the form of bird droppings, as large bird populations are often drawn by abundance of zooplankton (Davis, 1974; Javor, 2002; Oren, 2009). Not all salterns are operated the same way. Some local modifications to the ancient production technology are imposed by prevailing weather conditions and saltern geology (Pašić *et al.*, 2005; Moinier,

1999), and these differences likely affect the structure of the microbial community present. For example, in areas where salt production is limited to the arid part of the year the water-retention time in the unusually shallow crystallizer ponds can be reduced to less than 24 hours. One such example are the Sečovlje salterns in Slovenia, where halite is harvested as soon as it starts to precipitate, at about 26% (w/v) total salts (Pašić *et al.*, 2005). Quite the opposite is seen in the saltern of Torrevieja (Alicante, Spain), a saltern developed in a large natural hypersaline lagoon. This hypersaline environment works as a continuous system that is steadily replenished. Therefore, its highly concentrated brine (up to 32% of total salts) represents a constant environment.

The Bras del Port saltern, the subject of our studies, was established in 1900 in Santa Pola, a small town not far from Alicante, Spain. The climate in this area is mild Mediterranean with low average rainfall, and this allows the brine to be retained in individual ponds for months at a time. Consequently, it can reach astonishing 37% (w/v) of total salts in the terminal crystallizer ponds (Ghai *et al.*, 2011). The mode of operation in this saltern resembles that of a semi-continuous bioreactor, and the salterns provide a wide range of ecosystems spanning the salinities from that of seawater to halite saturation. However, once a year, at the end of the summer, crystallizers are completely emptied to harvest the salt, being afterwards replenished with water of lower salinity. A vast majority of scientific efforts regarding this habitat concentrated on samples originating from a single crystallizer pond, denominated CR30 (Fig. 1.1). Over the last 30 years the microbiota of CR30 were studied using a plethora of approaches ranging from cultivation attempts (Rodríguez-Valera *et al.*, 1985) to large-scale metagenomic studies (Ghai *et al.*, 2011).

The microbial species richness and community structure throughout the salinity gradient was recently revisited at a sequencing depth much higher than in any previous study (Ghai *et al.*, 2011). In contrast to the widely accepted paradigm that extreme environments support low microbial diversity, it was found that microbial communities were surprisingly

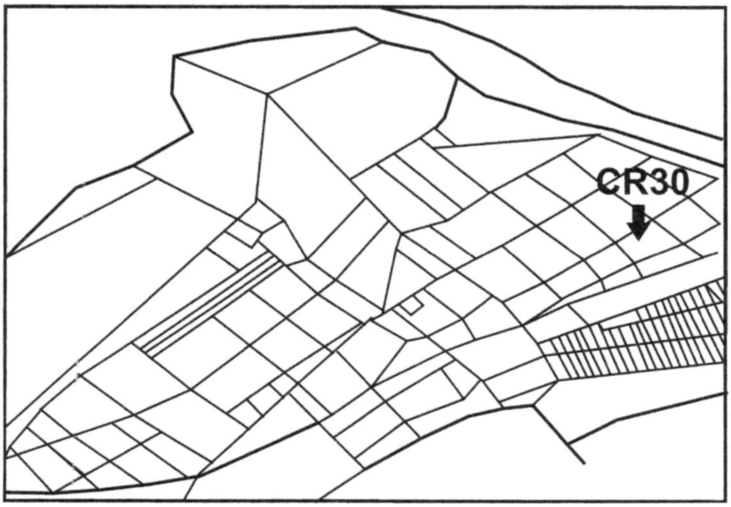

Figure 1.1 Bras del Port saltern in Santa Pola, near Alicante, Spain, and the location of the CR30 crystallizer.

diverse at all salinities. Their composition changed over the salinity gradient, and marine assemblages became replaced by heterotrophic microbial communities that were increasingly more halophilic in origin. Finally, at the high end of salinity gradient this community was composed almost exclusively of highly specialized halophiles (Fig. 1.2).

Over the years, natural populations of the square archaeon *Haloquadratum walsbyi* were studied in much detail. *Haloquadratum walsbyi* cells can represent up to two thirds of the crystallizer biomass and are followed in abundance by only one other microbe, *Salinibacter*

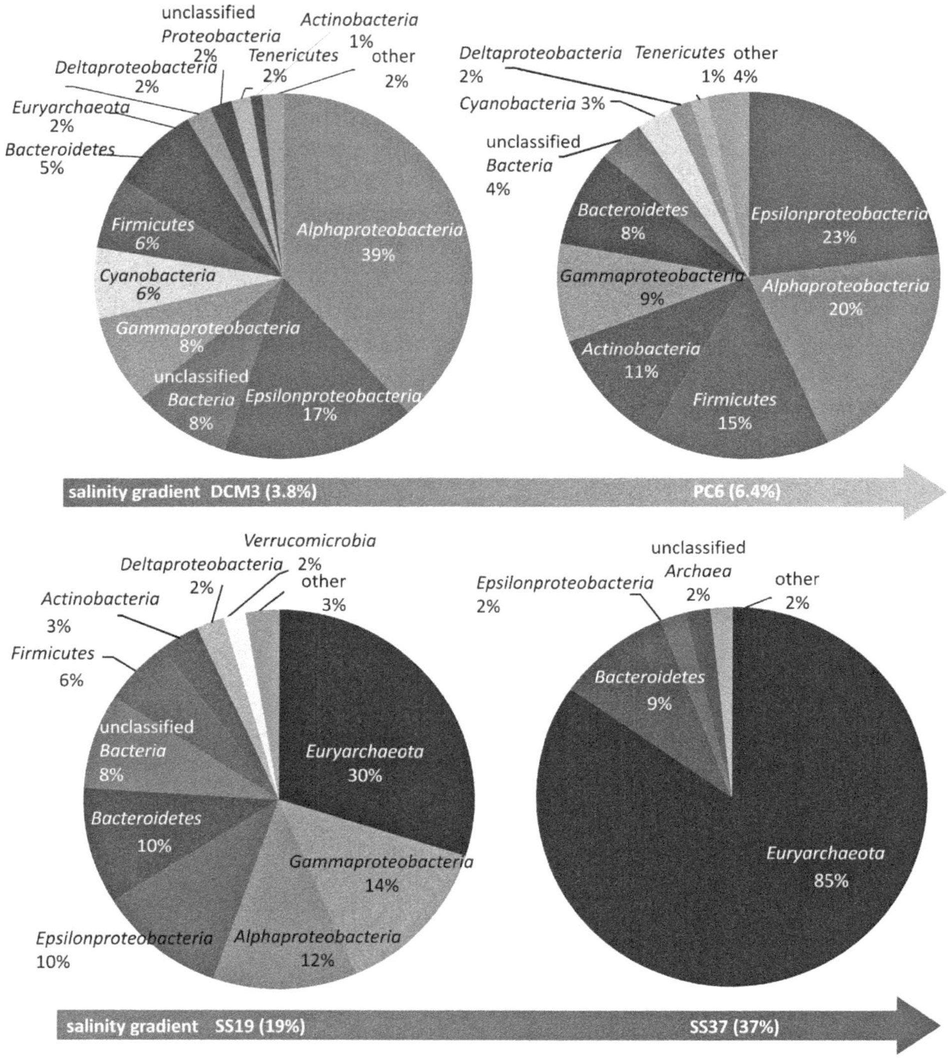

Figure 1.2 Taxonomic profiles using metagenomic 16S rRNA sequences across the salinity gradient. DCM3, 3% (w/v) salinity, PC6, 6% (w/v) salinity, SS19, 19% (w/v) salinity, SS37, CR30, 37% (w/v) salinity. (Adapted from Ghai *et al.*, 2011.)

ruber, a hyperhalophilic member of the Bacteroidetes (Antón *et al.*, 1999, 2000; Benlloch *et al.*, 2002; Casamayor *et al.*, 2002; Rosselló-Mora *et al.*, 2003, Ghai *et al.*, 2011). This bacterium can amount to one-third of the crystallizer biomass, and the two microbes appear to share much more than a common environment (Antón *et al.*, 2000). Both microbes are 'salt-in' strategists that compensate high extracellular sodium chloride concentrations by accumulating molar concentrations of potassium and chloride, a strategy that requires extensive metabolic adaptation. Additionally, they are aerobic heterotrophs and both contain retinal proton pumps in the membrane. The observed convergences might be a consequence of lateral gene transfer between the crystallizer species, as indicated by *Salinibacter ruber* genome analysis (Mongodin *et al.*, 2005).

There has been some debate as to what extent abiotic and biotic factors such as predation or competition control the relative abundance of taxa in the solar saltern crystallizer. Experimental data indicated that neither organic nutrient availability (Gasol *et al.*, 2004) nor grazing (Guixa-Boixareu *et al.*, 1996; Joint *et al.*, 2002) limit prokaryotic abundance, despite occasional presence of nanoflagellates in the crystallizer (Pedrós-Alió *et al.*, 2000). Thus, it appears that the cells present in the saltern crystallizer deal with predation pressure which is mainly viral in origin. Indeed, the salterns are well known to harbour unusually high numbers of virus-like particles, which reach up to 10^9 per ml in the crystallizer (Guixa-Boixareu *et al.*, 1996). However, the prokaryotic loss per day due to viruses was estimated to be lower than 5%, indicating that the viruses do not exert strong control over prokaryotic growth and abundance (Guixa-Boixareu *et al.*, 1996). Likewise, there has never been any report of massive lysis events collapsing the *Haloquadratum walsbyi* population.

Haloquadratum walsbyi: the uncultivable species

The cells of *Haloquadratum walsbyi* were first described in 1980. Anthony Edward Walsby observed them in a sample of brine that he had previously collected in a Sinai Peninsula hypersaline marsh (Walsby, 1980). The novel microbe presented a number of fascinating features. Its cells resembled squares with sharp edges and straight corners and formed transparent sheets that floated on the surface of the brine. The sides of square cells were about 2 μm long but the cells were only 0.25 μm thick, near the limit of resolution of a light microscope (Stoeckenius, 1981). In fact, they would probably go unnoticed had they not contained gas vacuoles, Walsby's scientific interest at the time (Walsby, 1980). Intriguingly, the cell thickness remained constant as the cells grew and divided. Besides gas vacuoles, square cells contained small dark granules, later identified as poly-β-hydroxybutyric acid (Stoeckenius, 1981; Kessel and Cohen, 1982; Burns *et al.*, 2004a). The cell surface was covered with a surface layer (S-layer) of hexagonal arrays of particles of ~20 nm in size (Parkes and Walsby, 1981). In the suspension of cells, researchers observed the presence of a pigment which resembled haloarchaeal bacteriorhodopsin, but no patches of purple membrane were observed (Stoeckenius, 1981).

Another interesting feature of square cells was their abundance. Where present, these cells often reached high densities, up to 10^7 cells per ml (Walsby, 1980; Antón *et al.*, 2000), and this prompted a series of cultivation attempts. Walsby was the first one to try to cultivate square cells, but failed within the three months that he had allowed himself for successful cultivation (Walsby, 2005). It soon became clear that this was not an easy task, a feature common to many microbes that are either ecologically relevant or particularly abundant in the ecosystem. Stoeckenius successfully obtained enrichment cultures by supplementing natural brine with peptone medium, but failed to obtain single colonies on a solid medium (Stoeckenius, 1981). In a study published in 1986, Torella reported that he grew the square shaped cells in low nutrient medium; however, this precious culture was lost, as it was not deposited in a culture collection (Torella, 1986). In the following two decades, the squares skillfully avoided cultivation regardless of methodology used. This was troubling, as cultivation is crucial in order to fully understand the physiology of a microbe and the extent of its ecological interactions. Therefore, in absence of an axenic culture, what was left was to study samples that are naturally enriched in square cells.

The first breakthrough in this respect was achieved with the development of cultivation-independent techniques. The first such study conducted on CR30 samples involved sequencing of 16S rRNA genes amplified from its environmental DNA with an aim to describe the extent of diversity present in this hypersaline environment (Benlloch *et al.*, 1995). The sequencing of the highly variable region corresponding to 16S rRNA gene nucleotides 100–300 showed that this part of gene sequence was identical in the first 12 sequenced clones and dissimilar to any known species. Furthermore, another variable region, corresponding to nucleotides 384–600 of the 16S rRNA gene, was identical in six clones, while two additional clones differed in 12 nucleotides. Thus, the two clones corresponding to the two subtypes detected were fully sequenced and their 16S rRNA gene sequences differed in only 15 nucleotides. This particular phylotype became known as SPhT, short for Sussana's phylotype, after its discoverer Sussana Benlloch (Benlloch *et al.*, 1995). Based on SPhT, a fluorescence *in situ* hybridization probe targeting nucleotides 1337–1354 of the 16S rRNA gene was developed. When FISH was performed using this probe it proved that the SPhT originated from square-shaped prokaryotic cells (Antón *et al.*, 1999). Then, in order to link phylogenetic affiliation and *in situ* metabolic activity, FISH-MAR was used. This approach combines fluorescence *in situ* hybridization and microautoradiography, a method that measures *in situ* uptake of specific radiolabelled substrates by individual cells (Okabe *et al.*, 2004). The squares recovered from samples taken from a crystallizer pond of a saltern in Mallorca, Spain, were found to metabolize acetate and amino acids, yet no uptake of glycerol was observed (Rosselló-Mora *et al.*, 2003).

With the availability of 16S rRNA sequences it became clear that Walsby's square archaeon is a true saltern dweller. Its 16S rRNA gene sequence was reported from geographically widely separated crystallizers of salterns in Spain (Benlloch *et al.*, 2002), Israel (Oren, 1994), Australia (Burns *et al.*, 2004b), Tunisia (Trigui *et al.*, 2011) and Peru (Maturrano *et al.*, 2006). In all these salterns, the phylotypes related to square microbes were the most abundant and represented up to 60% of the microbial population. It should be mentioned that some salterns were found devoid of this particular species (within the sampling effort conducted) and dominated by some other haloarchaeal group(s) such as *Halorubrum*, *Natrinema* or *Halobacterium* (e.g. Pašić *et al.*, 2005; Manikandan *et al.*, 2009). It was hypothesized that the slow-growing squares are unable to build up in numbers in those salterns where the solar salt production is limited to the arid part of the year (Pašić *et al.*, 2005). Indeed, the relative abundance of this phylotype was recently correlated with the concentration of sodium chloride and/or magnesium in the crystallizer brine, which is lower in salterns that harvest the salt as soon as it starts to precipitate (Boujelben *et al.*, 2012).

Isolation of *Haloquadratum walsbyi*

After years of following the population of Walsby's square archaeon in pond CR30 of the Bras del Port salterns, a sample retrieved from this pond in 2001 yielded the first isolate, denominated HBSQ001. Led by Henk Bolhuis, the Dutch group that retrieved the isolate reasoned that the squares might grow in pure culture if provided with amount of nutrients similar to that in their natural environment. Thus, the enrichment medium used contained low amount of nutrients (0.10% C-source and 0.01% yeast extract) and the colonies of isolate HBSQ001 were obtained on plates solidified with agarose instead of conventional agar (Bolhuis *et al.*, 2004). Phylogenetic and phenotypic characterization established that the square cells are unrelated to any of the known *Halobacteriaceae* (Bolhuis *et al.*, 2004).

Presently, only one more pure culture of this eccentric microbe is available and it was obtained from a geographically very distant location. It originates from samples taken from a crystallizer pond of a saltern in Geelong, Victoria, Australia, in 2002–2004. This scientific group, led by Mike Dyall-Smith, significantly contributed to the cultivability of halophilic Archaea and proved that the majority of microbial groups inhabiting the saltern they were studied could be obtained in pure culture (Burns *et al.*, 2004a). Similar to their Dutch colleagues, this scientific group used nutrient poor media, yet combined this approach with extinction–dilution culturing. Medium supplemented with mixed amino acids and pyruvate yielded isolate C23 upon 3 weeks of incubation (Burns *et al.*, 2004a). As it often is in science, a quarter of century of attempts resulted in concurrent cultivation of the two strains, and the two reports were published within a month of each other (Bolhuis *et al.*, 2004; Burns *et al.*, 2004a). In 2007, isolates C23 and HBSQ001 were formally described as a new species within the *Halobacteriaceae*. To honour its discoverer Anthony E. Walsby, the new species was named *Haloquadratum walsbyi* (Burns *et al.*, 2007).

The two genomes

In 2006, Bolhuis and co-workers took advantage of the expansion of sequencing techniques and reported the genomic sequence of strain HBSQ001 (Bolhuis *et al.*, 2006). The genomic sequence of type strain C23 followed and a detailed comparative study of these two genomes was reported in a 2011 paper (Dyall-Smith *et al.*, 2011). Both genomes showed traits regarded as genomic signatures of hypersaline adaptation (Paul *et al.*, 2008). Examples are a distinctive pattern of dinucleotide frequencies at the first and second codon position of genes, low average pI of proteins (5.1) and distinct usage of synonymous codons encoding the amino acids Arg, Val, Thr, Leu and Cys (Bolhuis *et al.*, 2006; Paul *et al.*, 2008). The genomic sequence of *Haloquadratum walsbyi* shows a number of other interesting features. Its coding density is unusually low (76%) compared to other halophilic Archaea (86–91%). Additionally, this genome contains a high number of pseudogenes and IS elements, which led to the assumption that it is undergoing genome shrinkage as a result of its specialization into a very restrictive niche (Bolhuis *et al.*, 2006). Perhaps most striking is its unusually low content of nucleotides guanine and cytosine (GC content), which averages at 47.9% and is in sharp contrast to the generally high GC content of other halophilic Archaea (60–70% GC) (Bolhuis *et al.*, 2006; Dyall-Smith *et al.*, 2011). Such a drift to an AT rich genome has been attributed to the extremely high concentration of $MgCl_2$ in saltern crystallizer. It was reasoned that if *Haloquadratum walsbyi* was to possess a GC-rich genome, the presence of excessive amount of magnesium ions might cause DNA rigidity that could interfere with

DNA replication and transcription (Bolhuis *et al.*, 2006). However unusual, the low GC content feature was essential for metagenomic studies of saturated brines and, together with dinucleotide frequency, was used to bin sequences related to *Haloquadratum walsbyi* (Legault *et al.*, 2006; Cuadros-Orellana *et al.*, 2007; Ghai *et al.*, 2011; Narasingarao *et al.*, 2012).

It recently became known that *Haloquadratum walsbyi* is not the only archaeon in CR30 with an unusually low GC-content. A study that combined metagenomics and single-cell genomics yielded the presence of *Candidatus* 'Haloredivivus' whose GC-content was only 42% (Ghai *et al.*, 2011). The sequencing of the genome amplified from a single cell of this organism suggested that *Candidatus* 'Haloredivivus' is a photoheterotroph capable of polysaccharide degradation (Ghai *et al.*, 2011). Phylogenetic analysis affiliated this species with the recently described class 'Nanohaloarchaea' that comprises Archaea with atypical metabolic pathways and unusually small cells that are on average 0.6 μm in diameter (Narasingarao *et al.*, 2012). Other members of this class are *Candidatus* 'Nanosalina' and *Candidatus* 'Nanosalinarum', that were found present in metagenomic datasets originating from a crystallizer pond of the acidic hypersaline lake Tyrrel, another neutral saltern in Australia, and crystallizers of the Chula Vista saltern in California, USA (Narasingarao *et al.*, 2012). However, sequences related to these organisms were not found in the CR30 datasets (Ghai *et al.*, 2011).

The giant protein

A peculiarity of this species is the presence of a gene that codes for halomucin, a giant protein that is similar to animal mucins. Halomucin is the largest protein reported to date in halophilic Archaea. In HBSQ001 it extends for 9159 amino acids, while its C23 orthologue is shorter (7836 amino acids), likely due to a series of independent deletions (Bolhuis *et al.*, 2006; Dyall-Smith *et al.*, 2011). Halomucin is predicted to have multiple glycosylation/sulfation sites. It is also considered to have structural potential to interact with sugar molecules, surface proteins and divalent cations, and it displays fractional homology to proteins involved in cellular adhesion (Rinck, 2009; Dyall-Smith *et al.*, 2011). Therefore, it was hypothesized that together with sialic acid and poly-γ-glutamate halomucin might form a rigid capsule which protects the cell against desiccation (Bolhuis *et al.*, 2006). Another bioinformatic study hypothesized that it might contribute to the spatial arrangement of *H. walsbyi* cells in the form of postage stamp sheets (Rinck, 2009). Some of these notions are supported by experimental data. These show that both strains of *Haloquadratum walsbyi* do invest in expression of this giant protein. Furthermore, halomucin could be microscopically observed as an external capsule and stained by a specific antibody coupled to fluorescein (Bolhuis *et al.*, 2006; Sublimi-Saponetti *et al.*, 2011; Dyall-Smith *et al.*, 2011).

Rhodopsin-related genes

There has been some debate about the intriguing flatness of *Haloquadratum walsbyi*. Bolhuis *et al.* (2006) postulated that the reason behind this phenomenon is the uptake of essential nutrients which are often complexed to cations in hypersaline environments. This way, the cells of this species can increase surface-to-volume ratio, retain a relatively large size and increase the number of membrane-dependent cellular processes without affecting other cellular functions (Bolhuis *et al.*, 2006). Indeed, the genomes of *Haloquadratum walsbyi* have a substantial number of transporter-related genes (Cai *et al.*, 2012) as well as other genes involved in membrane processes.

Perhaps the most intriguing of these proteins are the bacteriorhodopsins, integral membrane proteins discovered through studies about the purple membrane. Bacteriorhodopsins function as light-driven pumps that transfer protons from the cytoplasmic side to the extracellular surface of the membrane. This way, an electrochemical gradient is generated which then drives other metabolic processes such as ATP synthesis or active transport. In similar fashion, another group of proteins, known as halorhodopsins, transfer chloride ions from the extracellular surface of the membrane to the cytoplasmic side. The first experimental evidence that the membranes of *Haloquadratum walsbyi* might harbour these protein came from the work of Papke and co-workers on CR30 environmental DNA. The amplified and sequenced CR30 environmental bacteriorhodopsin sequences were later found to correspond to *Haloquadratum walsbyi* gene *bop* I (Papke *et al.*, 2003). The *Haloquadratum walsbyi* genome revealed other rhodopsin-related genes: bacteriorhodopsin II (*bop* II), a bacteriorhodopsin-like gene (*brp*), and a halorhodopsin (*hop*) gene.

Bop genes were studied in much detail and they encode functional proteins (Bolhuis *et al.*, 2006; Dyall-Smith *et al.*, 2011). Both native and purified recombinant Bop I showed characteristics typical of bacteriorhodopsin: in the dark it had an absorption maximum at 552 nm, had all-*trans* and 13-*cis*-retinal and had a fast photocycle (Lobasso *et al.*, 2012; Sudo *et al.*, 2011). In contrast, Bop II showed a number of intermediate properties between those of bacteriorhodopsin and sensory rhodopsin II. Its absorption maximum was at 485 nm, close to that of sensory rhodopsin II of *Natronomonas pharaonis* (498 nm), making it one of the most blue-shifted microbial rhodopsins identified to date. In addition, it displayed a retinal composition which included all-*trans* and 13-*cis*-retinal but also 11-cis-retinal, and displayed a fast photocycle (Sudo *et al.*, 2011). Interestingly, the unusual features of this protein coincide with its phylogenetic positioning. In phylogenetic reconstructions, the *bop* II gene did not cluster within any known functional groups but instead branched basal to the proton pumps (Bolhuis *et al.*, 2006; Sharma *et al.*, 2007). This led some authors to believe that the gene encoding this protein was acquired through later gene transfer (Bolhuis *et al.*, 2006; Sudo *et al.*, 2011). *Haloquadratum walsbyi* halorhodopsin gene also encodes a functional halorhodopsin with a unique tolerance to changes in chloride concentration, a feature attributed to the serine residue at position 262 (Bolhuis *et al.*, 2006; Fu *et al.*, 2012).

CRISPR

The CRISPR/Cas systems [clustered regularly interspaced short palindromic repeats/(CRISPR)-associated genes] provide a form of acquired immunity to exogenous genetic elements such as plasmids or bacteriophages. They consist of short direct tandem repeats (23–47 bp) that are separated by spacer sequences of variable length. The spacers originate from invading exogenous elements in which they are referred to as protospacers. Their presence in the genome provides a 'memory' of previous exposures to invasive elements. It has been demonstrated experimentally that spacers can recognize and silence plasmids or phages in a fashion analogous to that of RNAi in eukaryotic organisms (Barrangou *et al.*, 2007; Marraffini and Sontheimer, 2010).

The CRISPR/Cas systems are not equally represented in the two genomes sequenced. In the C23 genome there are three CRISPR/Cas loci, while the HBSQ001 genome has only a remnant of such a system. In C23 one such system is present in a divergent region which in HBSQ001 contains a transcription unit coding for a probable cell surface structure. Interestingly, two of the C23 CRISPR/Cas loci and the HBSQ001 remnant are present in the portion of genome that is strain-specific and in HBSQ001 is named metagenomic island 2 (see below). Common to both genomes are intact leader sequences and the first direct repeat. However, while the complete CRISPR system is present in C23, the syntenic region in HBSQ001 is followed only by a few remnant spacers (Dyall-Smith *et al.*, 2011). The spacers that are common to both genomes are not identical, but are highly similar. This led Dyall-Smith and co-workers to believe that *Haloquadratum walsbyi* strains experienced rapid global dispersal, possibly by airborne salt particles or migratory birds. They further considered that at some point of evolutionary history, the HBSQ001 chromosome suffered a deletion event which has removed most of this CRISPR system (Dyall-Smith *et al.*, 2011). Mutations or loss of CRISPR/Cas are not uncommon (Deveau *et al.*, 2008; Touchon *et al.*, 2010) and were recently attributed to the ability of this system to avoid autoimmunity, e.g. upon acquisition of foreign DNA targeted by a CRISPR system that carries a gene essential for cell growth (Gudbergsdottir *et al.*, 2011). CRISPRs are genomic signatures that are common to both hosts (CRISPR spacer) and viruses (proto-spacer) and typically differ in identity by less than 10% (Garcia Heredia *et al.*, 2012). Their presence can therefore be used to assign putative hosts to environmental viral sequences. This way, seven environmental viral fosmid sequences originating from the CR30 metavirome were recently found to contain proto-spacers that correspond to CRISPR spacers of *Haloquadratum walsbyi* strain C23. Additional six environmental phage sequences carried sequences with lower identity to spacers (75–89%). Thus, this species was assigned as putative host (Garcia-Heredia *et al.*, 2012). Even more importantly, the presence of spacers of strain C23, which originates from a location 16,000 km away from CR30, indicates global distribution of both *Haloquadratum walsbyi* lineages and its predators.

The metagenomic approach

The possibility to obtain unbiased samples of all genes from all the members of a sampled community was enormously appealing from the very early days of what is today known as metagenomics. Given that square cells are not readily obtained in culture media, this approach offered a way to circumvent this problem and to reveal the true extent of diversity of *Haloquadratum walsbyi* in its natural environment. Thus, from the early 2000s this new technology was used for CR30 samples, and a way to further enrich CR30 samples in square cells was devised. To this aim, some authors (Legault *et al.*, 2006) took advantage of the specific shape of *Haloquadratum walsbyi* (flat square cells of 2–5 μm sides) and its abundance in the solar saltern crystallizer. Specifically, prior to environmental DNA isolation, they passed the crystallizer brine through a 5 μm pore size filter in order to remove larger particles. Then, the flat square cells of *Haloquadratum walsbyi* were retained on a 2 μm pore size filter while rod and spiral shaped cells were passed into the filtrate. The efficiency of this methodology was tested by preparing clone libraries of the 16S rRNA gene (Legault *et al.*, 2006). This analysis revealed that the 2 μm fraction was indeed enriched in cells of *Haloquadratum walsbyi* which represented 66% of sequenced inserts, and this fraction was then used to obtain environmental DNA.

In a first set of experiments, conducted in 2003–2006, the environmental DNA was sheared and fragments in size range between 35 and 45 kb were cloned into fosmid vectors

(Legault *et al.*, 2006). Then, 2000 insert-positive clones were end-sequenced. This way, about two-thirds of 2.4 Mb of data were assigned to *Haloquadratum walsbyi*. Then, beginning in 2007 advantage was taken of the coverage and reliability of data obtained through massive parallel sequencing. Environmental DNA isolated from CR30 was 454 pyrosequenced in winter 2007 and summer 2008 (Ghai *et al.*, 2011). In 2010, the very same sample, taken in summer of 2008 was resequenced using Illumina technology obtaining ca. 5 Gb of sequence data (Martín-Cuadrado *et al.*, unpublished).

In addition, a metagenome enriched in viral DNA was also constructed. To this aim, microbial cells were removed by filtration and the filtrate concentrated about 250× through tangential flow filtration on a 100 kDa filter. Environmental viral DNA was then isolated by centrifugation in a caesium chloride gradient followed by formamide lysis and cetyltrimethylammonium bromide DNA extraction (Thurber *et al.*, 2009). This DNA was used to construct a fosmid metagenomic library. This way the sequences of 42 fosmids which correspond to 1.2 Mb of sequence data were obtained (Garcia-Heredia *et al.*, 2012).

Unveiling the extent of intragenomic diversity in *Haloquadratum walsbyi*

With the advent of genomics it became possible to evaluate to what extent the homogeneity of species in terms of 16S ribosomal RNA sequence could be extended to entire genomes. Some bacterial species appear to have very homogenous genomes. Metagenomic analysis of samples obtained in extreme environments reported that the majority of intraspecies variability concentrates on single-nucleotide polymorphism, a notion that has allowed reconstruction of entire genomes from environmental metagenomic libraries (Tyson *et al.*, 2004). On the other hand, complete genome sequences of *Escherichia coli* and cultivable pathogens revealed a wide variation in individual genome content (Bergthorsson and Ochman, 1998; Welch *et al.*, 2002; Lindsay and Holden, 2004; Thompson *et al.*, 2005). It appeared that while certain chromosomal regions were shared among highly related strains, others appeared to be highly variable. A detailed analysis of six genomes of the pathogen *Streptococcus agalactiae* led Tettelin and co-workers to propose the concept of the microbial pan-genome. They suggested that 'a bacterial species can be described by its 'pan-genome'... which includes a core genome, containing genes present in all strains, and a dispensable genome composed of genes absent from one or more strains and genes that are unique to each strain' (Tettelin *et al.*, 2005). This 'dispensable', 'accessory', 'flexible' or 'adaptive' portion of the species genome does not appear to be essential for cell survival, but instead it is considered to provide the possibility to adapt to fluctuating environmental parameters.

Metagenomic studies of aquatic environments indicated that abundant aquatic microbes likely have large pan-genomes that surpass several times in size the amount of genes found in their core genomes. For example, a group of coastal bacterioplankton with 16S rRNA sequence identity ≥ 99% to *Vibrio splendidus* was found to 'consist of at least a thousand distinct genotypes, each occurring at extremely low environmental concentrations (on average less than one cell per millilitre)' (Thompson *et al.*, 2005). Analysis of the Sargasso Sea environmental dataset found *Prochlorococcus marinus*, a photosynthetic cyanobacterium that is numerically dominant in the majority of the world's oceans, to be represented by a population of closely related strains that display considerable heterogeneity, not only at the nucleotide sequence level, but also at the gene content level (Venter *et al.*, 2004). Similar to the 16S rRNA genes of aquatic microbes mentioned above, the available data on the *Haloquadratum walsbyi* 16S rRNA gene indicate that it forms a very coherent species. The 16S rRNA genes of two described members of this genus, strains HBSQ001 (Bolhuis *et al.*, 2006) and C23 (Burns *et al.*, 2007), differ only in two nucleotides (Oh *et al.*, 2010). Besides, a number of independent studies conducted in geographically widely separated salterns reported that environmental 16S rRNA gene sequences affiliated with genus *Haloquadratum* do not diverge by more than 2% from those of the above mentioned isolates (Legault *et al.*, 2006; Oh *et al.*, 2010; Boujelben *et al.*, 2012; Zafrilla *et al.*, 2010). In similar fashion, in CR30 metagenomic libraries the vast majority of environmental fragments unambiguously identified as 16S rRNA of *H. walsbyi* varied by less than 2% (100% identity over the entire length of the environmental fragment). Furthermore, they were highly related (≥98.9% sequence identity) to sequences retrieved from the CR30 in studies conducted over the previous fifteen years (Benlloch *et al.*, 1995; Legault *et al.*, 2006; Ghai *et al.*, 2011). Even the sequences of two complete fosmids whose inserts overlapped this particular genomic region were nearly identical to the HBSQ001 genome, pointing towards the permanence and high conservation of this region in *Haloquadratum walsbyi* (Legault *et al.*, 2006).

The first metagenomic library published was a 2000 clones fosmid library constructed from environmental DNA enriched in *Haloquadratum walsbyi* cells. At the time, the sequencing of entire fosmid libraries using Sanger sequencing was financially straining for most laboratories. Thus, to evaluate the amount of diversity present in that library, only the ends of the 2000 fosmid clones were sequenced (Legault *et al.*, 2006). This way 2948 individual end sequences were obtained with an average size of 818 bp. This amounted to 2.4 Mb, a very small amount of data compared to the output obtained using second generation sequencing techniques. However, the size of this library was considered sufficient to cover the intragenomic diversity of *Haloquadratum walsbyi*, as the genomic diversity of the CR30 crystallizer calculated through DNA/DNA reassociation was estimated to be seven *E. coli* genome equivalents (Torsvik *et al.*, 2002). Two fosmids whose inserts overlapped the *Haloquadratum walsbyi* 16S rRNA gene were fully sequenced. Common to all environmental sequences, both fosmid-end sequences and the entire fosmid sequences, was the prevalence of sequences with GC content similar to that of *Haloquadratum walsbyi*. The environmental sequences were binned to *Haloquadratum walsbyi* based on their similarity to the HBSQ001 strain genome that was used as a reference. Three groups of sequences were analysed: those that shared more than 94% nucleotide identity to HBSQ001 (1029 sequences), those that shared 80–94% sequence similarity (261 sequences) and sequences that shared more than 80% sequence identity with *Haloquadratum walsbyi* on one fosmid end and were dissimilar to this species on the other fosmid end (173 sequences). The fosmid end sequences were first used to assess the level of synteny between environmental fosmid-end sequences and HBSQ001. This was achieved by assuming that their corresponding inserts are shorter than 60 kb and correspond to identical regions on the HBSQ001 genome. From this analysis it appeared that the genomes corresponding to individual lineages of *Haloquadratum walsbyi* are largely syntenic. In fact, end sequences of 457 fosmids had hits that were separated on average by 35.2 kb. If concatenated, these sequences would allow the reconstruction of 92% of the HBSQ001 chromosome with an average coverage of 5.7× (Legault *et al.*, 2006).

However, the analysis also identified 66 fosmid inserts in which the genomic synteny was largely broken and in which the fosmid end sequences corresponded to genomic regions that were on average separated by 589 kb. These genomic rearrangements were correlated with

DNA divergence. Together with fosmid sequences that shared similarity to the reference genome at only one fosmid end, these fosmids likely corresponded to accessory genomic regions. To confirm this, their location was plotted onto the reference genome (Fig. 1.3a). This plot identified that the intraspecies variation was by no means random. In fact, it was possible to delineate two regions that coincided with larger amplitude variance in the GC content (Legault *et al.*, 2006).

This first study also offered a way to characterize the accessory genome of *Haloquadratum walsbyi*. The fosmid-end sequences that were dissimilar to the *Haloquadratum walsbyi*

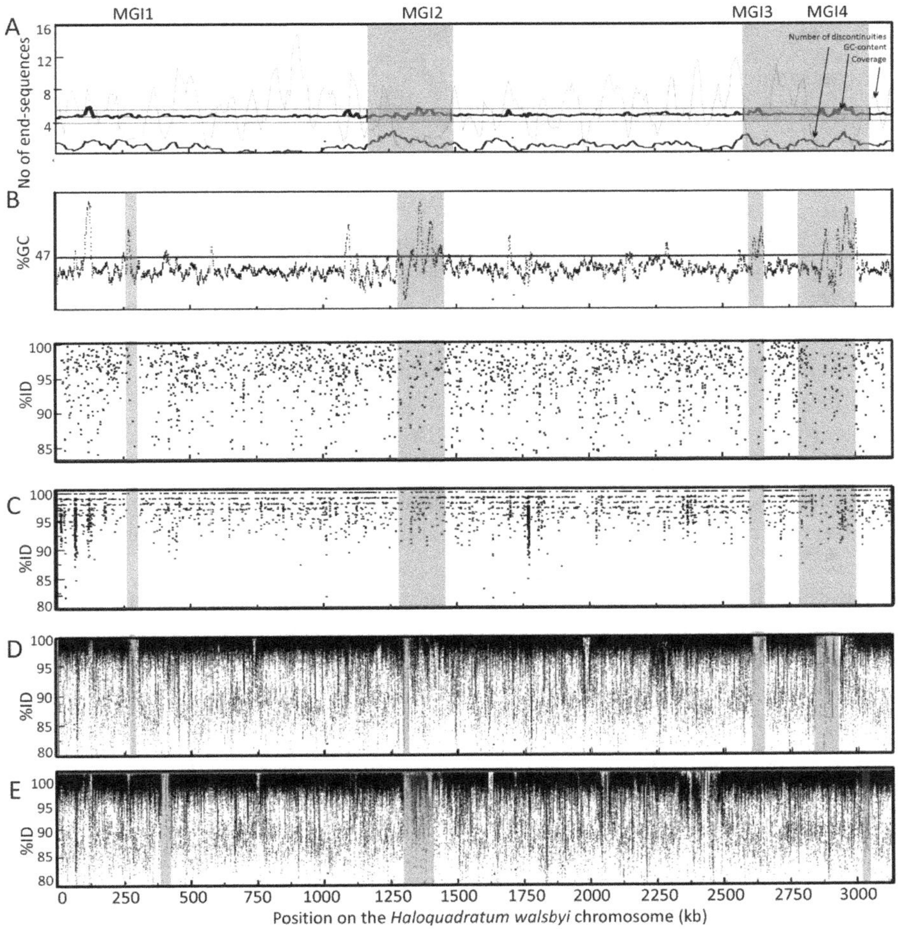

Figure 1.3 Metagenomic islands (MGI) are conserved features in *Haloquadratum walsbyi*. (A) GC-content and the distribution of the CR30 fosmid-end sequences along the HBSQ001 chromosome. (B) GC-content and genomic recruitment of individual CR30 fosmid-end sequences along the HBSQ001 genome. Genomic recruitment of (C) pyrosequenced metagenomic fragments from the Chula Vista solar saltern (California, USA) along HBSQ001 genome. (D) CR30 SS37 (SRP007685) dataset along HBSQ001 genome. (E) CR30 SS37 (SRP007685) dataset along C23 genome. Y axis reflects nucleotide percentage identity. (Adapted from Legault *et al.*, 2006; Cuadros-Orellana *et al.*, 2007; Pašić *et al.*, 2009.)

genome on one end were binned to the 'accessory genome' category. Another such group of sequences were fosmid-end sequences that did not share sequence similarity to HBSQ001, yet displayed signature GC content (≤ 55% GC). Combined, this database contained only 283 environmental sequences and yet its analysis revealed that their genes could be affiliated with a small number of functional categories. Particularly abundant were the genes involved in signal transduction and gene regulation and genes encoding components of cell envelope. This indicated the importance of environment sensing for the adaptation of individual lineages to specific heterotrophic niches with this habitat. Even more surprisingly, a large portion of intraspecies variability appeared to be at the level of the cellular envelope, as previously observed in natural populations of *Prochlorococcus marinus* (Coleman *et al.*, 2006).

Metagenomic islands

One way to discern between the core genome and variable genomic regions is to tile metagenomic reads of the environment from which the organism was isolated to its reference genome, a method known as genomic recruitment. The genes that are present in all strains will be well-represented in the environment and this will be reflected in a large number of metagenomic fragments recruited at high sequence similarity. Likewise, highly variable regions will be underrepresented, if at all present, in the metagenomic dataset. Although sceptics might find these gaps in recruitment an artefact of sequencing depth, this is highly unlikely as these are regularly observed in different environments whenever a microorganism is particularly abundant (Rodríguez-Valera *et al.*, 2009). As we will see below, the same variable regions are identified when geographically different metagenomic datasets are recruited to a reference genome. The presence of underrepresented genomic regions is observed regardless of the sequencing effort. However, larger sequencing efforts do contribute to clear delimitation of their borders. Besides, the gene content of these regions is often far from random and appears to be conserved in unrelated microorganisms.

Initially, variable genomic regions had been identified as genomic islands. The reason for this is historical, as similar regions were previously observed in highly related genomes of clinical isolates where they were defined as '... large chromosomal regions that are part of the flexible gene pool, previously transferred by other mobile genetic elements and are present in certain bacteria but are absent in most closely related variants' (Dobrindt *et al.*, 2004). Although these types of genomic island share numerous similarities with the aforementioned metagenomic islands, their overlap is not always complete. Some genomic islands are well represented in the metagenome while others are not. Furthermore, not all islands observed through comparisons with metagenomic datasets show clear signs of lateral gene transfer and thus the exact mechanisms of their generation remain to be fully understood. Thus, in order to distinguish between these two subtypes the term metagenomic island (MGI) was proposed to describe genome stretches that were identified through genome recruitment (Pašić *et al.*, 2009).

Initially, a genomic recruitment experiment was performed by using the fosmid-end sequence metagenomic library and these reads were tiled to the *Haloquadratum walsbyi* HBSQ001 genome that was available at the time (Cuadros-Orellana *et al.*, 2007). Although the metagenomic library contained less than 3000 sequences, this approach revealed a clear picture of the extent and the location of variable regions within this particular genome. Two additional lineage-specific regions were discerned this way, in addition to variable regions observed in the previous study detailed above (Legault *et al.*, 2006). Their location as bordered by environmental fosmid-end sequences highly similar to reference genome is depicted in Fig. 1.3b. This island pattern remained the same when the HBSQ001 genomic sequence was compared to metagenomic datasets of different geographic origins (Fig. 1.3c). Likewise, the location of islands was not affected by sequencing effort or sequencing technique used as it is evident from comparisons of HBSQ001 to a high sequencing coverage CR30 SS37 dataset (Fig. 1.3d). This indicates that metagenomic islands are a conserved genomic feature.

From the above plot, the genomic content of metagenomic islands of the reference strain HBSQ001 could be discerned, but not the amount of their natural variability with respect to gene synteny, gene content, copy number and their putative origin (i.e. prophage/plasmid/lateral gene transfer). Thus, the fosmids that targeted regions with broken synteny were recovered. They exhibited homology to the HSQ001 genome at distances much larger than the fosmid insert size (12 fosmids), exhibited homology only at one end of the insert (six fosmids) or had no sequence similarity to HBSQ001 genome but shared low GC content (five fosmids). The inserts of these fosmids were pyrosequenced and upon assembly, the size of sequenced inserts was between 10.5 and 38.2 kb. Twenty-one of these fosmids could be directly positioned along the reference genome and corresponded to individual lineages of *Haloquadratum walsbyi*. Most of these fosmids were either completely syntenic to the reference genome or had high sequence identity (≥ 98%). Only three of these fosmids overlapped metagenomic islands (Cuadros-Orellana *et al.*, 2007). Besides, when environmental gene fragments of CR30 crystallizer were compared to the genome of *Haloquadratum walsbyi* strain C23, isolated from a saltern in Geelong, Australia, 12 metagenomic islands could be discerned (Fig. 1.3e). Eight of these islands corresponded to classical genomic islands and were previously observed in comparative analysis of two genomes (Dyall-Smith *et al.*, 2011). Among these islands were the three metagenomic islands observed in HBSQ001. The remaining metagenomic and genomic islands were small in size and corresponded to highly divergent genes, insertion-coupled deletions and IS elements or reflected an insertion of prophage or integrative plasmid as detailed below.

Variability as a consequence of prophage/plasmid integration: metagenomic island 3

Prophages or plasmids frequently integrate into prokaryotic genomes and become associated with lineage-specific genomic material. Unsurprisingly, metagenomic islands of prophage/plasmid origin were reported from almost all microbes studied using approaches similar to the one presented here, namely *Salinibacter ruber* (Pašić *et al.*, 2009), *Prochlorococcus marinus* (Coleman *et al.*, 2006), *Candidatus* Accumulibacter phosphatis (García Martín *et al.*, 2006) and '*Ferroplasma acidarmanus*' (Allen *et al.*, 2007).

At least two such events are well-documented in sequenced genomes of *Haloquadratum walsbyi*. The first region corresponds to HBSQ001 genomic island 4, which had low coding density, was characterized by sharp GC-content changes and it had a high proportion of hypothetical genes, all tell-tale signs of highly unstable genomic regions. Adjacent to its left end, this metagenomic island contained a tRNA-His gene and a phage integrase similar to that of a halovirus HF2 that could be involved in the mobilization of this region (Dyall-Smith *et al.*, 2011). Other ORFs in the region were identified as homologues of Cdc6 and Cdc48 (HQ3268A, HQ3269A, HQ3297A) which could contribute to region replication, type IV secretory pathway related genes (HQ3289A, HQ3291A), genes involved in DNA modification and repair (HQ3274A-HQ3277A) and a gene related to bacterial conjugation (HQ3291A). The hypothetical proteins of ORFs HQ32902A and HQ32972A shared conserved domains with a metaviral attachment glycoprotein and HQ32950A was similar to a viral tail protein. Besides, this region also contained a partial copy of the tRNA-His sequence which could be used to target it (Cuadros-Orellana *et al.*, 2007; Dyall-Smith *et al.*, 2011). In HBSQ001, the phage-associated genomic island contained another set of genes. These genes (about 20 kbp) were identified as belonging to the *liv* cluster and as IS1341 transposases, which might have contributed to the spread of this genes or might act to promote their expression (Murai *et al.*, 1995). The genes of the *liv* cluster (*liv*FGMHJ) encode components of neutral and branched-chain amino acid transport system. Typically, these genes are organized in an operon which consists of two permease-coding genes, two genes that code for proteins that bind and hydrolyse ATP and a gene coding for substrate-binding protein. In *E. coli* this type of ABC transporter transports leucine, isoleucine and valine but also threonine, alanine, phenylalanine and tyrosine (Antonucci *et al.*, 1985) and other compounds such as 4-hydroxybenzoate (Egland *et al.*, 1997). It is this broad substrate specificity that has apparently contributed to the abundance of Liv family of proteins within bacteria (Larimer *et al.*, 2004). Indeed, the genomes of HBSQ001 and C23 were found to support a wide diversity of *liv* genes, containing both six such clusters, a characteristic typical for metabolically versatile microorganisms (Larimer *et al.*, 2004).

In contrast to HBSQ001, the C23 genome contained two other phage/integrative plasmid related elements. Similar to the GI3 in the HBSQ001 genome, the genes in the genomic region corresponding to nucleotides 161573–163558 of C23 genome included a phage integrase adjacent to a tRNA-Ala gene and genes related to DNA modification and repair. However, this region also contained a DNA polymerase similar to that of halophage HF1 and a region of likely co-transcribed genes. The second region, located at genomic positions 2411939–2454921 contained remnants of a phage integrase and was rich in transposons of which at least four appeared to have spread to other genomic regions (Dyall-Smith *et al.*, 2011).

Variability at the level of genes involved in uptake of nutrients: metagenomic islands 2 and 4

The crystallizer brine is endowed by heterotrophic microorganisms and both *Haloquadratum walsbyi* and *Salinibacter ruber* belong to this category. It is true that halophilic Archaea can derive energy from light by rhodopsins, but they can not fix CO_2 or N_2 (Falb *et al.*, 2008). To survive, these organisms exploit a wide diversity of organic compounds that are released at lower salinities by dense populations of *Dunaliella* sp. and other microbes (Pedrós-Alió *et al.*, 2000; Gasol *et al.*, 2004). To transport nutrients, microbes rely on specialized sets of proteins. The number of genes involved in nutrient transport is approximately proportional to the genome size (Saier *et al.*, 1998). For example, *Escherichia coli* has 78 ABC transporter systems, a number typical for this genome size. Bacteria that live in highly competitive soil environments, such as *Agrobacterium tumefaciens* (5.7 Mb) have up to 200 ABC transporter systems, while intracellular parasites and species that live in stable conditions have a lower number of ABC transporters. For example *Mycobacterium tuberculosis* (4.4 Mb) has only 38 systems (Davidson *et al.*, 2008). From this perspective, the number of transporters (39) in the 3.2 Mb genome of *Haloquadratum walsbyi* is within the range reported for Archaea

(Bolhuis *et al.*, 2006). The most frequent are transporters for branched-chain amino acids that are present in six copies. These sets of genes are followed in abundance by ABC transporters involved in transport of dipeptide/oligopeptide/nickel and sugar/*sn*-glycerol-3-phosphate, each present in three copies. Besides, *Haloquadratum walsbyi* possesses some systems that are unique to this archaeon such as a TRAP-type C4 dicarboxylate transport system, two different ABC-type sulfonate transport systems and a phosphonate transport system (Bolhuis *et al.*, 2006).

The presence of gene clusters involved in nutrient transport in variable genomic regions is another eye-catching feature of *Haloquadratum walsbyi*. A nice example of this is the metagenomic island 2 as evident from genome sequences of the strains HBSQ001 and C23 and the two environmental fosmids that overlap this region. In HBSQ001, metagenomic island 2 was found to contain the unique TRAP-type C4 dicarboxylate transport system, a ABC type branched amino acid transport system *liv*HMGF(J), a cluster involved in transport of nitrate/nitrite (*nar*K) and dissimilative nitrate reduction to ammonia (*nar*B and *nir*A) genes. In contrast, the corresponding region in the C23 genome contained genes for an ABC transporter involved in the transport of ribose. The sequence of fosmid eHwalsbyi485 that overlaps this region carried an ABC *sn*-glycerol-3-phosphate transport system, involved in catabolism of glycerol. Remarkably, despite the fact that they originate from widely separated geographical regions, the fosmid sequence of eHwalsbyi011 was 99% identical to the metagenomic island 2 of C23. It should be mentioned that the most striking difference in gene content of metagenomic island 2 in the two strains relates to other genomic features. In the C23 genome metagenomic island 2 contained two of the three CRISPR/Cas loci present and the only remnant of such a system was found in HBSQ001 genome. This is discussed in more detail in 'The two genomes' subsection above.

In the original study, when CR30 metagenomic fosmid-end reads were tiled to the HBSQ001 genome, metagenomic island 4 was found to contain genes involved in transport of nutrients (Cuadros-Orellana *et al.*, 2007). Likewise, the corresponding region in strain C23 also contained ABC transporters involved in transport of cobalt and urea/short-chain and other genes, but no genes involved in glycosylation of the cellular envelope. In the genome of HBSQ001, nutrient transport genes were located to the right end of this island and contained genes for ABC-type transporters involved in translocation of urea/short-chain amides and cobalt, respectively, as well as a cationic amino acid transporter and the urease operon. However, the genomic sequence of strain C23 revealed that this region is in fact syntenic in these two strains (Dyall-Smith *et al.*, 2011). Apart from the two ORFs (HQ3632A and HQ3633A) which correspond to the insertion element ISH2 and a hypothetical protein that are absent in C23 genome, the remaining genes shared ≥97% nucleotide identity. This indicates that in contrast to early observations, this part of the *Haloquadratum walsbyi* genome is in fact conserved in the two lineages. This is reflected in the recruitment studies against CR30 high-coverage SS37 dataset (Ghai *et al.*, 2011) which shows that larger sequencing efforts contribute to clear delimitation of metagenomic island borders (Fig. 1.2d).

Variability at the level of genes involved in cell envelope glycosylation: metagenomic island 4

As mentioned earlier, metagenomic island 4 in strain HBSQ001 was found to host genes putatively involved in the synthesis of sialic acid that could contribute to the formation of a capsule (Bolhuis *et al.*, 2006) and/or could facilitate viral attachment. Presence of genes involved in the synthesis of sialic acid is another feature that is common to metagenomic islands of several species, e.g. *Prochlorococcus marinus, Salinibacter ruber* and *Desulfovibrio vulgaris* (Schauer *et al.*, 2000; Coleman *et al.*, 2006; Pašić *et al.*, 2009; Walker *et al.*, 2009). Their putative involvement in viral attachment was recently demonstrated in haloarchaeal pleomorphic virus HRPV-1, the saltern phage that infects *Halorubrum* spp. Its major structural protein, termed VP4, protrudes from the membrane and was proposed to participate in host recognition (Pietilä *et al.*, 2009). When this virus is propagated in *Halorubrum* sp. strain PV6, VP4 is N-glycosylated by the host. The added pentasaccharide comprises glucose, glucuronic acid, mannose, sulfated glucuronic acid but also a terminal 5-N-formyl-legionaminic acid, a sialic acid-like monosaccharide. The importance of this terminal sugar in molecular recognition of the host cell was demonstrated in an infection inhibition assay in which using N-acetylneuraminic acid (closely resembling the sugar moiety) reduced progeny virus production by half four hours post infection. Even more interestingly, when VP4 was heterologously expressed in cells of *Haloferax volcanii*, it was also N-glycolysed, yet the pentasaccharide added was not the same as when virus was propagated in *Halorubrum* cells. Instead, the pentasaccharide added was the same as is N-linked to the S-layer glycoprotein in this species. This demonstrated that the N-glycosylation of the haloarchaeal viral protein involved in cell surface recognition is host-specific (Kandiba *et al.*, 2012).

In addition, the first part of genomic island 4 in HBSQ001 genome is marked by three glycoprotein-coding genes HQ3457, HQ3468 and HQ3469 that were paralogous (average nucleotide similarity ca. 40%) to HQ1197, HQ1196 and HQ1195 in genomic island 1. Given that these genes had no homologues among metagenomic sequences and that both regions are equidistant from the origin of replication, it was assumed that the variability of these paralogues is caused by intragenomic recombination. However no evidence of such events has been found *in silico*. To this date, we have little knowledge about this region in other lineages of *Haloquadratum walsbyi*. In strain C23, metagenomic island 4 contains genes related to nutrient transport. Unfortunately, besides the sequence of fosmid eHwalsbyi033 whose parts were syntenic to different regions of metagenomic island 4 in HBSQ001, none of the environmental fosmid sequences were found to overlap this particular region.

Variability at the level of cellular envelope: metagenomic island 1

Metagenomic island 1 was atypical in that its average GC-content was similar to that of the species genome and that it did not contain IS elements or putative phage genes. Metagenomic and cultivation efforts yielded three different versions of this island (Fig. 1.4a). In addition to genomic islands of strains HBSQ001 (Cuadros-Orellana *et al.*, 2007) and C23 (Dyall-Smith *et al.*, 2011), an environmental fosmid sequence corresponded to metagenomic island 1 of an uncultivated lineage of *Haloquadratum walsbyi* (Cuadros-Orellana *et al.*, 2007). Common to versions of metagenomic island 1 were numerous genes that code for cellular envelope components and that were annotated as cell surface glycoproteins. This fact is not surprising in itself since both sequenced *Haloquadratum walsbyi* strains contain 12 such genes (Bolhuis *et al.*, 2006; Dyall-Smith *et al.*, 2011). This has been considered a consequence of complex architecture associated with square cell morphology (Walsby, 2005). The recovered versions of metagenomic island 1 shared another striking feature, the presence of a gene that could be affiliated with a protein which forms the cellular wall of halophilic and other Archaea. This protein, known as the S-layer, is composed

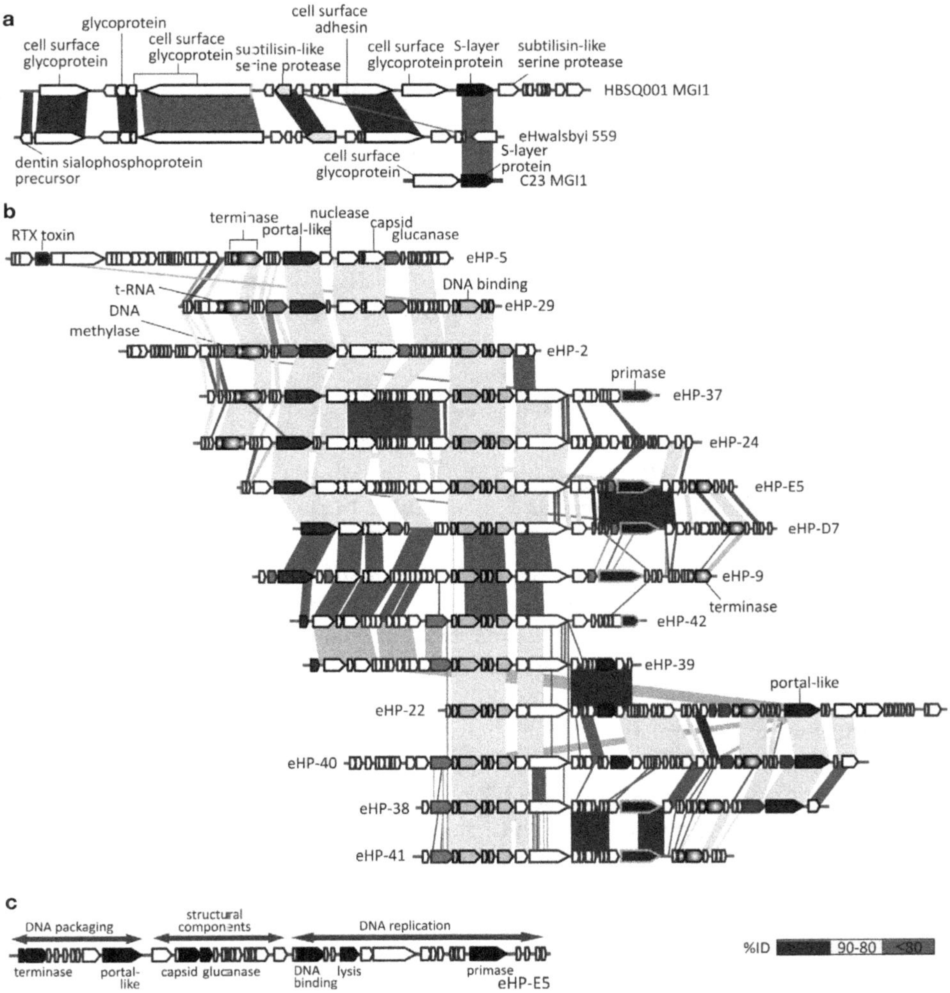

Figure 1.4 Comparative genomic organization of (a) metagenomic island 1 in strains HBSQ001, C23 and fosmid eHwalsbyi 599, (b) viral fosmids with *Haloquadratum walsbyi* as putative host. Conserved genomic regions are indicated by shading with intensity as a function of sequence similarity by BLASTN. (c) conserved modules in viral fosmids using eHP-E5 as a model. (Adapted from Cuadros-Orellana *et al.*, 2007; Garcia-Heredia *et al.*, 2012).

of oligomeric units that are anchored in the plasma membrane and which maintain the majority of S-layer above the membrane surface (Baumeister *et al.*, 1989, 1992). ORFs can be affiliated with the S-layer based on several features that are common to this group of proteins: (i) the high sequence identity that is in some species limited to the N-terminal region, which is responsible for subunit binding to the cellular envelope (Sleytr *et al.*, 1999); (ii) the variable middle and (iii) C-terminal protein parts that comprise domains involved in the self-assembly process, domains exposed inside the pores and on the surface of the S-layer. Furthermore, the S-layer proteins are heavily glycosylated (Mescher *et al.*, 1976; Sleytr *et al.*, 1976). While O-glycosylation sites are confined to the C-terminal protein region, the N-glycosylation sites are scattered throughout the sequence. In addition, these proteins are rich in acidic amino acid residues that help retain them soluble in high-salt conditions (Fukuchi *et al.*, 2003). Remarkably, in both genomes, the above criteria was met solely by orthologous genes located within metagenomic island 1- HQ1207A in HBSQ001 (988 aa) and Hqrw_1237 in C23 (822 aa) (Cuadros-Orellana *et al.*, 2007; Dyall-Smith *et al.*, 2011). The only other putative S-layer orthologues pair that shared about 40% sequence identity to the C-terminal residues of known S-layer proteins, was only about ¼ of length expected for such proteins (HQ1346 in HBSQ001, 220 aa; and Hqrw_1408 in C23, 222 aa; Dyall-Smith *et al.*, 2011). The above *in silico*-based assumptions were experimentally confirmed as the products of C23's ORF Hqrw_1237 and HBSQ001's ORF HQ1207A were identified as the major outer envelope proteins (Dyall-Smith *et al.*, 2011).

Such variability at the level of cellular envelope is considered responsible for the morphological differences between the cell walls of HBSQ001 (three-layered) and C23 (two-layered) (Burns *et al.*, 2007). It should also be mentioned that the S-layer is likely not the only cell wall component in *Haloquadratum walsbyi*. In the C23 strain three other proteins that correspond to ORFs Hqrw_1240, Hqrw_1641 and Hqrw_2184, were identified in the membrane fraction and their function remains unknown (Dyall-Smith *et al.*, 2011). Although the structure of the cell wall in HBSQ001 was not examined in that much detail, the cells of this strain were found to contain an additional cell layer, an external capsule probably formed by this strain's giant protein halomucin (Sublimi-Saponetti *et al.*, 2011). Thus, additional genes, some of which might as well be present in metagenomic island 1, might contribute to the formation of a functional cell wall.

Specific functional categories are highly overrepresented in metagenomic islands of marine microbes

The question that arose next regarded the extent to which above observations on the metagenomic islands of *Haloquadratum walsbyi* could be extended to other aquatic microbes. Indeed, the metagenomic islands of *Salinibacter ruber* were highly similar in gene content to those of *Haloquadratum walsbyi* (Pašić et al., 2009). Then, metagenomic islands larger than 10 kb in all sequenced marine bacterial species were systematically compared to marine metagenomes. This study revealed that all species studied had metagenomic islands which contained genes whose products are exposed extracellularly. Specifically, these most often corresponded to the O-chain of the lipopolysaccharide, a cellular component often recognized by viruses (Sharma *et al.*, 2008). Also found in metagenomic islands were genes that code for exopolysaccharide biosynthesis, genes involved in modifications of extracellular structures such as pilli and flagellar components and genes involved in transport of nutrients and environmental sensing (Rodríguez-Valera *et al.*, 2009). From this analysis it appears that in marine microbes the phage-interacting genes, nutrient transport genes and environmental-sensing genes are highly divergent and specific to individual clonal lineages.

On the origin of (meta)genomic islands in *Haloquadratum walsbyi*

The available hypotheses that try to explain the origin of (meta)genomic islands are all based on data obtained through the comparisons of available genomes. Besides the original description of metagenomic islands in *Haloquadratum walsbyi* (Cuadros-Orellana *et al.*, 2007), only one other study tackled this issue and reported meticulous comparison of the two available *Haloquadratum walsbyi* genomes (Dyall-Smith *et al.*, 2011). Below, we summarize their findings on the origin of the four genomic islands that are observed in comparative genomic and genomic recruitment studies.

The origin of metagenomic island 1 remains unclear. Early on it was proposed that its variation is caused by intragenomic recombination with metagenomic island 4, as both regions are equidistant from the origin of replication (Cuadros-Orellana *et al.*, 2007). However, this could not be supported *in silico*. Besides, in C23, this region shared no similarity with the rest of the genome and it was proposed that it diverged as a result of deletion-coupled insertion which resulted in what has been described as divergent genes in a conserved syntenic context (Cuadros-Orellana *et al.*, 2007; Dyall-Smith *et al.*, 2011). Metagenomic islands 2 and 4 had low coding density, were rich in transposases, putative phage-related genes and pseudogenes. In both genomes, these regions coincided with the deviation from average GC-content and differed dramatically in gene content. Metagenomic island 2 is considered to have developed as a result of a repeat-mediated deletion, which has eliminated just over 16 kb from the HBSQ001 genome (Dyall-Smith *et al.*, 2011). There is a particular direct repeat, ACATCAATTCT, that is present at the ends of the undeleted region in C23 genome and at the border of deletion in HBSQ001 genome. Besides, this metagenomic island's high GC-content (up to 57% in certain subregions) coincided with the presence of 14 pseudogenes, three functional IS1341 transposases, a probable phage integrase and several helicases that appeared to be phage related. In contrast, metagenomic island 4 likely evolved through a series of deletion-coupled insertions. Two such events, involving regions of 33.6 kb and 105 kb, were described in the HBSQ001 genome as well as a 15.3-kb insertion in the corresponding region of the C23 genome (Dyall-Smith *et al.*, 2011). Finally, some metagenomic islands simply reflect integration of prophages or plasmids as is the case with *Haloquadratum walsbyi* metagenomic island 3.

Interactions between *Haloquadratum walsbyi* and other species in the crystallizer pond

Synergistic interactions of *Haloquadratum walsbyi* and *Salinibacter ruber*

The ecological success of *Haloquadratum walsbyi* is likely to depend at least to some extent on its interactions with co-existing species. Synergistic interactions are not uncommon in bacteria and were previously observed in marine consortia of methane-oxidizing Archaea and sulfate-reducing bacteria (Boetius *et al.*, 2000; Caldwell *et al.*, 2008; Knittel and Boetius, 2009) and consortia in biofilms forming in acidic mine drainage (Tyson *et al.*, 2004). Typically, in such interaction one microbial partner consumes metabolites produced by the other partner. A number of studies aimed to investigate the possibility and the nature of putative association between *Haloquadratum walsbyi* and the second most abundant microbe in the crystallizer – the hyperhalophilic bacterium *Salinibacter ruber*. Both these microbes are heterotrophs and depend on nutrients produced at lower salinities, the most abundant being glycerol that is produced by blooms of *Dunaliella* algae. Experimental evidence pointed that glycerol is readily taken and rapidly turned over in the crystallizer (Oren, 2010), and together with its catabolism products dihydroxyacetone and pyruvate, glycerol was found to stimulate respiration of crystallizer community (Warketin *et al.*, 2009).

In sharp contrast, a microautoradiography and fluorescence *in situ* hybridization (FISH-MAR) study failed to identify glycerol as organic substrate readily used by *Salinibacter ruber* in its native environment (Rosselló-Mora *et al.*, 2003). Likewise, no growth enhancement was observed in pure cultures of *Salinibacter ruber* and *Haloquadratum walsbyi* when glycerol was used as a sole energy and carbon source (Antón *et al.*, 2002; Burns *et al.*, 2007). However, the uptake of glycerol could be induced if *Salinibacter ruber* cells were propagated in complex media with added glycerol (Sher *et al.*, 2004). This way, up to 25% of added glycerol was incorporated into biomass and respired, but also incompletely oxidized and excreted as dihydroxyacetone (Elevi Bardavid and Oren, 2008). Bolhuis *et al.* (2006) found genomic evidence that, unlike glycerol, this intermediate could be taken up by cells of *Haloquadratum walsbyi*. The square archaeon was found to possess a unique dihydroxyacetone transport system in which dihydroxyacetone is translocated over the membrane via facilitated diffusion and converted to dihydroxyacetonephosphate using a cytosolic phosphoenolpyruvate-dependent phosphotransferase system (Bolhuis *et al.*, 2006). Elevi Bardavid and Oren (2008) pursued this issue further and provided experimental evidence on dependence of *Haloquadratum walsbyi* on *Salinibacter ruber* for the supply of dihydroxyacetone. Although beneficial to both partners, this syntrophic-like relationship is not obligatory. Thus, it likely does not contribute to the relative abundance of taxa in the crystallizer but it could play an important role in the maintenance of crystallizer consortia and needs to be understood in more detail.

Metagenomic description of viruses that prey on *Haloquadratum walsbyi*

In the absence of grazing, dense populations of crystallizer microbes deal mainly with viral pressure. Viruses have a profound influence on prokaryotic populations (Angly, 2006). In the short term, viral induced lysis of a bacterial population will affect the flow of carbon and nutrients through different compartments of the food web (Weinbauer and Rassoulzadegan, 2004; Suttle, 2007). In the long term, the virus-mediated transfer of genetic material between host organisms can influence the evolution of the host population.

Viruses are often neglected in ecological studies, mainly because they are not readily visualized and isolated. Similar to other aquatic environments, in crystallizers the numbers of viral-like particles can surpass cell numbers by two orders of magnitude and reach 10^9 viral-like particles per ml (Guixa-Boixareu *et al.*, 1996). Under the microscope, four distinct viral morphologies, head and tail, round, spindle and pleomorphic, were repeatedly observed, with spindle-shaped particles being the most common morphotype in Bras del Port, and other hypersaline environments studied (Guixa-Boixerau *et al.*, 1996; Oren, 1997, Santos *et al.*, 2010). To date, about 100 halophilic viruses have been isolated in pure culture (Atanasova *et al.*, 2012; Senčilo *et al.*, 2013); they infect members of the haloarchaeal genera *Halobacterium*, *Halorubrum*, *Haloarcula*, *Haloferax*, *Natrialba* and *Natrinema* (Dyall-Smith *et al.*, 2003; Mei *et al.*, 2007; Pagaling *et al.*, 2007; Pietilä *et al.*, 2009; Atanasova *et al.*, 2012). Nearly all isolated halophilic viruses have been shown to have a double-stranded (dsDNA) and genomes in size range from 12 kb to 77 kb (Prangishvili *et al.*, 2006; www.haloarchaea.com). The one exception we are aware of is the ssDNA genome of HRPV-1, which is 7 kb in size (Pietilä *et al.*, 2009). The genomic sequences of halophilic viruses provided substantial novelty and were generally not similar to sequences in public databases. However, none of the isolated viruses infected either *Haloquadratum walsbyi* or *Salinibacter ruber*.

Environmental virology is hampered by the lack of a single phylogenetic marker with a signal similar to that of a 16S rRNA, which is commonly used for Archaea and Bacteria. Díez *et al.* (2000) were first to develop a protocol that allowed them to fingerprint the viral community using pulsed-field gel electrophoresis (PFGE). Specifically, a protocol was designated to concentrate viral particles. These concentrates were then used for environmental DNA isolation and intact viral genomes were size-separated using PFGE. This analysis showed that in Bras del Port salterns metagenomic assemblages changed from low to high salinities with populations in genomic ranges from 25 to 300 kb, and revealed that only four to five viral subpopulations were present in the crystallizer (Díez *et al.*, 2003; Sandaa *et al.*, 2003). Santos *et al.* (2007) took this type of analysis one step further. This time the fraction containing viruses was concentrated and mixed with low-melting agarose. These agarose plugs were treated to remove non-viral DNA and capsid proteins, and viral genomes were separated using PFGE. This way the PFGE analysis of samples taken from the CR30 pond showed a dominant band of around 37 kb. This individual band was excised from the gel, cloned into a fosmid vector and sequenced. This sequence became the first genomic sequence of an environmental halophage (Santos *et al.*, 2007). The sequenced genome was 35 kb in size and had a GC-content of 51%. It contained 40 predicted ORFs, among which were genes that putatively code for proteins involved in DNA replication, normally found in lytic viruses. However, the putative host to this environmental virus could not be inferred.

Viruses are known to exhibit genomic signatures that (with exceptions) mimic those of their primary hosts (Sharp and Li, 1987; Carbone, 2003). For example, unless the virus carries its own tRNA genes, it usually requires host translational mechanisms (Limor-Waisberg *et al.*, 2011). Thus, optimal expression of viral genes occurs when the usage of codons in viruses match that of a host (Cardinale and Duffy, 2011). Furthermore, viruses display unique and distinct patterns of nucleotide usage both at the dinucleotide and the tetranucleotide level (Burge *et al.*, 1992; Pride *et al.*, 2003), except in regions hypothesized to be acquired through lateral gene transfer (Karlin, 2001). Again, nucleotide usage patterns of viruses and their hosts are frequently similar, which is likely a consequence of their co-evolution (Pride *et al.*, 2006). A study on environmental halophilic viruses took advantage of genomic signatures described above (Santos *et al.*, 2010). Here, the viral concentrates preserved in agarose plugs were used to extract environmental DNA and prepare fosmid libraries and yielded 1.35 Mbp of sequenced environmental fosmid DNA. Like in previous studies, viral sequences revealed substantial novelty as most ORFs coded for hypothetical proteins (75.3–87.5%) and only a small fraction of ORFs could be associated with known viral genes. However, when individual fosmid sequences were binned into groups based on dinucleotide frequency and GC-content, five groups termed HVS-1 through HVS-5 were observed. Viral genomes of cluster HVS-4 shared dinucleotide frequency and GC-content with the genome of *Haloquadratum walsbyi* and according to this grouping schema those authors suggested that this cluster would include viruses that infect lineages of *Haloquadratum walsbyi* (Santos *et al.*, 2010).

A further insight into this particular group of viruses was recently obtained by Garcia-Heredia *et al.* (2012), who cloned and subsequently sequenced 42 viral fosmids originating from CR30 (ca. 1.2 Mb). The sequenced inserts were in size (20.–43.6 kb) and GC-content range (43.9–60.8%) characteristic for complete viral dsDNA genomes of hypersaline origin. Based on sequence similarity (≥75% over 3000 bp), frequency of tetranucleotides and codon usage the majority of fosmids (14) was found to correspond to viruses that prey on *Haloquadratum walsbyi* (Fig. 1.4b). The environmental fosmids showed several genomic hallmarks of head-tail viruses belonging to the Caudoviridae. Their genes were organized in three separate modules that correspond to DNA packaging, structural components and DNA replication module (Krupovic *et al.*, 2010). On the level of individual genes, the fosmid inserts showed the presence of a gene coding for a terminase large subunit (TerL), together with a small protein that corresponded to a small terminase subunit (TerS). Likewise, the capsid genes shared similarities with capsid genes of head-tail phages. Besides, the environmental fosmids contained putative portal proteins that are essential for development and infection of tailed viruses (Krupovic *et al.*, 2010).

The synteny of fosmids was largely undisrupted although some regions were rearranged in a way reminiscent of the circularly permuted gene order seen in some phages that replicate by the rolling circle mechanism (Bath *et al.*, 2006). The environmental fosmids displayed high level of mosaic diversity in local populations. This was evident as nearly identical sequences were followed by sequences that are unique to each genome (Fig. 1.4b).

However, the group of genes which is of utmost importance for the purpose of this chapter were located in variable regions just downstream from the capsid gene. These genes were underrepresented in metaviromes and corresponded to a viral ORF annotated as glucanase. Phage glucanases are involved in degradation of host cell wall either during virus entry or release (Xu *et al.*, 2001; Cornelissen *et al.*, 2011) and could act either on glycoproteins of haloarchaeal S-layer or on expolysaccharides present in cell envelopes. Similar to genomic island genes involved in cell surface decoration, these genes could be specific to individual phage lineages. In this case, the high diversity achieved through this approach would be essential to recognize the diversity of cell envelope polysaccharides of the putative host.

We also know something about the expression of viral genes under different conditions in CR30. Santos *et al.* (2011) addressed this important question and quantified the expression of viral genes using microarrays containing clones from previously characterized metavirome. The level of expression of genes associated with putative viruses of *Haloquadratum walsbyi* was not the highest detected. Instead, viral groups with the highest expression levels were those related to high GC-content haloarchaea and *Salinibacter ruber* that are minor components in the environment. Under UV and osmotic stress the gene expression in some viral groups increased and the number of viruses that were found in the sample changed. This type of strong response was highest in contigs that were associated with Archaea, in particular with *Haloquadratum walsbyi*. This was not unexpected since UV radiation inactivates viruses, and reduced salinity makes archaeal cells unstable, which stimulates viruses to reproduce and quickly exit cells (Dyall-Smith *et al.*, 2003; Santos *et al.*, 2011). Putative archaeal viruses showed more activity under stress leading to the drop in numbers of their hosts. However, neither of these factors affected the population structure at the species level at the level of resolution provided by the analysis of DGGE patterns (Santos *et al.*, 2011).

Viral sequences point towards global distribution of *Haloquadratum walsbyi* lineages

Although the methodologies used above to infer the phylogeny among environmental viruses are well-suited, they can be failable if used as sole merits to affiliate viruses to their putative hosts. An example from hypersaline environments are viruses His1 and His2 that prey on *Haloarcula hispanica* and differ significantly in GC-content (Bath *et al.*, 2006). Alternatively, environmental virus sequences can be matched to their putative hosts through analysis of virus-derived sequences present in host genomes. A typical example are the CRISPR/Cas systems (Clustered Regularly Interspaced Short Palindromic Repeats/CRISPR associated) that consist of short direct repeats (23–47 bp) that are separated by spacer sequences of variable length. The spacer sequences frequently match sequences found in phages (i.e. protospacers) and are considered to be copied or transferred to host genome during an unsuccessful phage attack. While their presence in a host genome is an indication that the strain has been a host of the infectious element, their presence in an environmental sequence can be used to assign hosts to putative viruses.

The two sequenced genomes of *Haloquadratum walsbyi* differ in CRISPR/Cas systems – the genome of HBSQ001 contains remnants of the CRISPR/Cas subtype I-B and the genome of strain C23 contains two CRISPR/Cas systems, that belong to the subtypes I-D and I-B, respectively. Interestingly, although the strains originate from geographically widely separated locations, the spacer sequences of HBSQ001 were present in genome of C23 and matched Cas genes and a predicted endonuclease and vice versa, one C23 spacer was an exact match to a sequence within a HBSQ001 ORF gene (Dyall-Smith *et al.*, 2011). Likewise, 12 spacers of Australian C23 strain were either highly similar or completely identical to environmental viral sequences recovered from the CR30 crystallizer in Spain (Santos *et al.*, 2010). This suggested not only that the environmental sequences belonged to *Haloquadratum*-specific viruses, but also that the viruses inhabiting Spanish and Australian salterns are highly related, indicative of their global distribution.

The metagenomic insight into predator–prey interplay in aquatic environments

Metagenomics proved to be a useful tool and contributed significantly to our understanding of ecology of the CR30 crystallizer pond. We now know that this simple system is dominated by clonal lineages of a single species, i.e. *Haloquadratum walsbyi*, whose cells amount to 50–80% of cells in the crystallizer (Antón *et al.*, 2002; Ghai *et al.*, 2011). We know that this species has a large pan-genome (Legault *et al.*, 2006) and that the individual clonal lineages differ mainly in genes that are involved in nutrient uptake and genes that code for components of cellular envelope that include S-layer (Cuadros-Orellana *et al.*, 2007). This amount of intraspecies genetic diversity is highly unexpected in asexual populations. It seems likely that rather than just being accumulated junk, this DNA is important to the environmental adaptation of different lineages. In the absence of grazing, the cells of *Haloquadratum walsbyi* deal mainly with viral pressure (Guixa-Boixerau *et al.*, 1995). However, it remained undetermined to what extent if any, the observed diversity reflects the adaptation of these cells to their predators. It is evident from other ecosystems that the viruses are major players that exert a profound influence on host population size and the ecosystem food web dynamics (Wren *et al.*, 2006; Suttle, 2007). The viruses alter the behaviour and reduce the fitness of host populations and induce mortality rates comparable to those of grazing (Suttle, 1994; Fuhrman and Noble, 1995). However, the fine mechanisms of virus–host interplay and their role in host evolution remain to be fully understood.

It has been proposed that natural populations of bacterial species are composed of distinctive lineages that arise by 'Periodic selection' (Atwood, 1951). This process of diversification is selectively neutral. Periodically, the selection that favours an adaptive mutation leads to a sweep that eliminates nearly all the diversity within natural population. On the other hand, in marine environments, where nutrients are not the limiting resource (Wilcox and Fuhrman, 1994; Sano *et al.*, 2004), the relative abundance of host cells is controlled by predating viruses. They prevent the best bacterial competitors from building up a high biomass in a 'Kill-the-winner' fashion (Thingstad, 2000). More specifically, predator–prey pairs are thought to oscillate in time in a periodic out-of-phase motion in which the bloom of a host is followed by that of its predating virus (Thingstad and Lignell, 1997; Thingstad, 2000). This way, the populations of predator and prey evolve in time according to a pair of first-order, non-linear, differential equations known as equations of Lotka–Volterra (Hofbauer and Sigmund, 1992). Accordingly, viral dynamics closely tracked that of a host in experimental microbial systems (chemostats) with a limited number of predator–prey pairs (Bull *et al.*, 2006). In such systems, the intragenomic microbial diversity is expected to be low and the entire population is expected to be dominated by a single clonal lineage. This proved to be the case and besides, the host populations appeared to experience rapid evolution as the resistance to predating virus arose in as little as seven days following their introduction (Lennon and Martiny, 2008). Based on previous adsorption kinetic assays with *Synechococcus* WH7803 and its virus, these authors hypothesized that viral resistance most likely arose from the loss or modification of cell surface receptors used for viral attachment (Stoddard *et al.*, 2007; Lennon and Martiny, 2008), a phenomenon similar to what was observed in genome recruitment studies (see above).

Analysis of metagenomic data from a study of marine viruses suggested that the 'Kill the winner' dynamics should be modified to encompass the hypothesis of neutral evolution. This was achieved by applying power law distribution to Lotka–Volterra equations (Hoffmann *et al.*, 2007). From the biological point of view, this modification assumes that in a marine system all predator–prey pairs follow identical 'Kill the winner' dynamics, but bloom at independent times. Thus, in a typical virus–host interaction, the number of host cells remains low for extended periods of time. Any increase in abundance of a particular host is followed by blooms of its predating virus. It also predicts that upon the bloom, the viral populations experience an abundance decrease which is initially rapid and decreases in time, thus resulting in long-term population viability (Hoffmann *et al.*, 2007). The modified 'Kill the winner' dynamics was experimentally tested in 29 viral and prokaryotic metagenomes that were obtained from freshwater and San Diego saltern samples (California, USA) (Rodríguez-Brito *et al.*, 2010). The four environments tested included a crystallizer pond and were sampled at irregular intervals ranging from days to over a year. The dynamics of predator–prey interaction was investigated at taxonomic levels of species and at a level of viral genotype/microbial strain (Rodríguez-Brito *et al.*, 2010). The microbial community composition and its metabolic potential remained stable throughout the time series. The changes in relative abundance of taxa were restricted to relatively less abundant taxa or taxa considered to be transient. Thus, at the species level, the dynamics of virus–host interplay failed to follow 'Periodic selection' model as no perturbations of dominant species were observed throughout the time-series. In these terms, the community response was similar to that observed in the CR30 crystallizer pond upon UV or salinity stress (Santos *et al.*, 2010). However, this temporal stability of microbial communities failed to persist below the species level. Here again, a significant amount of intraspecies diversity was observed, indicating that the population of dominant species and its predator is likely composed of several co-existing lineages. The time-dependent shuffling of the most successful lineage was observed in all environments and even in samples collected just one day apart. However, even at this fine-grained level, the dominant microbial taxa were not periodically reduced in numbers as it is assumed by 'Periodic selection' model. Instead, an increase in abundance of individual microbial strain was met by increased viral predation and the subsequent decline in host abundance was quickly followed by a decline of that particular viral genotype. Thus, it appeared that the power law modification of 'Kill-the-winner' dynamics operates at the level of individual microbial clonal lineages/viral genotypes but not at the level of microbial species.

The 'Constant diversity' dynamics model

To encompass the phenomenon of intraspecies diversity observed through metagenomics in hypersaline and aquatic environments and the absence of 'Periodic selection' at this taxonomic level, the 'Constant diversity' model of population dynamics was introduced (Rodríguez-Valera *et al.*, 2009). The 'Constant diversity' model is based on 'Kill the winner' dynamics and it assumes that the metabolically superior clonal lineages will be selected against by predating viruses in density-dependent fashion. However, under this model, the viral population guards prokaryotic intraspecies diversity and sudden drops in cell numbers are not observed. In a nutshell, while 'Periodic selection' predicts that the intragenomic variability is driven by niche exploitation, 'Constant diversity' assumes that it is driven by phage-host interplay. In more detail, 'Constant diversity' is considered to occur in those aquatic ecosystems in which bacterial populations compete with each other and predating viruses have a similar chance of infecting any cell within the community. In such systems, a large diversity of clonal lineages that differ in viral susceptibility is a prerequisite to avoid losses in cell numbers imposed by viral lysis. As indicated by metagenomics, the concurrent clonal lineages differ in genome stretches (metagenomic islands) that code for a number of extracellular components recognized by predating viruses as well as in genes responsible for exploitation of available nutrients. Under the 'Periodic selection' model, the appearance of a mutant or recombinant that is ecologically superior to its peers will lead to its expansion and eventual domination over ecologically less successful lineages. 'Constant diversity' differs from this model in that it predicts that this situation can occur only transiently. Any increase in numbers of cells corresponding to the ecologically superior lineage will be met by an increase in numbers of viruses that target its cell-surface receptor. This way, the invasive lineage will be selected against and eventually replaced by original 'normal fitness' lineages. Therefore, under this model of population dynamics, the high level of clonal lineage diversity is constantly maintained and results in a large pan-genome within the bacterial population, as indicated by metagenomic data and mathematical modelling (Rodríguez-Valera *et al.*, 2009).

The 'Constant diversity' model has another important consequence. Given that the ecological success of individual lineages is influenced by their ability to efficiently exploit available nutrients, the co-existing lineages are expected to differ in this respect. Therefore, the ensemble of lineages that individually differ in transporter genes could exploit all available nutrients. This way the 'Constant diversity' population dynamics not only preserves intraspecies diversity but also leads towards a more efficient exploitation of niche resources and better ecosystem functioning as illustrated in Fig. 1.5. Finally, the two models of bacterial

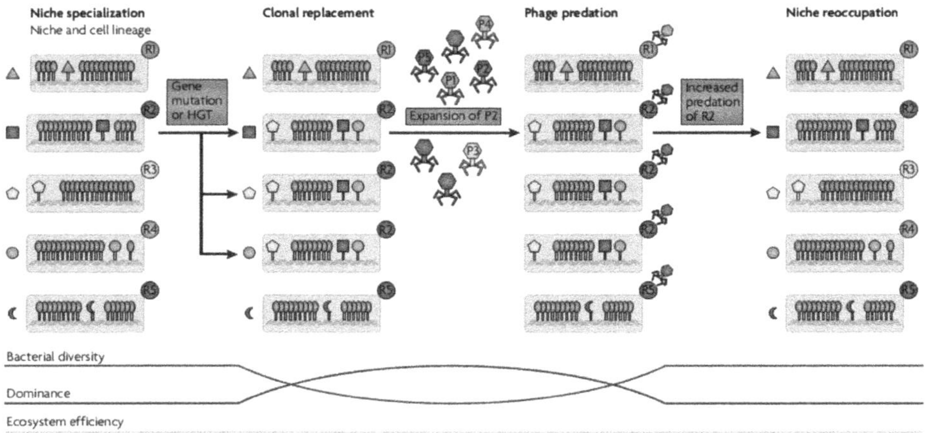

Figure 1.5 Population dynamics under constant-diversity dynamics. Clonal lineages that co-exist in a natural population exploit different niche resources (geometric symbols) and differ in the set of receptors (R1–R5) to viruses (P1–P5) that they carry. The major viral receptor is lineage specific. An increase in fitness gained by a lineage would unsettle the predator-prey equilibrium and select against the invasive clone which is eventually replaced by the original 'normal fitness' lineages (adapted from Rodríguez-Valera *et al*., 2009).

population dynamics are not mutually exclusive and each is considered to prevail under certain environmental conditions. The influence of phage predation on bacterial diversity requires that bacterial populations interact with each other. Thus, 'Constant diversity' is not expected to occur in physically constrained environments such as biofilms. Likewise, pathogens are likely to follow 'Periodic selection' unless they become very numerous or persist for a long time.

Evidence for each of the two alternatives can be found in natural populations of bacteria. For example, if a bacterial population follows 'Periodic selection', the number of co-existing lineages found in a population would be small. Besides, a single dominant lineage would be observed and this dominant lineage will eventually be replaced by another at different time and/or space points. Adversely, under 'Constant diversity' many clonal lineages will co-exist at each time and space point and the very same lineages will be retrieved in each sampling effort. Very recently, attempts to discern between these two alternatives through analysis of metagenomic datasets was submitted for consideration (Gonzaga *et al.*, 2012; López-Pérez *et al.*, 2013; Martín-Cuadrado *et al.*, unpublished). These studies concentrated on two environments and model organisms, the marine environment and *Alteromonas macleodii* and hypersaline crystallizer brine and *Haloquadratum walsbyi*. The hypersaline aquatic model is considered particularly well-suited for this type of study for a number of reasons: the crystallizer is a fairly stable environment which experiences only low-amplitude fluctuations in environmental parameters, it is endowed with low microbial diversity and it supports high densities of *Haloquadratum walsbyi* cells. Today, a wealth of data is available from this particular habitat: two large metagenomes sequenced using 454 and Illumina technologies and obtained from the CR30 crystallizer in winter 2007 and summer 2008. The two datasets originate from the same location but different water bodies, as the salterns are emptied between two salt-harvesting season. A third metagenome used in this study was obtained from nearby saltern of Torrevieja, built in a natural hypersaline lagoon that is never replenished and thus originates from a constant water body.

Analysis of the metagenomic datasets showed that *Haloquadratum walsbyi* represented 64% of the cells (Ghia *et al.*, 2011). If the CR30 *Haloquadratum walsbyi* population is to follow 'Periodic selection', then the number of co-existing lineages will be low and it will be possible to reconstruct large portions of the genome from environmental DNA.

The second model organism, *Alteromonas macleodii,* follows quite a different lifestyle. Compared with *Haloquadratum walsbyi,* which is a slow-growing K strategist, *Alteromonas macleodii* is an *r* strategist whose populations experience rapid growth (and subsequent decay) in nutrient rich temperate marine waters, often near or on particulate material. Four *Alteromonas macleodii* genomes, originating from the same location in the deep Mediterranean, have been sequenced and compared to a fosmid library originating from the same environment (Gonzaga *et al.*, 2012). This analysis indicated that the genomes were well-represented in the environment, and even metagenomic islands were preserved in the metagenomic dataset. Out of 160 fosmid inserts sequenced, two additional inserts corresponded to the O-chain polysaccharide metagenomic island and indicated that at least two clonal lineages of *Alteromonas macleodii* were present in metagenomic libraries. This data indicated that a limited number of *Alteromonas macleodii* clonal lineages was present in this environment (Gonzaga *et al.*, 2012) yet no dominant lineage was observed.

The question that arose next regarded the preservation of clonal lineages in space and time and whether it could be observed in metagenomic datasets. In the case of *Alteromonas macleodii* strains, AltDE and AltDE1 were recovered in a dataset originating from a different location in the Eastern Mediterranean (López-Pérez *et al.*, 2013).

Conclusion

Extensive metagenomic surveys on natural populations of *Haloquadratum walsbyi* in the CR30 crystallizer of the Santa Pola saltern led to a number of conclusions. It is now clear that this organism lives in polyclonal populations in which individual clonal lineages differ by less than 5% in their core genome sequence. The variable genomic content of this species is concentrated in regions we call metagenomic islands. The gene content of these regions is not random and is in fact preserved in different aquatic species. In natural populations of *Haloquadratum walsbyi,* metagenomic islands code for genes whose products transport different nutrients and genes whose products are exposed on the cellular surface and could be recognized by predatory viruses. The predator–prey interaction in crystallizer follows the 'Constant diversity' model of dynamics in which a bacterial species is characterized by a large pan-genome (i.e. numerous clonal lineages) that is preserved in space and time as a consequence of viral predation. Although the cell-surface targets populations and the exact mechanisms of viral infection of *Haloquadratum walsbyi* cells are only beginning to be revealed, metagenomic data indicates that similar to host populations, viral assemblages are also composed of individual subpopulations of genetically almost identical viruses, likely specific to individual lineages. This might be a universal feature of marine ecosystems as 'Constant diversity'-like dynamics and variability at the genomic level were recently reported in marine Cyanobacteria and their viruses (Marston *et al.*, 2012; Avrani *et al.*, 2012). The above studies yielded another important aspect that should be addressed in more detail

– the existence of global clones beyond clinical settings. In these terms it would be interesting to gain insight into the evolution of different clonal lineages of this species. The versatile metagenomic studies outlined above demonstrate that this approach can yield answers that rise above a census and metabolic potential of microbial community studied.

References

Acinas, S.G., Klepac-Ceraj, V., Hunt, D.E., Pharino, C., Ceraj, I., Distel, D.L., and Polz, M.F. (2004). Fine-scale phylogenetic architecture of a complex bacterial community. Nature *430*, 551–554.

Allen, E.E., Tyson, G.W., Whitaker, R.J., Detter, J.C., Richardson, P.M., and Banfield, J.F. (2007). Genome dynamics in a natural archaeal population. Proc. Natl. Acad. Sci. U.S.A. *104*, 1883–1888.

Angly, F.E., Felts, B., Breitbart, M., Salamon, P., Edwards, R.A., Carlson, C., Chan, A.M., Haynes, M., Kelley, S., Liu, H., *et al.* (2006). The marine viromes of four oceanic regions. PLoS Biol. *4*, e368.

Antón, J., Llobet-Brossa E., Rodríguez-Valera, F., and Amman, R. (1999). Fluorescence *in situ* hybridization analysis of the prokaryote community inhabiting crystallizer ponds. Environ. Microbiol. *1*, 517–523.

Antón, J., Rosselló-Mora, R., Rodríguez-Valera, F., and Amann, R. (2000). Extremely halophilic bacteria in crystallizer ponds from solar salterns. Appl. Environ. Microbiol. *66*, 3052–3057.

Antón, J., Oren, A., Benlloch, S., Rodríguez-Valera, F., Amman, R., and Rosselló-Mora, R. (2002). *Salinibacter ruber*, gen. nov., sp. nov., a novel extremely halophilic member of the Bacteria from the saltern crystallizer ponds. Int. J. Syst. Evol. Microbiol. *52*, 485–491.

Antonucci, T.K., Landick, R., and Oxender, D.L. (1985). The leucine binding proteins of *Escherichia coli* as models for studying the relationships between protein structure and function. J. Cell. Biochem. *29*, 209–216.

Atanasova, N.S., Roine, E., Oren, A., Bamford, D.H., and Oksanen, H.M. (2011). Global network of specific virus–host interactions in hypersaline environments. Environ. Microbiol. *14*, 426–440.

Atwood, K.C., Schneider, L.K., and Ryan, F.J. (1951). Periodic selection in *Escherichia coli*. Proc. Natl. Acad. Sci. U.S.A. *37*, 146–155.

Avrani, S., Schwartz, D.A., and Lindell, D. (2012). Virus-host swinging party in the oceans: incorporating biological complexity into paradigms of antagonistic coexistence. Mob. Genet. Elements *2*, 88–95.

Bapteste, E., O'Malley, M.A., Beiko, R.G., Ereshefsky, M., Gogarten, J.P., Franklin-Hall, L., Lapointe, F.J., Dupré, J., Dagan, T., Boucher, Y., *et al.* (2009). Prokaryotic evolution and the tree of life are two different things. Biol. Direct *4*, 34.

Barrangou, R., Fremaux, C., Deveau, H., Richards, M., Boyaval, P., Moineau, S., Romero, D.A., and Horvath, P. (2007). CRISPR provides acquired resistance against viruses in prokaryotes. Science *315*, 1709–1712.

Bath, C., Cukalac, T., Porter, K., and Dyall-Smith, M. (2006). His1 and His2 are distantly related, spindle-shaped haloviruses belonging to the novel virus group, Salterprovirus. Virology *350*, 228–239.

Baumeister, W., and Lembcke, G. (1992). Structural features of archaebacterial cell envelopes. J. Bioenerg. Biomembr. *24*, 567–575.

Baumeister, W., Wildhaber, I., and Phipps, B.M. (1989). Principles of organization in eubacterial and archaebacterial surface-proteins. Can. J. Microbiol. *35*, 215–227.

Benlloch, S., Rodríguez-Valera, F., and Martinez-Murcia, A.J. (1995). Bacterial diversity in two coastal lagoons deduced from 16S rDNA PCR amplification and partial sequencing. FEMS Microbiol. Ecol. *18*, 267–279.

Benlloch, S., López-López, A., Casamayor, E.O., Øvreås, L., Goddard, V., Daae, F.L., Smerdon, G., Massana, R., Joint, I., Thingstad, F., *et al.* (2002). Prokaryotic genetic diversity throughout the salinity gradient of a coastal solar saltern. Environ. Microbiol. *4*, 349–360.

Bergthorsson, U., and Ochman, H. (1998). Distribution of chromosome length variation in natural isolates of *Escherichia coli*. Mol. Biol. Evol. *15*, 6–16.

Boetius, A., Ravenschlag, K., Schubert, C.J., Rickert, D., Widdel, F., Gieseke, A., Amman, R., Widdel, F., Jørgensen, B.B., Witte, U., *et al.* (2000). A marine microbial consortium apparently mediating anaerobic oxidation of methane. Nature *407*, 623–626.

Bolhuis, H., te Poele E.M., and Rodríguez-Valera, F. (2004). Isolation and cultivation of Walsby's square archaeon. Environ. Microbiol. *6*, 1287–1291.

Bolhuis, H., Palm, P., Wende, A., Falb, M., Rampp, M., Rodríguez-Valera, F., Pfeiffer, F., and Oesterhelt, D. (2006). The genome of the square archaeon 'Haloquadratum walsbyi': life at the limits of water activity. BMC Genomics. *7*, 169.

Boujelben, I., Gomariz, M., Martínez-García, M., Santos, F., Peña, A., López, C., Antón, J., and Maalej, S. (2012). Spatial and seasonal prokaryotic community dynamics in ponds of increasing salinity of Sfax solar saltern in Tunisia. Antonie van Leeuwenhoek *4*, 845–857.

Bull, J.J., Millstein, J., Orcutt, J., and Wichman, H.A. (2006). Evolutionary feedback mediated through population density, illustrated with viruses in chemostats. Am. Nat. *167*, E39–51.

Burge, C., Campbell, A.M., and Karlin, S. (1992). Over- and under-representation of short oligonucleotides in DNA sequences. Proc. Natl. Acad. Sci. U.S.A. *89*, 1358–1362.

Burns, D.G., Camakaris, H.M., Janssen, P.H., and Dyall-Smith, M.L. (2004a). Cultivation of Walsby's square haloarchaeon. FEMS Microbiol. Lett. *238*, 469–473.

Burns, D.G., Camakaris, H.M., Janssen, P.H., and Dyall-Smith, M. (2004b). Combined use of cultivation-dependent and cultivation-independent methods indicates that members of most haloarchaeal groups in an Australian crystalliser pond are cultivable. Appl. Environ. Microbiol. *70*, 5258–5265.

Burns, D.G., Janssen, P.H., Itoh, T., Kamekura, M., Li, Z., Rodríguez-Valera, F., Bolhuis, H., and Dyall-Smith, M.L. (2007). *Haloquadratum walsbyi* gen. nov., sp. nov., the square haloarchaeon of Walsby, isolated from saltern crystallizers in Australia and Spain. Int. J. Syst. Evol. Microbiol. *57*, 387–392.

Cai, L., Zhao, D., Hou, J., Wu, J., Cai, S., DasSarma, P., and Xiang, H. (2012). Cellular and organellar membrane-associated proteins in haloarchaea: Perspectives on the physiological significance and biotechnological applications. Sci. China Life Sci. *55*, 404–414.

Casamayor, E.O., Massana, R., Benlloch, S., Øvreås., L., Díez, B., Goddard, V.J., Gasol, J.M., Joint, I., Rodríguez-Valera, F., and Pedrós-Alió, C. (2002). Changes in archaeal, bacterial and eukaryal assemblages along a salinity gradient by comparison of genetic fingerprinting methods in a multipond solar saltern. Environ. Microbiol. *4*, 338–348.

Caldwell, S.L., Laidler, J.R., Brewer, E.A., Eberly, J.O., Sandborgh, S.C., and Colwell, F.S. (2008). Anaerobic oxidation of methane: mechanisms, bioenergetics, and the ecology of associated microorganisms. Environ. Sci. Technol. *42*, 6791–6799.

Carbone, A. (2003). Codon adaptation index as a measure of dominating codon bias. Bioinformatics *19*, 2005–2015.

Cardinale, A., and Duffy, S. (2011). Single-stranded genomic architecture constrains optimal codon usage. Bacteriophage *1*, 219–224.

Coleman, M.L., Sullivan, M.B., Martiny, A.C., Steglich, C., Barry, K., DeLong, E.F., and Chisholm, S.W. (2006). Genomic islands and the ecology and evolution of *Prochlorococcus*. Science *311*, 1768–1770.

Cornelissen, A., Ceyssens, P.J., T'Syen, J., Van Praet, H., Noben J.P., Shaburova, O.V., Krylov, V.N., Volckaert, G., and Lavigne, R. (2011). The T7-related *Pseudomonas putida* phage φ15 displays virion-associated biofilm degradation properties. PLoS One *6*, e18597.

Cottrell, M.T., and Kirchman, D.L. (2003). Contribution of major bacterial groups to bacterial biomass production (thymidine and leucine incorporation) in the Delaware estuary. Limnol. Oceanogr. *48*, 168–178.

Cuadros-Orellana, S., Martín-Cuadrado, A.B., Legault, B., D'Auria, G., Zhaxybayeva, O., Papke, R.T., and Rodríguez-Valera, F. (2007). Genomic plasticity in prokaryotes: the case of the square haloarchaeon. ISME J. *1*, 235–245.

Davis, J.S. (1974). Importance of microorganisms in solar salt production. In 4th Symposium on Salt, A.L. Coogan, ed. (Northern Ohio Geological Society, Cleveland), vol. 1, pp. 369–372.

Davidson, A.L., Dassa, E., Orelle, C., and Chen, J. (2008). Structure, function and evolution of bacterial ATP-binding cassette systems. Microbiol. Mol. Biol. Rev. 72, 317–364.

Deveau, H., Barrangou, R., Garneau, J.E., Labonte, J., Fremaux, C., Boyaval, P., Romero, D.A., Horvath, P., and Moineau, S. (2008). Phage response to CRISPR-encoded resistance in *Streptococcus thermophilus*. J. Bacteriol. *190*, 1390–1400.

Díez, B., Antón, J., Guixa-Boixaereu, N., Pedrós-Alió, C., and Rodríguez-Valera, F. (2003). Pulsed-field gel electrophoresis analysis of virus assemblages present in a hypersaline environment. Int. Microbiol. 3, 159–164.

Dobrindt, U., Hochhut, B., Hentschel, U., and Hacker, J. (2004). Genomic islands in pathogenic and environmental microorganisms. Nat. Rev. Microbiol. 2, 414–424.

Dyall-Smith, M., Tang, S.L., and Bath, C. (2003). Haloarchaeal viruses, how diverse are they? Res. Microbiol. *154*, 309–313.

Dyall-Smith, M.L., Pfeiffer, F., Klee, K., Palm, P., Gross, K., Schuster, S.C., Rampp, M., and Oesterhelt, D. (2011). *Haloquadratum walsbyi*: limited diversity in a global pond. PLoS One *6*, e20968.

Egland, P.G., Pelletier, D.A., Dispensa, M., Gibson J., and Harwood, C.S. (1997). A cluster of bacterial genes for anaerobic benzene ring biodegradation. Proc. Natl. Acad. Sci. U.S.A. *94*, 6484–6489.

Elevi Bardavid, R., and Oren, A. (2008). Dihydroxyacetone metabolism in *Salinibacter ruber* and in *Haloquadratum walsbyi*. Extremophiles *12*, 125–131.

Falb, M., Müller, K., Königsmaier, L., Oberwinkler, T., Horn, P., von Gronau, S., Gonzalez, O., Pfeiffer, F., Bornberg-Bauer, E., and Oesterhelt, D. (2008). Metabolism of halophilic archaea. Extremophiles *12*, 177–196.

Fu, H.Y., Chang, Y.N., Jheng, M.J., and Yang, C.S. (2012). Serine 262 determines the chloride-dependent color tuning of a new halorhodopsin from *Haloquadratum walsbyi*. Biosci. Rep. *32*, 501–509.

Fuhrman, J.A., and Noble, R.T. (1995). Viruses and protists cause similar bacterial mortality in coastal seawater. Limnol. Oceanogr. *40*, 1236–1242.

Fukuchi, S., Yoshimune, K., Wakayama M., Moriguchi, M., and Nishikawa, K. (2003). Unique amino acid composition of proteins in halophilic bacteria. J. Mol. Biol. *327*, 347–357.

Garcia-Heredia, I., Martín-Cuadrado, A.B., Mojica, F.J., Santos, F., Mira, A., Antón, J., and Rodríguez-Valera, F. (2012). Reconstructing viral genomes from the environment using fosmid clones: the case of haloviruses. PLoS One *7*, e33802.

García Martín, H., Ivanova, N., Kunin, V., Warnecke, F., Barry, K.W., McHardy, A.C., Yeates, C., He, S., Salamov, A.A., Szeto, E., *et al.* (2006). Metagenomic analysis of two enhanced biological phosphorus removal (EBPR) sludge communities. Nat. Biotechnol. *24*, 1263–1269.

Gasol, J.M., Casamayor, E.O., Joint, I., Garde, K., Gustavson, K., Benlloch, S., Díez, B., Schauer, M., Massana, R., and Pedrós-Alió, C. (2004). Control of heterotrophic prokaryotic abundance and growth rate in hypersaline planktonic environments. Aquat. Microb. Ecol. *34*, 193–206.

Ghai, R., Pašić, L., Fernández, A.B., Martín-Cuadrado, A.B., Mizuno, C.M., McMahon, K.D., Papke, R.T., Stepanauskas, R., Rodríguez-Brito, B., Rohwer, F., *et al.* (2011). New abundant microbial groups in aquatic hypersaline environments. Sci. Rep. *1*, 135.

Gonzaga, A., Martín-Cuadrado, A.B., López-Pérez, M., Mizuno, C.M., García-Heredia, I., Kimes, N.E., Lopez-García, P., Moreira, D., Ussery, D., Zaballos, M., *et al.* (2012). Polyclonality of concurrent natural populations of Alteromonas macleodii. Genome Biol. Evol. *4*, 1360–1374.

Gudbergsdottir, S., Deng, L., Chen, Z., Jensen, J.V., Jensen, L.R., She, Q., and Garrett, R.A. (2011). Dynamic properties of the *Sulfolobus* CRISPR/Cas and CRISPR/Cmr systems when challenged with vector-borne viral and plasmid genes and protospacers. Mol. Microbiol. *79*, 35–49.

Guixa-Boixareu, N., Calderón-Paz, J.I., Heldal, M., Bratbak, G., and Pedrós-Alió, C. (1996). Viral lysis and bacterivory as prokaryotic loss factors along a salinity gradient. Aquat. Microb. Ecol. *11*, 215–227.

Hofbauer, J., and Sigmund, K. (1992). The Theory of Evolution and Dynamical Systems (Cambridge University Press, Cambridge, UK).

Hoffmann, K.H., Rodríguez-Brito, B., Breitbart, M., Bangor, D., Angly, F., Felts, B., Nulton, J., Rohwer, F., and Salamon, P. (2007). Power law rank-abundance models for marine phage communities. FEMS Microbiol. Lett. *273*, 224–228.

Javor, B.J. (2002). Industrial microbiology of solar salt production. J. Ind. Microbiol. Biotechnol. *28*, 42–27.

Joint, I., Henriksen, P., Garde, K., and Riemann, B. (2002). Primary production, nutrient assimilation and microzooplankton grazing along a hypersaline gradient. FEMS Microbiol. Ecol. *39*, 245–257.

Kandiba, L., Aitio, O., Helin, J., Guan, Z., Permi, P., Bamford, D.H., Eichler, J., and Roine, E. (2012). Diversity in prokaryotic glycosylation: an archaeal-derived N-linked glycan contains legionaminic acid. Mol. Microbiol. *84*, 578–593.

Karlin, S. (2001). Detecting anomalous gene clusters and pathogenicity islands in diverse bacterial genomes. Trends Microbiol. *9*, 335–343.

Kessel, M., and Cohen, Y. (1982). Ultrastructure of square bacteria from a brine pool in southern Sinai. J. Bacteriol. *150*, 851–860.

Knittel, K., and Boetius, A. (2009). Anaerobic oxidation of methane: progress with an unknown process. Annu. Rev. Microbiol. *63*, 311–334.

Krupovic, M., Forterre, P., and Bamford, D.H. (2010). Comparative analysis of the mosaic genomes of tailed archaeal viruses and proviruses suggests common themes for virion architecture and assembly with tailed viruses of bacteria. J. Mol. Biol. *397*, 144–160.

Larimer, F.W., Chain, P., Hauser, L., Lamerdin, J., Malfatti, S., Do, L., Land, M.L., Pelletier, D.A., Beatty, J.T., Lang, A.S., *et al.* (2004). Complete genome sequence of the metabolically versatile photosynthetic bacterium *Rhodopseudomonas palustris*. Nat. Biotechnol. *22*, 55–61.

Legault, B.A., López-López, A., Alba-Casado, J.C., Doolittle, W.F., Bolhuis, H., Rodríguez-Valera, F., and Papke, R. (2006). Environmental genomics of "*Haloquadratum walsbyi*" in a saltern crystallizer indicates a large pool of accessory genes in an otherwise coherent species. BMC Genomics *7*, 171.

Lennon, J.T., and Martiny, J.B.H. (2008). Rapid evolution buffers ecosystem impacts of viruses in a microbial food web. Ecol. Lett. *11*, 1178–1188.

Limor-Waisberg, K., Carmi, A., Scherz, A., Pilpel, Y., and Furman, I. (2011). Specialization versus adaptation: two strategies employed by cyanophages to enhance their translation efficiences. Nucleic Acids Res. *39*, 6016–6028.

Lindsay, J.A., and Holden, M.T. (2004). *Staphylococcus aureus*: superbug, super genome? Trends. Microbiol. *12*, 378–385.

Lobasso, S., Lopalco, P., Vitale, R., Saponetti, M.S., Capitanio, G., Mangini, V., Milano, F., Trotta, M., and Corcelli, A. (2012). The light-activated proton pump Bop I of the archaeon *Haloquadratum walsbyi*. Photochem Photobiol. *88*, 690–700.

López-Pérez, M., Gonzaga, A., and Rodriguez-Valera F. (2013). Genomic diversity of 'deep ecotype' *Alteromonas macleodii* isolates: evidence for Pan-Mediterranean clonal frames. Genome Biol. Evol. *5*, 1220–1232.

Malmstrom, R.R., Cottrell, M.T., Elifantz, H., and Kirchman, D.L. (2005). Biomass production and assimilation of dissolved organic matter by SAR11 bacteria in the Northwest Atlantic Ocean. Appl. Environ. Microbiol. *71*, 2979–2986.

Manikandan, M., Kannan, V., and Pašić, L. (2009). Diversity of microorganisms in solar salterns of Tamil Nadu, India. World J. Microbiol. *25*, 1007–1017.

Marraffini, L.A., and Sontheimer, E.J. (2010). CRISPR interference: RNA-directed adaptive immunity in bacteria and archaea. Nat. Rev. Genet. *11*, 181–190.

Marston, M.F., Pierciey, F.J. Jr, Shepard, A., Gearin, G., Qi, J., Yandava, C., Schuster, S.C., Henn, M.R., and Martiny, J.B. (2012). Rapid diversification of coevolving marine *Synechococcus* and a virus. Proc. Natl. Acad. Sci. U.S.A. *109*, 4544–4549.

Maturrano, L., Santos, F., Rosselló-Mora, R., and Antón, J. (2006). Microbial diversity in Maras salterns, a hypersaline enviornment in the Peruvian Andes. Appl. Environ. Microbiol. *72*, 3887–3895.

Medini, D., Donati, C., Tettelin, H., Masignani, V., and Rappuoli, R. (2005). The microbial pan-genome. Curr. Opin. Genet. Dev. *15*, 589–594.

Medini, D., Serruto, D., Parkhill, J., Relman, D.A., Donati, C., Moxon, R., Falkow, S., and Rappuoli, R. (2008). Microbiology in the post-genomic era. Nat. Rev. Microbiol. *6*, 419–430.

Mei, Y., Chen, J., Sun, D., Chen, D., Yang, Y., Shen, P., and Chen, X. (2007). Induction and preliminary characterization of a novel halophage SNJ1 from lysogenic *Natrinema* sp. F5. Can. J. Microbiol. *53*, 1106–1110.

Mescher, M.F., Hansen, U., and Strominger, J.L. (1976). Formation of lipid-linked sugar compounds in *Halobacterium salinarium* – presumed intermediates in glycoprotein synthesis. J. Biol. Chem. *251*, 7289–7294.

Moinier, B. (1999). The appropriate size of saltworks to meet environmental and production requirements. Proceedings of the Post Conference Symposium Saltworks: Preserving Saline Coastal Ecosystems. Global NEST Samos, 49–65.

Mongodin, E.F., Nelson, K.E., Daugherty, S., Deboy, R.T., Wister, J., Khouri, H., Weidman, J., Walsh, D.A., Papke, R.T., Sanchez Perez, G., *et al.* (2005). The genome of *Salinibacter ruber*: convergence and gene exchange among hyperhalophilic bacteria and archaea. Proc. Natl. Acad. Sci. U.S.A. *102*, 18147–18152.

Murai, N., Kamata, H., Nagashima, Y., Yagisawa, H., and Hirata, H. (1995). A novel insertion sequence (IS)-like element of the thermophilic bacterium PS3 promotes expression of the alanine carrier protein-encoding gene. Gene *163*, 103–107.

Narasingarao, P., Podell, S., Ugalde, J.A., Brochier-Armanet, C., Emerson, J.B., Brocks, J.J., Heidelberg, K.B., Banfield, J.F., and Allen, E.E. (2012). De novo metagenomic assembly reveals abundant novel major lineage of Archaea in hypersaline microbial communities. ISME J. *6*, 81–93.

Oh, D., Porter, K., Russ, B., Burns, D.G., and Dyall-Smith, M. (2010). Diversity of *Haloquadratum* and other haloarchaea in three, geographically distant, Australian saltern crystallizer ponds. Extremophiles *14*, 161–169.

Okabe, S., Kindaichi, T., and Ito, T. (2004). MAR-FISH-an ecophysiological approach to link phylogenetic affiliation and *in situ* metabolic activity of microorganism at a single-cell resolution. Microbes Environ. *19*, 83–98.

Oren, A. (1994). Characterization of the halophilic archaeal community in saltern crystallizer ponds by means of polar lipid analysis. Int. J. Salt Lake Res. *3*, 15–29.

Oren, A. (2009) Saltern evaporation ponds as model systems for the study of primary production processes under hypersaline conditions. Aquat. Microb. Ecol. *56*, 193–204.

Oren, A. (2010). Thoughts on the 'missing link' between saltworks biology and solar salt quality. Global NEST J. *12*, 417–425.

Pagaling, E., Haigh, R.D., Grant, W.D., Cowan, D.A., Jones, B.E., Ma, Y., Ventosa, A., and Heaphy, S. (2007). Sequence analysis of an archaeal virus isolated from a hypersaline lake in Inner Mongolia, China. BMC Genomics *8*, 410.

Palenik, B. (1994). Cyanobacterial community structure as seen from RNA polymerase gene sequence analysis. Appl. Environ. Microbiol. *60*, 3212–3219.

Palys, T., Nakamura, L.K., and Cohan, F.M. (1997). Discovery and classification of ecological diversity in the bacterial world: the role of DNA sequence data. Int. J. Syst. Bacteriol. *47*, 1145–1156.

Papke, R.T., Douady, C.J., Doolittle, W.F., and Rodríguez-Valera, F. (2003). Diversity of bacteriorhodopsins in different hypersaline waters from a single Spanish saltern. Environ. Microbiol. *5*, 1039–1045.

Parkes, K., and Walsby, A.E. (1981) Ultrastructure of a gas-vacuolate square bacterium. J. Gen. Microbiol. *126*, 503–506.

Pašić, L., Galan-Bartual, S., Poklar-Ulrih, N., Grabnar, M., and Herzog-Velikonja, B. (2005). Diversity of halophilic archaea in the crystallizers of an Adriatic solar saltern. FEMS Microbiol. Ecol. *54*, 491–498.

Pašić, L., Rodríguez-Mueller, B., Martín-Cuadrado Ana-Belen, Mira, A., Rohwer, F., and Rodríguez-Valera, F. (2009). Metagenomic islands of hyperhalophiles: the case of *Salinibacter ruber*. BMC Genomics *10*, 570.

Paul, S., Bag, S.K., Das, S., Harvill, E.T., and Dutta, C. (2008). Molecular signature of hypersaline adaptation: insights from genome and proteome composition of halophilic prokaryotes. Genome Biol. *9*, R70.

Pedrós-Alió, C. (2006). Marine microbial diversity: can it be determined? Trends Microbiol. *14*, 257–263.

Pedrós-Alió, C., Calderón-Paz, J.I., MacLean, M.H., Medina, G., Marrasé, C., Gasol, J.M., and Guixa-Boixareu, N. (2000). The microbial food web along salinity gradients. FEMS Microbiol. Ecol. *32*, 143–155.

Pietilä, M.K., Laurinavicius, S., Sund, J., Roine, E., and Bamford, D.H. (2010). The single-stranded DNA genome of novel archaeal virus *Halorubrum* pleomorphic virus 1 is enclosed in the envelope decorated with glycoprotein spikes. J. Virol. *84*, 788–798.

Prangishvili, D., Forterre, P., and Garrett, R.A. (2006). Viruses of the Archaea: a unifying view. Nat. Rev. Microbiol. *4*, 837–848.

Pride, D.T., Meinersmann, R.J., Wassenaar, T.M., and Blaser, M.J. (2003). Evolutionary implications of nucleotide usage patterns in prokaryotes. Genome Res. *13*, 145–155.

Pride, D.T., Wassenaar, T.M., Ghose, C., and Blaser, M.J. (2006). Evidence of host-virus co-evolution in tetranucleotide usage patterns of bacteriophages and eukaryotic viruses. BMC Genomics *7*, 8.

de Queiroz, K. (2005). Ernst Mayr and the modern concept of species. Proc. Natl. Acad. Sci. U.S.A. *102*, 6600–6607.

Rinck, A. (2009). Bioinformatical analysis of extraordinarily large, prokaryotic proteins. Master thesis (Technische Universität München, Munich, Germany). Available at: http://www.mosaic.ethz.ch/research/docs/Rinck2009.pdf

Rodríguez-Brito, B., Li, L., Wegley, L., Furlan, M., Angly, F., Breitbart, M., Buchanan, J., Desnues, C., Dinsdale, E., Edwards, R., *et al.* (2010). Viral and microbial community dynamics in four aquatic environments. ISME J. *4*, 739–751.

Rodríguez-Valera, F., Ventosa, A., Juez, G., and Imhoff, J.F. (1985). Variation of environmental features and microbial populations with salt concentrations in a multi-pond saltern. Microb. Ecol. *11*, 107–115.

Rodríguez-Valera, F., Martín-Cuadrado, A.B., Rodríguez-Brito, B., Pašić, L., Thingstad, T.F., Rohwer, F., and Mira, A. (2009). Explaining microbial population genomics through phage predation. Nat. Rev. Microbiol. *7*, 828–836.

Rosselló-Mora, R., Lee, N., Antón, J., and Wagner, M. (2003) Substrate uptake in extremely halophilic microbial communities revealed by microautoradiography and fluorescence *in situ* hybridization. Extremophiles *7*, 409–413.

Saier, M.H. Jr, Paulsen, I.T., Sliwinski, M.K., Pao, S.S., Skurray, R.A., and Nikaido, H. (1998). Evolutionary origins of multidrug and drug-specific efflux pumps in bacteria. FASEB J. *12*, 265–274.

Sandaa, R.A., Foss Skjoldal, E., and Bratbak, G. (2003). Virioplankton community structure along a salinity gradient in a solar saltern. Extremophiles *7*, 347–351.

Sano, E., Carlson, S., Wegley, L., and Rohwer, F. (2004). Movement of viruses between biomes. Appl. Environ. Microbiol. *70*, 5842–5846.

Santos, F., Meyerdierks, A., Peña, A., Rosselló-Mora, R., Amann, R., and Antón, J. (2007). Metagenomic approach to the study of halophages: the environmental halophage 1. Environ. Microbiol. *9*, 1711–1723.

Santos, F., Yarza, P., Parro, V., Briones, C., and Antón, J. (2010). The metavirome of a hypersaline environment. Environ. Microbiol. *12*, 2965–2976.

Santos, F., Moreno-Paz, M., Meseguer, I., López, C., Rosselló-Mora, R., Parro, V., and Antón, J. (2011). Metatranscriptomic analysis of extremely halophilic viral communities. ISME J. *5*, 1621–1633.

Schauer, R. (2000). Achievements and challenges of sialic acid research. Glycoconj. J. *17*, 485–499.

Schleper, C., DeLong, E.F., Preston, C.M., Feldman, R.A., Wu, K.-Y., and Swanson, R.V. (1998). Genomic analysis reveals chromosomal variation in natural populations of the uncultured psychrophilic archaeon *Cenarchaeum symbiosum*. J. Bacteriol. *180*, 5003–5009.

Senčilo, A., Jacobs-Sera D., Russel, D.A., Ko, C.C., Bowman, C.A., Atanasova, N.S., Osterlund, E., Oksanen, H.M., Bamford, D.H., Hatfull, G.F., *et al.* (2013). Snapshot of haloarchaeal tailed virus genomes. RNA Biol. *10*, 803–816.

Sharma, A.K., Walsh, D.A., Bapteste, E., Rodríguez-Valera, F., Doolittle, W.F., and Papke, R.T. (2007). Evolution of rhodopsin ion pumps in haloarchaea. BMC Evol. Biol. *7*, 79.

Sharma, R.S., Mishra, V., Mohmmed, A., and Babu, C.R. (2008). Phage specificity and lipopolysaccarides of stem- and root-nodulating bacteria (*Azorhizobium caulinodans, Sinorhizobium* spp., and *Rhizobium* spp.) of *Sesbania* spp. Arch. Microbiol. *189*, 411–418.

Sharp, P.M., and Li, W.H. (1987). The codon adaptation index – a measure of directional synonymous codon usage bias, and its potential applications. Nucleic Acids Res. *15*, 1281–1295.

Sher, J., Elevi, R., Mana, L., and Oren, A. (2004). Glycerol metabolism in the extremely halophilic bacterium *Salinibacter ruber*. FEMS Microbiol. Lett. *19*, 211–215.

Sleytr, U.B., and Thorne, K.J. (1976). Chemical characterization of the regularly arranged surface layers of *Clostridium thermosaccharolyticum* and *Clostridium thermohydrosulfuricum*. J. Bacteriol. *126*, 377–383.

Sleytr, U.B., Messner, P., Pum, D., and Sára, M. (1999). Crystalline bacterial cell surface layers (S-layers): from supraMol. Cell structure to biomimetics and nanotechnology. Angew. Chem. Int. Ed. *38*, 1034–1054.

Stoddard, L.I., Martiny, J.B.H., and Marston, M.F. (2007). Selection and characterization of cyanophage resistance in marine *Synechococcus* strains. Appl. Environ. Microbiol. *73*, 5516–5522.

Stoeckenius, W. (1981). Walsby's square bacterium: fine structure of an orthogonal procaryote. J. Bacteriol. *148*, 352–360.

Sublimi-Saponetti, M., Bobba, F., Salerno, G., Scarfato, A., Corcelli, A., and Cucolo, A. (2011). Morphological and structural aspects of the extremely halophilic archaeon *Haloquadratum walsbyi*. PLoS One *6*, e18653.

Sudo, Y., Ihara, K., Kobayashi, S., Suzuki, D., Irieda, H., Kikukawa, T., Kandori, H., and Homma, M. (2011). A microbial rhodopsin with a unique retinal composition shows both sensory rhodopsin II and bacteriorhodopsin-like properties. J. Biol. Chem. *286*, 5967–5976.

Suttle, C.A. (1994). The significance of viruses to mortality in aquatic microbial communities. Microbial Ecol. *28*, 237–243.

Suttle, C.A. (2007). Marine viruses – major players in the global ecosystem. Nat. Rev. Microbiol. *5*, 801–812.

Tettelin, H., Masignani, V., Cieslewicz, M.J., Donati, C., Medini, D., Ward, N.L., Angiuoli, S.V., Crabtree, J., Jones, A.L., Durkin, A.S., *et al.* (2005). Genome analysis of multiple pathogenic isolates of *Streptococcus agalactiae*: implications for the microbial 'pan-genome'. Proc. Natl. Acad. Sci. U.S.A. *102*, 13950–13955.

Thingstad, T.F. (2000). Elements of a theory for the mechanisms controlling abundance, diversity, and biogeochemical role of lytic bacterial viruses in aquatic systems. Limnol. Oceanogr. *45*, 1320–1328.

Thingstad, T.F., and Lignell, R. (1997). Theoretical models for the control of bacterial growth rate, abundance, diversity and carbon demand. Aquat. Microb. Ecol. *13*, 19–27.

Thompson, J.R., Pacocha, S., Chanathip, P., Klepac-Ceraj, V., Hunt, D.E., Benoit, J., Sarma-Rupavtarm, R., Distel, D.L., and Polz, M.F. (2005). Genotypic diversity within a natural coastal bacterioplankton population. Science *307*, 1311–1313.

Thurber, R.V., Haynes, M., Breitbart M., Wegley, L., and Rohwer, F. (2009). Laboratory procedures to generate viral metagenomes. Nat. Protoc. *4*, 470–483.

Torrella, F. (1986). Isolation and adaptive strategies of Haloarculae to extreme hypersaline habitats. Abstracts 4th Internat. Symp. Microbiol. Ecol. p. *59*, C19–3.

Torsvik, V., Øvreås, L., and Thingstad, T.F. (2002). Prokaryotic diversity – magnitude, dynamics, and controlling factors. Science *296*, 1064–1066.

Touchon, M., and Rocha, E. (2010). The small, slow and specialized CRISPR and anti-CRISPR of *Escherichia* and *Salmonella*. PLoS One *5*, e11126.

Trigui, H., Masmoudi, S., Brochier-Armanet, C., Barani, A., Grégori, G., Denis, M., Dukan, S., and Maalej, S. (2011). Characterization of heterotrophic prokaryote subgroups in the Sfax coastal solar salterns by combining flow cytometry cell sorting and phylogenetic analysis. Extremophiles *15*, 347–358.

Tyson, G.W., Chapman, J., Hugenholtz, P., Allen, E.E., Ram, R.J., Richardson, P.M., Solovyev, V.V., Rubin, E.M., Rokhsar, D.S., and Banfield, J.F. (2004). Community structure and metabolism through reconstruction of microbial genomes from the environment. Nature *428*, 37–43.

Venter, J.C., Remington, K., Heidelberg, J.F., Halpern, A.L., Rusch, D., Eisen, J.A., Wu, D., Paulsen, I., Nelson, K.E., Nelson, W., *et al.* (2004). Environmental genome shotgun sequencing of the Sargasso Sea. Science *304*, 66–74.

Walker, C.B., Stolyar, S., Chivian, D., Pinel, N., Gabster, J.A., Dehal, P.S., He, Z., Yang, Z.K., Yen, H.C., Zhou, J., *et al.* (2009). Contribution of mobile genetic elements to *Desulfovibrio vulgaris* genome plasticity. Environ. Microbiol. *11*, 2244–2252.

Walsby, A.E. (1980). A square bacterium. Nature *283*, 69–71.

Walsby, A.E. (2005). Archaea with square cells. Trends Microbiol. *13*, 193–195.

Warkentin, M., Schumann, R., and Oren, A. (2009). Community respiration studies in saltern crystallizer ponds. Aquat. Microb. Ecol. *56*, 255–261.

Weinbauer, M.G., and Rassoulzadegan, F. (2004). Are viruses driving microbial diversification and diversity? Environ. Microbiol. *6*, 1–11.

Welch, R.A., Burland, V., Plunkett, G., Redford, P., Roesch, P., Rasko, D., Buckles, E.L., Liou, S.R., Boutin, A., Hackett, J., *et al.* (2002). Extensive mosaic structure revealed by the complete genome sequence of uropathogenic *Escherichia coli*. Proc. Natl. Acad. Sci. U.S.A. *99*, 17020–17024.

Whitaker, R.J., Grogan, D.W., and Taylor, J.W. (2003). Geographic barriers isolate endemic populations of hyperthermophilic Archaea. Science *301*, 976–978.

Wilcox, R.M., and Fuhrman, J.A. (1994). Bacterial viruses in coastal seawater: lytic rather than lysogenic production. Mar. Ecol. Prog. Ser. *114*, 35–45.

Wren, J.D., Roossinck, M.J., Nelson, R.S., Scheets, K., Palmer, M.W., and Melcher, U. (2006). Plant virus biodiversity and ecology. PLoS Biol. *4*, 314–315.

Xu, Y., Wang, Z., Xue, Y., Zhou, P., Ma, Y., Ventosa, A., and Grant, W.D. (2001). *Natrialba hulunbeirensis* sp. nov. and *Natrialba chahannaoensis* sp. nov., novel haloalkaliphilic archaea from soda lakes in Inner Mongolia Autonomous Region, China. Int. J. Syst. Evol. Microbiol. *51*, 1693–1698.

Zafrilla, B., Martínez-Espinosa, R.M., Alonso, M.A., and Bonete, M.J. (2010). Biodiversity of Archaea and floral of two inland saltern ecosystems in the Alto Vinalopó Valley, Spain. Saline Syst. *6*, 10.

Salinibacter ruber: The Never Ending Microdiversity? 2

Arantxa Peña, María Gomariz, Marianna Lucio, Pedro González-Torres, Jaime Huertas-Cepa, Manuel Martínez-García, Fernando Santos, Phillippe Schmitt-Kopplin, Toni Gabaldón, Ramon Rosselló-Móra and Josefa Antón

Abstract

Salinibacter ruber is an extremely halophilic bacterium of the Bacteroidetes phylum that thrives in hypersaline environments. This bacterium shares the environment, as well as many phenotypic traits, with extremely halophilic Archaea. The study of the wide collection of strains of *S. ruber* isolated from around the world has shown that the species is very homogeneous from a phylogenetic point of view although it shows a very wide genomic microdiversity. In this chapter, we provide stat-of-the art data on abundance, distribution, metabolomic and genomic microdiversity of *S. ruber* and discuss the contribution of recombination and lateral gene transfer to the shaping of this species.

Introduction

Salinibacter ruber was discovered serendipitously when trying to link molecular and microscopy data drawn from hypersaline environments. On one side, light and electron microscopy analysis of crystallizers indicated the conspicuous presence of Walsby's square 'bacteria' while 16S rRNA gene clone libraries from these environments (Rodríguez-Valera *et al.*, 1999) repeatedly retrieved a sequence, then known as 'Susana's phylotype' SPhT (Benlloch *et al.*, 1995), that was only distantly related to the closest cultured haloarchaea. The obvious question arising from these two observations was whether the squares corresponded to SPhT, and the obvious way to answer it was by using fluorescence *in situ* hybridization (FISH) with SPhT-specific probes. Among the different FISH controls, Archaea and Bacteria specific probes were used. Unexpectedly, Bacteria probes hybridized with a relatively high number of rods present in the hypersaline samples. DGGE and 16S rRNA gene clone libraries unveiled the presence of two very close bacterial phylotypes (EHB-1 and EHB-2, from 'Extremely Halophilic Bacteria'). Extremely halophilic bacteria corresponding to phylotype EHB-1 were soon isolated from crystallizer water samples and led to the description of the species *S. ruber*. Shortly after the discovery of *Salinibacter* it became evident that it was widespread in hypersaline environments. Studies carried out during these years finally allowed the isolation of the square archaeon, named as *Haloquadratum walsbyi*, as well as considerable advances in the characterization of its genome and metagenome in different saline systems (Bolhuis *et al.*, 2004; Burns *et al.*, 2004, Cuadros Orellana *et al.*, 2007; Dyall-Smith *et al.*, 2011; Legault *et al.*, 2006). The description of crystallizers as 'monospecific cultures of Archaea', that preceded the discovery of *Salinibacter* and *Haloquadratum*, was substituted by (a rather simplistic) description according to which crystallizer were dominated by the square Archaea and *Salinibacter ruber*.

Abundance and distribution revisited

Although indeed *Haloquadratum* and *Salinibacter* are the dominant Archaea and Bacteria, respectively, in crystallizer ponds in many coastal solar salterns at some points of the year, there are relevant deviations to this pattern. The application of new techniques and the study of new hypersaline environments have allowed a more precise description of their microbiota (Antón *et al.*, 2012). For instance, *Haloquadratum* is not always the clearly dominant archaeon, such in the Maras salterns discussed below, or can even be outnumbered by *Halorubrum* spp. as in crystallizer ponds from some Adriatic salterns (Pašić *et al.*, 2005). The Maras salterns (Maturrano *et al.*, 2006a), located at 3500 m above sea level in the Peruvian Andes, also harbour a peculiar bacterial community, dominated by gammaproteobacteria now classified as *Salicola marasensis* (Maturrano *et al.*, 2006b).

The microbiota of Turkish Tuz Lake, which harbours one of the highest proportions of Bacteria reported for close-to-saturation environments, also has some unique traits (Mutlu *et al.*, 2008), mainly regarding the bacterial assemblage. While *Haloquadratum* phylotypes retrieved from the lake were very closely related to those repeatedly found in salterns around the world (Australia, Peruvian Andes and both sides of the Mediterranean), the *Salinibacter* assemblage was quite site specific. Indeed, only a small fraction of the sequences could be affiliated to *S. ruber* and a 'Tuz Lake *Salinibacter* spp. group' was found to be clearly different from the previously *Salinibacter* described phylotypes EHB-1 and EHB-2. In the Tunisian coastal solar salterns of Sfax (Baati *et al.*, 2008, 2010, 2011; Boujelben *et al.*, 2012) Bacteroidetes was also the most frequently retrieved bacterial group, but they included a high representation of new sequences not previously found in nature, of which many are distantly related to *S. ruber*.

The use of new approaches such as metagenomics and single cell isolation followed by amplification and sequencing of the individual genomes (i.e. SCG or single cell genomics) has also contributed to the knowledge of crystallizer microbiota, the discovery of the low GC content Nanohaloarchaea (Antón *et al.*, 2012; Ghai *et al.*, 2011; Narasingarao *et al.*, 2011) being an outstanding example. Very recently (Zhaxybayeva *et al.*, 2013) SCG has been used to compare the microbiota of crystallizers from the South Bay Salt Works in Chula Vista in California (USA) and the Santa Pola salterns (the same CR30 where *S. ruber* was first detected in 1998). The authors concluded that although there was some overlap in community composition, such as the presence in both of Nanohaloarchaea, both communities were significantly different. Most specifically, the bacterial assemblage in Chula Vista consisted of diverse Bacteroidetes and Proteobacteria while in Santa Pola the bacterial fraction consisted of the previously known *Salinibacter* lineages. The authors argued 'that observed differences in community composition reflect restricted dispersal between sites, a likely mechanism for diversification of halophilic microorganisms'.

Thus, although there are hypersaline environments where *Salinibacter* and relatives have not been detected, during these last years many sequences affiliating with the phylum Bacteroidetes have been retrieved from hypersaline environments around the world, both thalassohaline and athalassohaline (for a review see Antón *et al.*, 2008).

Salinibacter and other Bacteroidetes seem to be thus the 'bacterial masters' although, so far, only a few extremely halophilic Bacteroidetes have been brought into pure culture: *Salisaeta longa* (Vaisman *et al.*, 2009), *S. ruber*, and two new species of *Salinibacter* (*S. luteus* and *S. iranicus*) isolated from an Iranian salt lake (Makhdoumi-Kakhki *et al.*, 2012). *S. ruber* and *S. longa* can be differentiated based on their morphology, 16S rRNA gene sequences and salt optimum needed for growth, among other characteristics. However, a 16S rRNA primer originally designed as specific for *S. ruber* (Antón *et al.*, 2002) is no longer specific. Thus, there is an urgent need for re-designing specific probes for *Salinibacter* and Bacteroidetes in hypersaline environments.

In addition to the differences that can be found among different solar salterns, the microbiota of a given pond can undergo seasonal changes. This has been observed using a metagenomic approach in San Diego salterns (Rodríguez-Brito *et al.*, 2010), that were analysed by taking samples at various time points separated by one day to more than one year. The authors found that most abundant microbial 'species' persisted over time but underwent changes in their relative abundances and a variation of their microdiversity.

Fig. 2.1 shows a summary of the presence/absence of different phylotypes related to *Salinibacter* and other Bacteroidetes in five ponds from a multipond solar saltern in Santa Pola (SE Spain), an environment that has been extensively studied during the years (Antón *et al.*, 2000, 2005, 2012; Benlloch *et al.*, 1996, 2001, 2002; Casamayor *et al.*, 2002; Rodriguez-Valera *et al.*, 1999; Oren and Rodriguez Valera, 2001; Peña *et al.*, 2005, 2010). The data shown in Fig. 2.1 were obtained in the course of an annual study (Gomariz *et al.*, submitted) that included the sampling of the ponds at nine different time points. The five ponds included three adjacent concentrators (CM1, CM2 and CCAB) of increasing salinity through which the saline water was circulating to finally feed two different crystallizer ponds (CR30 and CR41), both fed with water from the same concentrator (CCAB). The microbiota was analysed by different 16S rRNA gene targeting techniques that unveiled

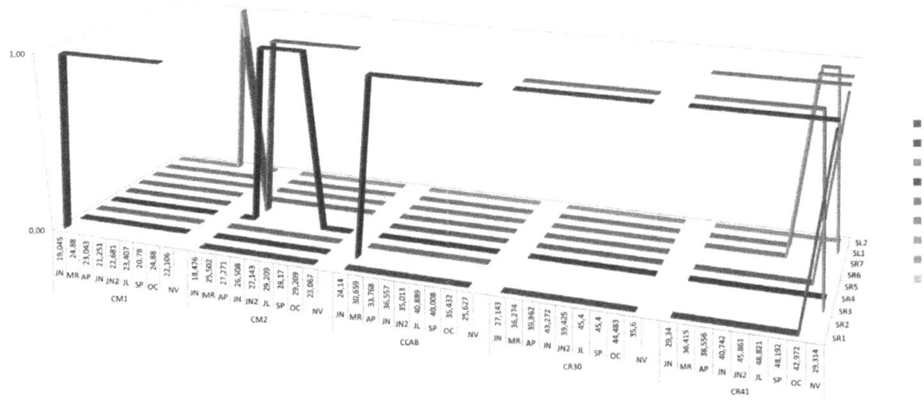

Figure 2.1 Presence (1 in the vertical axis) or absence (0) of several Bacteroidetes phylotypes in five different ponds (CM1, CM2, CCAB, CR30, and CR41) of Santa Pola solar salterns at nine time points during year 2009. For every pond and sampling date, the corresponding salinity values (numbers above months labelling) expressed in% are indicated. Phylotypes related to *Salinibacter ruber* and to *Salisaeta longa* are labelled, respectively, as SR and SL. A colour version of this figure is located in the plate section at the back of the book.

the presence of nine different phylotypes belonging to the Bacteroidetes showing different levels of similarity to *S. ruber* or *S. longa* (labelled in Fig. 2.1 as SR and SL, respectively). SR1 was the phylotype including a higher number of sequences and had a similarity above 99% with *S. ruber* (corresponding to the originally discovered EHB-1). The rest of SR phylotypes included less sequences and were only distantly related to *S. ruber* (always less than 93%) and in some cases matched very closely environmental clones form Tunisian salterns (Baati *et al.*, 2010). Phylotypes SL1 and 2 had similarities of around 93% with *S. longa*.

As shown in Fig. 2.1, some of these phylotypes can be detected only at some time points, underscoring the relevance of temporal sampling. In addition, it is also noteworthy that two crystallizer ponds fed with the same water can harbour different *Salinibacter* assemblages as shown by phylotypes SR1 and SR7. Furthermore, phylotype fluctuation changes from pond to pond: while CM ponds underwent rather drastic changes, crystallizer CR30 was the most stable pond, which is worth mentioning considering that most studies on Santa Pola salterns were carried out with samples taken from this very pond (Antón *et al.*, 2000, 2002; Rodríguez-Valera *et al.*, 1999; Santos *et al.*, 2007, 2010, 2011, 2012). Multivariate analyses have shown that seasonal variations in the prokaryotic community harboured in the different pond ponds are correlated both with changes in the ionic composition of the water and environmental factors (Gomariz *et al.*, submitted). Remarkably, crystallizer CR30 was the pond in which *S. ruber* was first detected. At that time (summer 1998), only two bacterial phylotypes (EHB-1 and EHB-2) were detected in this pond by DGGE and 16S rRNA gene clone libraries, although our current data indicates a wider diversity.

Microdiversity and recombination

One of the most noticeable traits of *S. ruber* isolates is their homogeneity at the 16S rRNA gene level. Indeed, all the strains isolated so far belong to the phylotype EHB-1, most of them with very high similarities, often very close to 100%. However, genomic comparisons using pulsed-field gel electrophoresis (PFGE) have shown a relatively high genomic microdiversity, according to the high variability of their genomic restriction patterns (Peña *et al.*, 2005). Throughout the following discussion we must keep in mind that different PFGE restriction patterns do not always imply different gene content, since intra-chromosomic rearrangements could also alter the relative position among the restriction enzyme targets, thus affecting the corresponding restriction pattern.

New PFGE patterns are observed in every new round of *S. ruber* isolations from water samples while in no single instance have the patterns of the original strains (used for the species description) been retrieved again. Fig. 2.2 shows the PFGE patterns obtained for some of the strains included in Table 2.1 and the resulting dendrogram obtained from the data of presence/absence of PFGE bands in the different patterns (modified from Antón *et al.*, 2013). Although all the genomes are resolved using the same electrophoretic conditions, which indicates their similarity, each strain has a distinct restriction pattern. This has always been the observed trend for *S. ruber* isolates, with the exception of some strains isolated from Peruvian salterns in which this bacterium is below FISH detection limit, and the PFGE of the isolates were identical (Antón *et al.*, 2005). This intraspecific diversity has also been found when comparing *S. ruber* plasmid patterns (Peña *et al.*, unpublished results).

Although the initial comparisons of PFGE patterns rendered some biogeographic clustering of the strains (Peña *et al.*, 2005), this trend was lost when a larger collection of strains was

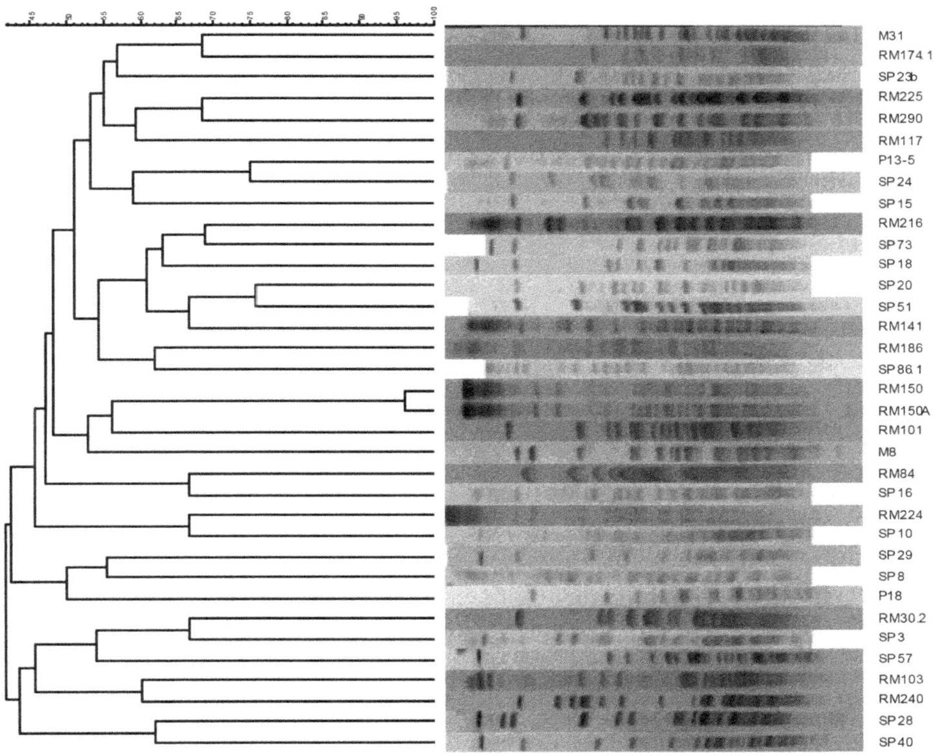

Figure 2.2 Similarity dendrogram (left) comparing the analysed *Salinibacter ruber* strains based on unweighted pair group method and arithmetric mean analysis of their *Xba*I genomic restriction products separated by pulsed-field gel electrophoresis (right).

analysed. In addition, and contrary to observations for fluorescent pseudomonads (Cho and Tiedje, 2000), internal transcribed spacer (ITS) sequences were not suitable for studying biogeographical segregation due to their high sequence similarity (Peña *et al.*, 2005). Thus, in order to investigate the biogeographical differences among *S. ruber* strains, a multilocus sequence analysis (MLSA) was undertaken (Rosselló-Mora *et al.*, 2008). This approach had proven successful for thermophilic Archaea (Whitaker *et al.*, 2003) and extremely halophilic Archaea of the genus *Halorubrum* (Papke *et al.*, 2004; Papke and Gogarten, 2012). Ten strains (labelled with crosses in Table 2.1) were selected and for each of them the 16S rRNA gene plus seven additional housekeeping genes were sequenced. These seven housekeeping genes had been previously shown to be phylogenetically informative (Soria-Carrasco *et al.*, 2007) and were: *eno* F (coding for an enolase), the gene coding for the ribosomal protein S5, *gap* (coding for glyceraldehyde-3-phosphate dehydrogenase), *pyr*G (coding for CTP synthase), *rps*C (coding for ribosomal protein S3), *tuf* (coding for a translation elongation factor G), and *gly*A (coding for a serine hydroxymethyl transferase). The concatenated DNA stretch rendered an alignment of 7995 homologous sites. Phylogenetic analyses were performed by including and excluding indels, and 16S rRNA genes. Altogether, the reconstructions (Fig. 2.3A) did not show clear geographical segregation of the selected strains, in contrast to observations made with other extremophiles (Whitaker *et al.*, 2003).

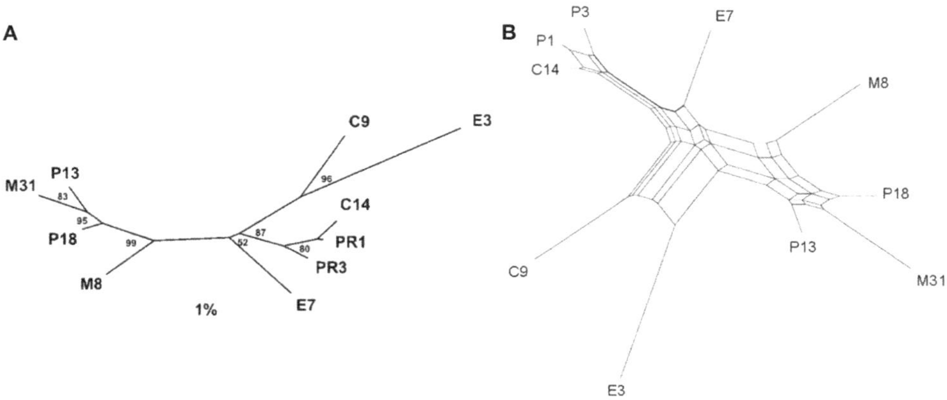

Figure 2.3 (A) Phylogenetic reconstruction based on a PHYML algorithm of a 7995 nucleotide alignment corresponding to eight housekeeping genes, including SSU rRNA. The bar indicates 1% sequence divergence. From Rosselló-Mora *et al.* (2008). (B) Split decomposition of the data shown on the left panel.

Phylogenetic trees for the ten strains were also reconstructed individually for each of the seven analysed genes (some examples are shown in Fig. 2.4). Such phylogenies turned out to be incongruent, which would be expected if recombination had occurred during the divergence of the strains (Papke *et al.*, 2004). To explore further the likelihood of recombination as a force driving the evolution of these *S. ruber* strains, a split decomposition was carried out (Huson, 2003). This method can be used to assess the support of the data for a strictly bifurcating tree (such as the one shown in Fig. 2.3A). When a split decomposition produces a network with many pairs of incompatible splits (in a split network, every edge is associated with a split of the taxa, with a number of parallel edges associated with each split) there is an indication that the dataset contains conflicting signals (that could be due to recombination). On the other hand, in the absence of these conflicting signals, the split-decomposition would produce a fairly resolved tree (Huson, 2003; Huson and Bryant, 2006). As shown in Fig. 2.3B, the analysis of the eight housekeeping genes in the ten analysed *S. ruber* strains produced a network rather than a clearly resolved tree, which, together with the incongruent phylogenies derived from the individual genes, supports the role of recombination as driving force in the evolution of the species. Whether this force is driving diversification or homogenization remains an open question since these two different outcomes can derive from the action of recombination (Papke *et al.*, 2007).

It has been suggested that recombination would be enhanced in extremophile environments, hypersaline and thermophilic, due to high-intensity UV light that would induce expression of the recombination machinery (Papke *et al.*, 2007) and of type IV pili (Ajon *et al.*, 2011). Two different approaches have been used to study recombination in extremophiles: whole genome comparison, such in the work with *Sulfolobus islandicus* (Cadillo-Quiroz *et al.*, 2012) and multilocus strain analysis (MLSA) of a large collection of strains, such in *Halorubrum* species (Papke *et al.*, 2004, 2007). The last, which share their habitat with *Salinibacter*, have shown such a high degree of homologous recombination that they would be close to panmixis (Papke *et al.*, 2004, 2007).

In 2010 we undertook the genomic comparison of two very closely related *S. ruber* strains

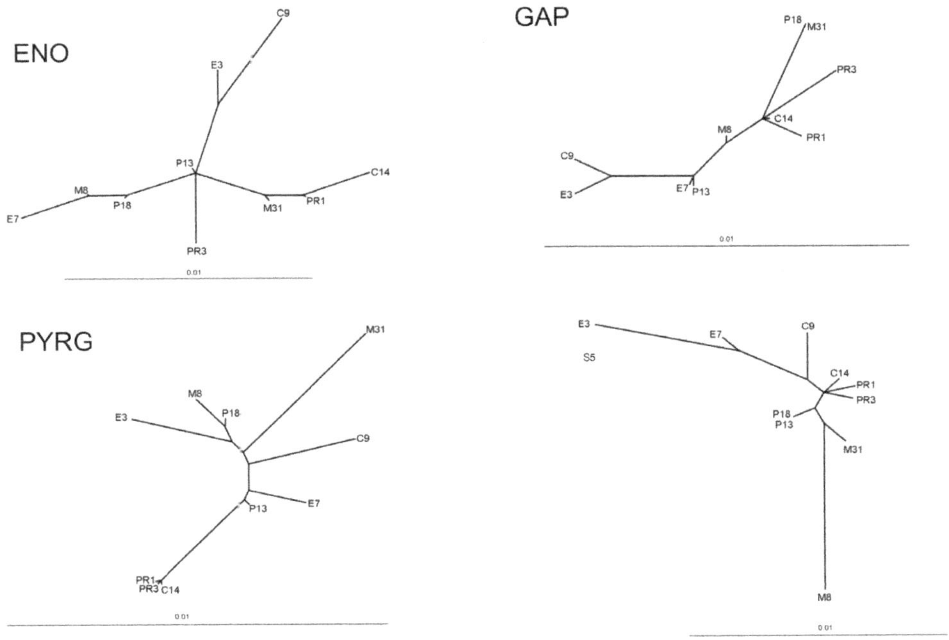

Figure 2.4 Phylogenetic trees of four of the eight genes used for MLSA of ten *S. ruber* strains isolated from different locations.

(M8 and M31, both simultaneously isolated from Mallorca salterns in 1999 and used for the species description) in order to elucidate the extent of the genomic differences among these strains, and explore the ecological consequences of these differences. The results of this study have been extensively discussed elsewhere (Peña *et al.*, 2010, 2011) although there are some issues relevant in the context of recombination that are worth presenting here. The M8 and M31 genomes, as expected, showed a considerable degree of synteny although there were some regions of the genome where differences among the two strains accumulated. These regions were enriched in transposases, genes related to surface properties, strain-specific genes and highly divergent orthologous. Overall, differences between M8 and M31 included strain-specific genes (around 10% of the total), insertions and deletions outside of the hypervariable regions (HVRs), divergence among homologues, and specific gene duplications in every genome.

Inter-domain (Archaea–Bacteria) lateral gene transfer: archaeal genes in *Salinibacter ruber* genomes

Salinibacter and extremely halophilic Archaea have a high phenotypic similarity (summarized in Oren, 2013). In the words of Professor Oren 'phenotypically *Salinibacter* is thus a prokaryote with a 'split personality': phylogenetically it belongs to the Bacteroidetes, but physiologically it behaves like the extremely halophilic Archaea'. This phenotypic similarity, together with the fact that *S. ruber* shares its habitat with extremely halophilic Archaea, suggested that *S. ruber* was a good candidate for intra-domain lateral gene transfer (LGT).

The analysis of the *S. ruber* M31 genome suggested that this was indeed the case, although the number of genes likely involved in LGT was found to be lower than expected in view of the high phenotypic similarities (Mongodin *et al.*, 2005). During the analysis of strain M8 genome, we increased the stringency of these analyses and found 40 candidate LGT genes (Peña *et al.*, 2005).

The 40 (Table 2.2) genes identified as putatively involved in inter-domain transfers had GC contents and tetranucleotide frequencies that did not identify them as deviated from the background values in the genome. In addition, most of them had high codon adaptation index values, suggesting that they were transcribed at average or high levels compared to the rest of the genes. A recent transcriptomic analysis of M8 and M31 genomes indicated that all these genes were indeed expressed in M8 and M31 during normal growth (González-Torres *et al.*, in preparation). In addition, 34 of these genes were part of the core genome shared by the strains M8 and M31, for which a detailed genomic comparison was carried out (Peña *et al.*, 2010). Thus, these putative LGT genes would have been incorporated from Archaea into the *S. ruber* genome at a very early stage in the evolution of the species. This hypothesis was also supported by the analysis of intergenic regions adjacent to the putative LGT genes. According to Hooper *et al.*, 2009, conservation of these regions would be indicating that these genes were transferred from a common ancestor. Similarities of intergenic regions for LGT genes present in both strains M8 and M31 was above 90% in 37 of the 40 cases, which indicated that these genes probably did not derive from a recent transfer. Among the LGT genes (Table 2.2) there was an overrepresentation of gene products involved in ion transport, rhodopsin-related proteins and signal transduction systems. The genes were scattered along the M8 chromosome, most of them located outside both variable and conserved regions (Peña *et al.*, 2010).

A collection of 92 *S. ruber* strains from different origins (Table 2.1) was checked for the presence of these genes to ascertain to which extent they conformed part of the species core genome when isolates from different geographical origins were tested. Three groups of strains were used: (i) our old collection of 35 strains isolated from Mediterranean, Peruvian and Atlantic salterns, that included the isolates used for the species description (Antón *et al.*, 2002) as well as the ones used for the first microdiversity (Peña *et al.*, 2005) and metabolomic (Rosselló-Mora *et al.*, 2008) studies carried out with *Salinibacter*, (ii) a set of 22 co-occurring strains isolated in 2006 from a Mallorca crystallizer and (iii) a set of 35 co-occurring strains isolated in 2008 from a Santa Pola crystallizer. These two last groups of strains, together with some control strains from the 'old' collection, have been recently analysed by a metabolomic approach (Antón *et al.*, submitted), as discussed below. Putative LGT were searched for by using PCR specific primers (Peña *et al.*, 2010, 2011) designed based on the M8 genomic sequence. They were checked with strain M31 and all the results were as expected according to its genome sequence. A given strain was considered to harbour a given putative LGT when a PCR product of the expected size was obtained. Obviously, a LGT which had undergone changes in the primer target sequenced would not be detected by this approach.

Using this approach, 13 out of 40 putative LGT genes were present in all strains. For RM and SP sets, the number of LGT genes present in all the co-occurring strains was 17 and 16, respectively. Both strain sets shared 16 common genes (the 13 genes present in all strains plus three conserved hypothetical proteins).

Table 2.1 *Salinibacter ruber* strains analysed from 1999 to 2013

Geographical area	Strains	Place/date of isolation	Analysed by	References
Mediterranean	M1, M8,[a,b] M31[a,b]	Campos salterns, Mallorca, 1999	PFGE, LGT, metabolomics, genomic, MLST, phylogeny	Antón *et al.* (2002), Peña *et al.* (2005, 2010), Rosselló-Mora *et al.* (2008), Soria-Carrasco *et al.* (2008), Antón *et al.*, (2013), this work
	RM30, RM84, RM101, RM103, RM117, RM129, RM131, RM141, RM150, RM158, RM159, RM172, RM174, RM179, RM186, RM216, RM224, RM225, RM240, RM272, RM20	Campos salterns, Mallorca, 2006	PFGE, metabolomics, LGT	Antón *et al.*, (2013), this work
	Pola 13 [a,b] Pola 18[a,b]	Bras del Port salterns, Alicante, 1999	PFGE, MLST, LGT, metabolomics	Antón *et al.* (2002), Peña *et al.* (2005), Rosselló-Mora *et al.* (2008), Antón *et al.*, (2013), this work
	SP3, SP7, SP8, SP10, SP11, SP15–24, SP26, SP28–30, SP32, SP35, SP36, SP38–40, SP51, SP57, SP73, SP79, SP84, SP86, SP99, SP100	Bras del Port salterns, Alicante, 2007	PFGE, MLST, LGT, metabolomics	Antón *et al.*, (2013), this work
	E1,[a] E3,[a,b] E7,[a,b] E11, E12[a]	San Carles de la Ràpita, Tarragona, 2001	PFGE, MLST, metabolomics, LGT	Peña *et al.* (2005), Rosselló-Mora *et al.* (2008), Antón *et al.*, (2013), this work
	IL3[a]	Ses Salines, Ibiza, 2001	PFGE, MLST, LGT, metabolomics	Peña *et al.* (2005), Rosselló-Mora *et al.* (2008), Antón *et al.*, (2013), this work
	A1	S'Avall salterns, Mallorca, 2001	PFGE, MLST, LGT	Peña *et al.* (2005)
	ES4[a]	Eilat saltern, Israel, 2001	PFGE, MLST, LGT	Peña *et al.* (2005), Rosselló-Mora *et al.* (2008)
Atlantic	C3,[a] C4,[a] C5, C6,[a] C7, C9,[a,b] C12,[a] C14,[a,b] C15,[a] C16, C17,[a] C18, C22,[a] C24, C25A,[a] C26,[a] C27,[a] C28, C29[a]	Gran Canaria, Canary Islands, 2001	PFGE, MLST, LGT, metabolomics	Peña *et al.* (2005), Rosselló-Mora *et al.* (2008), Antón *et al.*, (2013), this work
Peruvian	PR1,[a,b] PR2, PR3,[a,b] PR4, PR6,[a] PR8[a]	Maras salterns, Peru, 2001/2003	PFGE, MLST, LGT, metabolomics	Peña *et al.* (2005), Rosselló-Mora *et al.* (2008), Antón *et al.*, (2013), this work

[a]Strains analysed in Peña *et al.* (2005). [b]Strains analysed by MLST in Rosselló-Mora *et al.* (2008).
PFGE, pulsed-field gel electrophoresis; LGT, lateral gene transfer; MLST, multilocus sequence typing.

Table 2.2 Genes putatively in Archaea–Bacteria lateral gene transfer present in the *Salinibacter ruber* M8 genome

	Gene name	Orf in M8	% strains[a]
1	Signal-transducing histidine kinase	94	65
2	**Putative light-and oxygen-sensing transcription regulator**	**150**	**100**
3	Adaptive-response sensory-kinase	169	42
4	Conserved hypothetical protein containing PIN domain (putative nucleic acid-binding protein)	226	3,3
5	Conserved hypothetical protein containing DUF0175 domain	227	50
6	Conserved hypothetical protein	274	99
7	Cytochrome c oxidase subunit II	393	61
8	Deoxycytidine triphosphate deaminase	538	82
9	Glycosyl transferase, group 1	655	33
10	Conserved hypothetical protein containing PIN domain (putative nucleic acid-binding protein)	831	61
11	Probable exodeoxyribonuclease VII small subunit	911	82
12	Ferredoxin	917	74
13	**Conserved hypothetical protein, membrane**	**918**	**100**
14	**Probable beta-carotene 15,15′-dioxygenase Brp**	**919**	**100**
15	Phytoene dehydrogenase	920	98
16	Probable cell division protein	964	80
17	**Translation initiation factor eIF-2B alpha subunit**	**995**	**100**
18	Conserved hypothetical protein, secreted	1011	64
19	Uncharacterized ACR	1308	55
20	Trk system potassium uptake protein trkA	1444	15
21	**d-lactate dehydrogenase, putative**	**1448**	**100**
22	Conserved hypothetical protein	1707	88
23	Putative zinc-binding dehydrogenase	1709	85
24	**Anion permease**	**1859**	**100**
25	**Phosphate transporter**	**1860**	**100**
26	Conserved hypothetical protein	1903	83
27	Multiantimicrobial extrusion protein MatE	1924	100
28	Sodium/calcium exchanger protein, membrane	2026	93
29	Conserved hypothetical protein containing UPF0047	2049	62
30	Bacterio-opsin-linked product	2365	89
31	Haem exporter protein C	2415	65
32	**Trk potassium uptake system protein**	**2493**	**100**
33	**Glycerol-3-phosphate dehydrogenase subunit A**	**2657**	**100**
34	**Conserved hypothetical protein, membrane**	**2720**	**100**
35	Sensory rhodopsin I (SR-I)	2733	68
36	**Putative small solutes transporter**	**2780**	**100**
37	Sensory rhodopsin I (SR-I)	2798	76
38	**Sodium:proline symporter**	**2993**	100

Table 2.2 continued

	Gene name	Orf in M8	% strains[a]
39	Halorhodopsin	2994	67
40	Conserved hypothetical protein	3079	38

[a]Percentage of the analysed strains containing each gene. Genes in bold are those present in all of the strains. Shadowed in grey are the lateral gene transfer candidates 'specific' of M8 (when compared with M31, Peña *et al.*, 2010).

The *S. ruber* genome harbours four rhodopsins: the proton pump xanthorhodopsin (XR), the chloride pump halorhodopsin (HR) and two sensorial rhodopsins (SR) (Mongodin *et al.*, 2005; Peña *et al.*, 2005). Three of them (HR and SRs) have a clear archaeal phylogenetic signal. In addition, two bacteriorhodopsin related genes (encoded by orfs 919 and 2365) are included in the archaeal gene list (Table 2.2). Thus, five putative retinal-binding proteins have an archaeal origin according to our phylogenetic approach (trees are shown in Fig. 2.5). Out of these five phototransduction archaeal genes, only one is conserved in all the analysed genomes (a bacteriorhodopsin related protein probable beta-carotene 15,15'-dioxygenase). The rest show a patchy distribution in the analysed strains, in a similar way of what has been observed in haloarchaea (Sharma *et al.*, 2007). *Salinibacter* HR is the only known (putative) light-driven chloride pump present in the domain Bacteria. In a previous analysis (Peña *et al.*, 2005) we looked for HR genes in the genomes of *S. ruber* strains included in what we call here the 'old' collection using a Southern hybridization approach, and we found a clustering of HR-containing strains according to the place of isolation. Indeed, when the same strains were checked for HR genes by PCR, we found the same results. However, the inclusion of new strains presented here shows that strains isolated from the same place at the same time

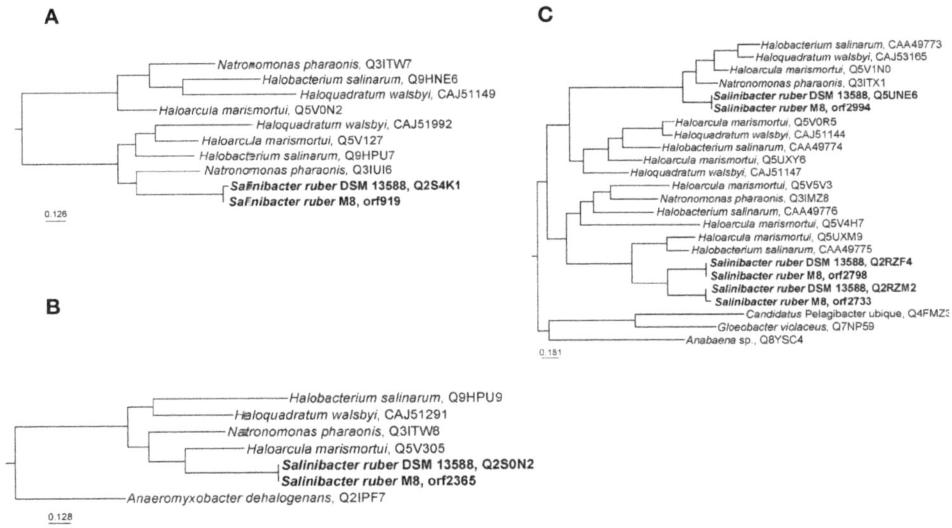

Figure 2.5 Phylogenetic trees of selected examples of putative lateral gene transfer genes present in the genome of strains *S. ruber* M8 and M31 (DSM13588).

can show different patterns. For instance, 4 out of 22 and 7 out of 35 new Mallorca and Santa Pola strains, respectively, did not harbour HR in their genomes. This again, resembles the situation found in some species of haloarchaea such as *Halorubrum* spp. (Papke *et al.*, 2007), in which not all members have the same patterns of genes coding for retinal-binding proteins.

One of the reasons of undertaking the analysis of LGT in the strain collections was to check for biogeographic trends. Hypersaline environments are patchily distributed on Earth, in an inland-like fashion. As pointed out in Zhaxybayeva *et al.* (2013), 'the evidence for endemic microbial populations in isolated environments suggests that allopatric speciation might be common there'. So far, the use of genetic markers, such as MLSA shown above, has not unveiled clear cut grouping of geographically related *S. ruber* strains, although metabolomic analysis showed some trends (see below).

When the presence/absence patterns of LGT genes were computed by calculating similarity matrices and the corresponding dendrograms, no clustering related with the origin of isolation was found. However, a correspondence analysis unveiled some relationships between these patterns and the origin of the strains. Correspondence analysis (CA) is a multivariate ordination method that summarizes a set of data in a two-dimensional biplot, treating rows (genes) and columns (strains) equivalently. This technique has been used in microbial ecology as a way to address, given a multidimensional table of sites (or samples) by species, whether certain species occur at specific sites (Ramette, 2007). The overall aim of the method is to compare the correspondence between samples (strains in our case) and species (here, LGT genes) from a table of counted data (or any dimensionally homogenous table) and to represent it in a reduced ordination space. Noticeably, instead of maximizing the amount of variance explained by the ordination, CA maximizes the correspondence between genes scores and strains scores.

As shown in Fig. 2.6, CA showed a distribution of strains related to their origin of isolation, which indicated a similar history of the LGT genes carried in the genomes of the geographically related strains. In a similar way of what metabolomic analysis of the old collection showed (discussed below), Atlantic strains (those isolated from the Canary Islands) showed an intermediate position between Mediterranean (those from Santa Pola, Mallorca and Ebro) and Peruvian strains. In addition, there was a striking differentiation between strains isolated from the same place at different time points: (M1, M8 and M31 isolated in 1999 and RM strains isolated in 2006; P13 and P18 and SP strains). Although, in both cases, the number of old strains is much lower than the new strains, old strains (labelled with arrows in Fig. 2.6) always sit in the outside boundaries of the space delimiting the new strains. This is particularly noteworthy, for instance, for strain M31, which is the type strain of the species. These results indicate not only the high genomic microdiversity of the species, but also the temporal genomic variation it undergoes even for a relatively short period of time (less than ten years).

Some of the putative LGT genes (shadowed in grey in Table 2.2) analysed in the strain collection were originally labelled as specific of strain M8 because they were present in M8 and absent from M31. These genes cannot longer be considered M8 specific since all of them were present at least in one additional *S. ruber* strain. In some instances (orfs 274 and 911), these genes were indeed very widespread in the collection, which could be indicating their ancestral nature. These data, together with the patchiness of the presence of LGT genes in the collection, point to recombination as a strong driver of *S. ruber* evolution.

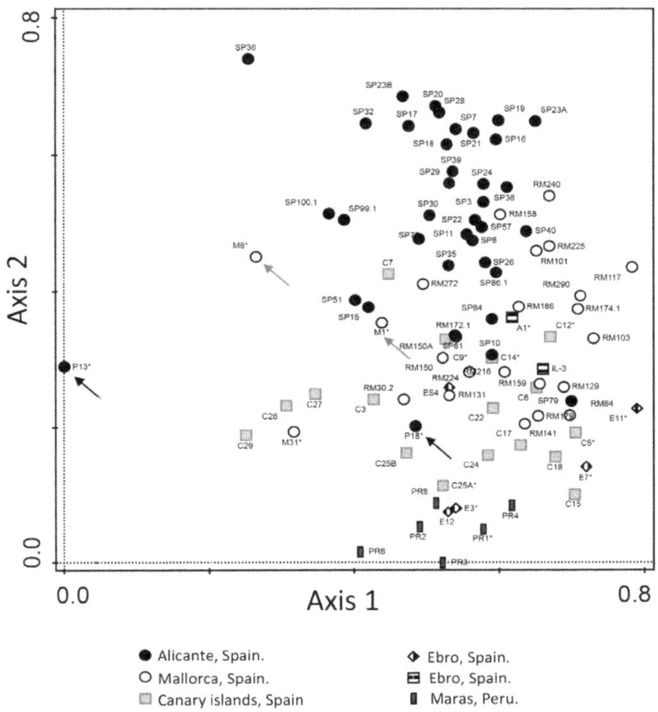

Figure 2.6 Redundancy analysis of the data on absence/presence of 40 lateral gene transfer genes in the genomes of 92 *S. ruber* strains. Arrows point to strains isolated in 1999 from Santa Pola (black circles) and Mallorca (white circles) salterns.

The power of metabolomics to explore phenotypic microdiversity

As mentioned above, the MLSA of *S. ruber* strains isolated from three different geographical locations (Mediterranean, Atlantic and Peruvian salterns) did not render any geographical pattern, contrary to what had been observed in other extremophilic organisms. For this reason, we investigated the suitability of high resolution mass spectrometry as a tool to differentiate among strains of different origin (Rosselló-Mora *et al.*, 2008). As explained below, this approach turned out to be very useful in the characterization of the species *S. ruber*.

The technique chosen for studying *S. ruber* was high-field ion cyclotron resonance Fourier transform mass spectrometry (12 T ICR-FT/MS) that enables the assignment of thousands of elemental compositions of metabolites in a mass range from 120 to 800 kDa directly out of complex mixtures. This technique provides ultrahigh resolved profiles with thousands of accurate mass values (m/z) that can be transformed into real elementary compositions. This assignment is the crucial initial step in subsequent metabolite annotation, that is carried out using different databases (such as KEGG, Kyoto Encyclopedia of Genes and Genomes database), in a similar way as genomic annotation. This technique, that is becoming very outstanding in 'metabolomics', has shown to have the highest resolution among all spectrometric methods in revealing fine-scale diversity in complex mixtures (Marshall, 2004; Want *et al.*, 2007).

The idea was to look for patterns of geographically isolated organisms at the level of the direct interaction with the environment (phenotype) rather than at the genotype level. In this study, we used a collection of 28 strains isolated from five different locations in the world (see Table 2.1). The isolates were obtained from five different locations in the Mediterranean area, the Atlantic Canary archipelago and from the 3500-m-high salterns in the Peruvian Andes (Maras). Ten of these isolates were the ones used for the MLSA discussed above. For each strain, three fractions were analysed: extracellular, cellular soluble (cytoplasm) and cellular insoluble (membrane and envelopes). The first step was to screen for the metabolites present in every sample using a non-targeted approach in order to look for their amount and diversity. Then, in a second step, a targeted approach was undertaken to obtain a detailed analysis of specific metabolites following a specific chemical structure hypothesis (for instance, the different sulfonolipids present in each samples were analysed).

The core metabolome (peaks common for all extracts) consisted of 2550 single masses, whereas the discriminative metabolome (peaks not common to all extracts) consisted of 6323 single metabolites. Comparative analyses were then carried out with qualitative data coded as presence or absence of single metabolites and expressed in a binary matrix. Neither cladistic (parsimony) nor phenetic (UPGMA) treatment of the data revealed clear geographical trends. However, when the relative intensity of each individual peak was weighted and the data treated by using multivariate analyses, statistically significant differences were found between the different samples. In other words, clear-cut allopatric differences for the analysed strains could be obtained by metabolome comparisons focusing on geographically discriminative data. A detailed analysis of the discriminative metabolomes indicated that those components generally associated with cell membranes, such as fatty acids and terpenoids, could be responsible for the geographic differentiation among the strains. Furthermore, it seemed that the Atlantic isolates had a position in the 'metabolomic space' that was intermediate between Mediterranean and Peruvian strains.

The targeted approach unveiled a set of conspicuous compounds belonging to a sulfonolipid family whose members are major components of the cell envelope of *Cytophaga* (Godchaux and Leadbetter, 1984), a genus included in the Bacteroidetes phylum (like *Salinibacter*). Indeed, one of the members of this sulfonolipid family accounts for 10% of *S. ruber* lipids and has been proposed as signature for this bacterium (Corcelli *et al.*, 2004). Our metabolomic study indicated that *S. ruber* contained at least nine additional sulfonolipids that differed from the sulfonolipid originally described in their elementary composition, with variations in their side chain length, unsaturation or hydroxylation degree. Two of these sulfonolipids seemed to be exclusive of the Atlantic strains.

During the course of this metabolomic study a comparison at smaller scale between four replicates of five Mediterranean strains (P13 and P18 from Alicante, M8 and M31 from Mallorca and IL3 from Ibiza) was also undertaken. Results indicated that the ICR-FT/MS approach was also useful for finding discriminative phenotypes at a more reduced geographical scale. The main discriminative metabolomic features between these Mediterranean strains were different from those giving resolution at a larger geographical scale and were involved primarily in the core metabolism (carbohydrate, amino acid and fatty acid biosynthesis and metabolism).

Thus, ICR-FT/MS not only was useful for finding biogeographical traits but also for exploring phenotypic differences between closely related strains. We took advantage of this fact in order to correlate the genomic differences found between strains M8 and M31 with their phenotypic traits (Peña *et al.*, 2010). As one of the most apparent differences between both genomes was the abundance of genes coding for sulfotransferases and glycosyltransferases, the metabolomes of M8 and M31 were compared for the presence of sulfonated/S-containing and glycosylated metabolites. Results indicated a consistent increase in sulfonated and glycosylated metabolites in M8 compared with M31, increase that was considerably higher in the extracellular fraction of the cultures. Thus, metabolomics proved also useful for understanding the genomic differences between these two very closely related strains. The metabolomic differences found were mostly related to molecules released to the medium or loosely attached to the cell surface, which are very good candidates for mediating interaction with phages.

In order to explore whether this metabolomic diversity was a general trend in the species and as a proxy of its 'meta-metabolome' in the environment, recently (Antón *et al.*, submitted) we have undertaken a detailed metabolomic analysis of the strains included in the new collections of isolates from Santa Pola and Mallorca. In total, 22 Mallorca and 35 Santa Pola isolates were analysed together with a set of old isolates used as controls (M8, M3, P13, P18, and IL3; see Table 2.1). In parallel to the metabolomic study, strains were characterized by means of two well-established taxonomic tools: MALDI-TOF and PFGE. All these strains were included in the LGT analysis discussed above.

The first outcome of the metabolic characterization of the 57 new strains and the five controls was an amazingly high diversity of metabolites in the three analysed fractions (extracellular, cytoplasmatic and cellular insoluble). Most outstanding was the metabolomic diversity in the extracellular fraction since it was so high that practically every strain behaved in a different manner. Among the thousands of metabolites that could be identified by ICR-FT/MS, only from 10 to 20%, depending on the considered fraction, could be annotated to previously known metabolites and only a small fraction of them could be assigned to metabolic pathways, underscoring the width of the unexplored metabolomic diversity in the species. This low number of annotable metabolites is reflecting the limitations of databases, in a similar way to what was the case for genome annotations 20 years ago. Such a wide metabolomic diversity has also been observed by means of nanospray desorption electrospray ionization mass spectrometry when sampling living colonies of different bacteria, including *Bacillus subtilis, Streptomyces coelicolor, Mycobacterium smegmatis* and *Pseudomonas aeruginosa*. (Waltrous *et al.*, 2012). As pointed out by Traxler and Kolter (2012), 'the chemical landscape inhabited and manipulated by bacteria is vastly more complex and sophisticated that previously thought'.

Strains from Mallorca and Santa Pola were analysed in two independent experiments, both of which included the same controls. In both cases, old (control) strains presented homogeneity in their profiles that distinguished them from the rest of strains. In other words, the old strains isolated from a given pond (either in the Santa Pola or the Mallorca salterns) resembled more the old strains from different origin than the new strains isolated from the same pond eight years later. One could argue that old and new strains can have different genomes (as indeed we have shown using the LGT approach shown above) but, in any case, there is a consistent cluster of old strains based on their metabolomic profiles and independent of the place and date of isolation. This is most likely pointing to a metabolomic shift that may result from the storage under lab conditions for years.

The third main outcome from this ICR-FT/MS analysis was that strains co-occurring in a same water sample had different metabolomic profiles (Fig. 2.7) that allowed the clustering

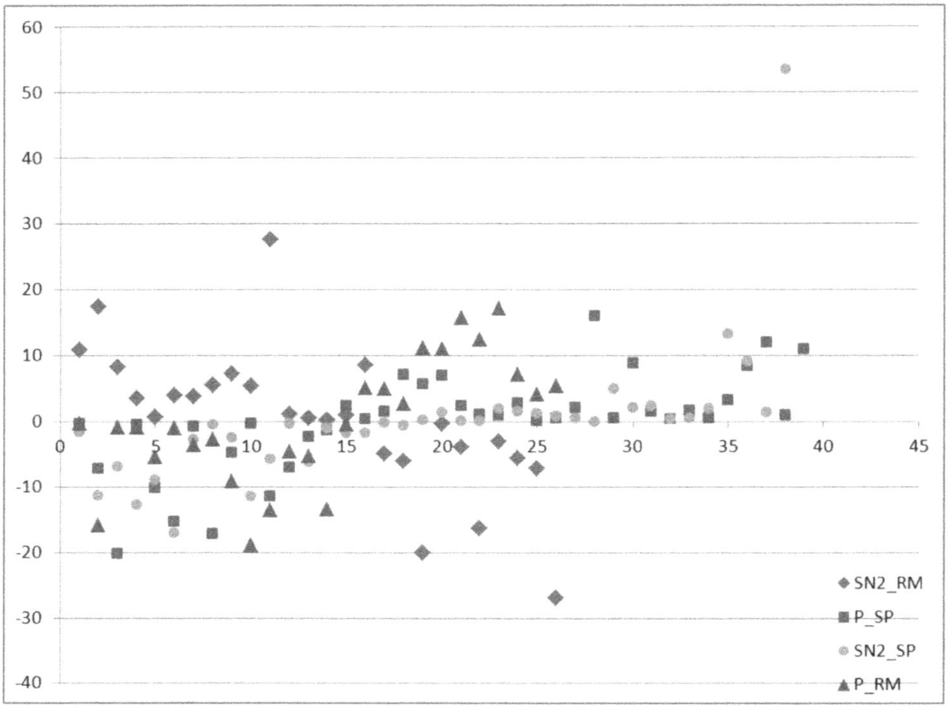

Figure 2.7 Comparison of intracellular (SN2) and pellet (P) samples for the 22 Mallorca (RM) and the 35 Santa Pola (SP) strains. Each dot represent a sample. The closer two dots are, the more similar the corresponding samples within the experiment. For instance the datasets with the highest variability are SN2_RM and P_RM.

of the isolates into statistically supported groups. We called these clusters '*metabotype* based OTUs': or *m*-OTUs. As with the overall metabolome, a large proportion (from 79 to 93%) of the metabolites responsible for the differentiation among the strains could not be annotated with the current databases. Among the annotable metabolites, only 22–36% could be assigned to metabolic pathways.

Both for Mallorca and Santa Pola strains, the metabolic classes including the highest number of metabolites were, in this order, lipid metabolism, metabolism of other amino acids, biosynthesis of other secondary metabolites and metabolism of terpenoids and polyketides. Within the lipid metabolism class, the pathways ether lipid metabolism, fatty acid biosynthesis and sphingolipid metabolism accumulated the highest number of metabolites responsible for the differences among strain clusters. Lipid metabolism seemed in fact to be the most versatile metabolic network in *S. ruber*. It is also worth mentioning the high numbers of metabolites responsible for differences among *m*-OTUs belonging to the biosynthetic pathways of antibiotics such as novobiocin, puromycin and penicillin, and cephalosporin, included in the class 'Biosynthesis of other secondary metabolites'.

Finally, we took an even closer look at the metabolomic diversity within *S. ruber* by looking at the metabolomic profiles of five Mallorca strains included within a given *m*-OTU. These strains could not be distinguished by statistical tools but still displayed different metabolic profiles. For instance, there were 88 annotable metabolites that were present in one of the strains but absent form the rest. One-third of these metabolites were lipids (the most variable pool in *S. ruber* metabolome) and the rest encompassed a wide variety of compounds, including antibiotic related products and alkaloid-analogues among others.

All these examples illustrate the power of high-resolution metabolomics to resolve differences between *S. ruber* strains and, in some cases, to correlate them with environmental data. Metabolomics is definitely a powerful tool for exploring intraspecific diversity, as already observed when studying different strains of the myxobacterium *Myxococcus xanthus* (Krug *et al.*, 2008). Furthermore, this technique has allowed an insight into the wide metabolomic diversity of bacteria in nature, a diversity that can change even among extremely closely related strains, a fact that should be taking into account when looking from new bioactive substances.

Future trends

In some of our previous reviews on *Salinibacter* (Antón *et al.*, 2005, 2008; Peña *et al.*, 2011) several points were included in our 'future research' sections, from the description of intra-specific genomic diversity to understanding the relationships between phages and prokaryotes in hypersaline environments. While some of these questions have been answered, other remain open and new questions have enlarged the list.

One of the new questions is how many different *S. ruber* (or any other bacterium) are present in a given environmental sample small enough to allow the interaction among them. Here, the key point is how to define 'different'. Different strains would be, for instance, the ones used for the LGT and metabolomics analyses discussed above. They all come from a small sample of water (0.1 ml) but they are isolates and therefore could not be representing the diversity encountered in nature for members of the species. Thus, ideally, all the individual *S. ruber* present in a sample should be separated and their genomes sequenced. Single cell technologies followed by amplification and sequence of the individual genomes could then provide the answer to our question. However, this would require the separation of all the members of the species present in the sample (independently or their nucleic acid content) and the correct amplification of all the whole individual genomes, which at this moment is not technically feasible. On the other hand, metagenomics can not answer to this question since the environmental metagenome of a given organisms is the average of the individual genomes but does not differentiate between them.

A third line of future research will need to address the interactions between *S. ruber* and the 'living entities' with which it shares the environment: Bacteria, Archaea and viruses. While we have made some progress understanding the role of phages in controlling *S. ruber* numbers (Santos *et al.*, 2010, 2011, 2012), we would like to know the role of viruses in shaping *S. ruber* genomes, both as vectors of LGT and selective forces. Regarding the interaction among different co-occurring strain of *S. ruber* and between *S. ruber* and extremely halophilic Archaea there are many questions that we find central to understand how bacteria interact in nature: Do different strains of a given species interact among them in nature? Do they perceive the differences among them? Do they react accordingly or they just behave as if the were in a pure culture? Are Archaea directly affecting the growth of *Salinibacter*? Which are the mechanisms or interaction? Our goal for the next future years of research is to answer these questions.

Web resources

As result of our studies, we produced a number of phylogenetic trees that were compiled in the form of full-genome catalogues of individual gene evolutionary histories (phylomes). *S. ruber* M8 and M31 phylomes are publicly available through the phylomeDB database (Huerta-Cepas *et al.*, 2011), under the identifiers 20 and 21 respectively. *Salinibacter* phylomes at phylomeDB provide precomputed maximum likelihood trees, multiple sequence alignments and automatic orthology and paralogy predictions for every gene in their genomes. In addition, PhylomeDB interface allows precomputed phylogenies and alignments to be searched and iteratively explored.

References

Ajon, M., Fröls, S., van Wolferen, M., Stoecker, K., Teichmann, D., Driessen, A.J., Grogan, D.W., Albers, S.V., and Schleper, C. (2011). UV-inducible DNA exchange in hyperthermophilic archaea mediated by type IV pili. Mol. Microbiol. *82*, 807–817.

Antón, J., Oren, A., Benlloch, S., Rodríguez-Valera, F., Amann, R., and Rosselló-Mora, R. (2002). *Salinibacter ruber* gen. nov., sp. nov., a novel, extreme halophilic member of the Bacteria from saltern crystallizer ponds. Int. J. Syst. Evol. Microbiol. *52*, 485–491.

Antón, J., Peña, A., Valens, M., Santos, F., Glöckner, F.-O., Bauer, M., Dopazo, J., Herrero, J., Rosselló-Mora, R., and Amann, R. (2005). *Salinibacter ruber*: genomics and biogeography. In Adaptation to Life at High Salt Concentrations in Archaea, Bacteria, and Eukarya, Gunde-Cimerman, N., Oren, A., Plemenitaš, A., eds. (Springer, New York), pp. 255–266.

Antón, J., Peña, A., Santos, F., Martínez-García, M., Schmitt-Kopplin, P., and Rosselló-Mora, R. (2008). Distribution, abundance and diversity of the extremely halophilic bacterium *Salinibacter ruber*. Saline Syst. *4*, 15.

Antón, J., Martínez-García, M., and Santos, F. (2012). Metagenomics of brine pools. In Encyclopedia of Metagenomics, Nelson, K., ed. (Springer, New York), Available at: http://www.springerreference.com/docs/html/chapterdbid/303289.html

Antón, J., Lucio M., Peña, A., Cifuentes, A., Brito-Echeverría, J., Moritz, F., Tziotis, D., López, C., Urdiain, M., Schmitt-Kopplin, P., *et al.* (2013). High metabolomic microdiversity within co-occurring isolates of the extremely halophilic bacterium *Salinibacter ruber*. PLoS One *8*, 5.

Baati, H., Guermazi, S., Amdouni, R., Gharsallah, N., Sghir, A., and Ammar, E. (2008). Prokaryotic diversity of a Tunisian multipond solar saltern. Extremophiles *12*, 505–518.

Baati, H., Guermazi, S., Gharsallah, N., Sghir, A., and Ammar, E. (2010). Microbial community of salt crystals processed from Mediterranean seawater based on 16S rRNA analysis. Can. J. Microbiol. *56*, 44–51.

Baati, H., Jarboui, R., Gharsallah, N., Sghir, A., and Ammar, E. (2011). Molecular community analysis of magnesium-rich bittern brine recovered from a Tunisian solar saltern. Can. J. Microbiol. *57*, 975–981.

Benlloch, S., Martínez-Murcia, A.J., and Rodríguez-Valera, F. (1995). Sequencing of bacterial and archaeal 16S rRNA genes directly amplified from a hypersaline environment. Syst. Appl. Microbiol. *18*, 574–581.

Benlloch, S., Acinas, S.G., Martínez-Murcia, A.J., and Rodríguez-Valera, F. (1996). Description of prokaryotic biodiversity along the salinity gradient of a multipond solar saltern by direct PCR amplification of 16S rDNA. Hydrobiologia *329*, 19–31.

Benlloch, S., Acinas, S.G., Antón, J., López-López, A., Luz, S.P., and Rodríguez-Valera, F. (2001). Archaeal biodiversity in crystallizer ponds from a solar saltern: culture versus PCR. Microb. Ecol. *41*, 12–19.

Benlloch, S., López-López, A., Casamayor, E.O., Øvreås, L., Goddard, V., Daae, F.L., Smerdon, G., Massana, R., Joint, I., Thingstad, F., *et al.* (2002). Prokaryotic genetic diversity throughout the salinity gradient of a coastal solar saltern. Environ. Microbiol. *4*, 349–360.

Bolhuis, H., te Poele, E.M., and Rodriguez-Valera, F. (2004). Isolation and cultivation of Walsby's square archaeon. Environ. Microbiol. *6*, 1287–1291.

Boujelben, I., Gomariz, M., Martínez-García, M., Santos, F., Peña, A., López, C., Antón, J., and Maalej, S. (2012). Spatial and seasonal prokaryotic community dynamics in ponds of increasing salinity of Sfax solar saltern in Tunisia. Antonie van Leeuwenhoek *101*, 845–857.

Burns, D.G., Camakaris, H.M., Janssen, P.H., and Dyall-Smith, M.L. (2004). Cultivation of Walsby´s square haloarchaeon. FEMS Microbiol. Lett. *238*, 469–473.

Cadillo-Quiroz, H., Didelot, X., Held, N.L., Herrera, A., Darling, A., Reno, M.., Krause, D.J., and Whitaker, R.J. (2012). Patterns of gene flow define species of thermophilic Archaea. PLoS Biol. *10*, e1001265.

Casamayor, E.O., Massana, R., Benlloch, S., Øvreås, L., Díez, B., Goddard, V.J., Gasol, J.M., Joint, I., Rodríguez-Valera, F., and Pedrós-Alió, C. (2002). Changes in archaeal, bacterial and eukaryal assemblages along a salinity gradient by comparison of genetic fingerprinting methods in a multipond solar saltern. Environ. Microbiol. *4*, 338–348.

Cho, J.C., and Tiedje, J.M. (2000). Biogeography and degree of endemicity of fluorescent *Pseudomonas* strains in soil. Appl. Environ. Microbiol. *66*, 5448–5456.

Corcelli, A., Lattanzio, V.M.T., Mascolo, G., Babudri, F., Oren, A., and Kates, M. (2004). Novel sulfonolipid in the extremely halophilic bacterium *Salinibacter ruber*. Appl. Environ. Microbiol. *70*, 6678–6685.

Cuadros-Orellana, S., Martin-Cuadrado, A.-B., Legault, B., D'Auria, G., Zhaxybayeva, O., Papke, R.T., and Rodriguez-Valera, F. (2007). Genomic plasticity in prokaryotes: the case of the square haloarchaeon. ISME J. *1*, 235–245.

Dyall-Smith, M.L., Pfeiffer, F., Klee, K., Palm, P., Gross, K., Schuster, S.C., Rampp, M., and Oesterhelt, D. (2011). *Haloquadratum walsbyi*: Limited diversity in a global pond. PLoS One *6*, e20968.

Ghai, R., Pašic, L., Fernández, A.B., Martin-Cuadrado, A.-B., Mizuno, C.M., McMahon, K.D., Papke, R.T., Stepanauskas, R., Rodriguez-Brito, B., Rohwer, F., *et al.* (2011). New abundant microbial groups in aquatic hypersaline environments. Sci. Rep. *1*, 135.

Godchaux, W., and Leadbetter, E.R. (1984). Sulfonolipids of gliding bacteria. Structure of the N-acylaminosulfonates. J. Biol. Chem. *259*, 2982–2990.

Hooper, S.D., Mavromatis, K., and Kypirides, N.C. (2009). A molecular study of microbe transfer between distant environments. PLoS One *3*, e2607.

Huerta-Cepas, J., Capella-Gutierrez, S., Pryszcz, L.O., Kormes, I., Marcet-Houben, M., and Gabaldón, T. (2011). PhylomeDB v3.0: an expanding repository of genome-wide collections of trees, alignments and phylogeny-based orthology and paralogy predictions. Nucleic Acids Res. *39*, D556–D560.

Huson, D.H. (2003). What if I don't have a tree?: split decomposition and related models. In Current Protocols in Bioinformatics (Wiley, Chichester), pp. 6.7.1–6.7.13.

Huson, D.H., and Bryant, D. (2006). Application of phylogenetic networks in evolutionary studies. Mol. Biol. Evol. *23*, 254–267.

Krug, D., Zurek, G., Revermann, O., Vos, M., Velicer, G.J., and Müller, R. (2008). Discovering the hidden secondary metabolome of *Myxococcus xanthus*: a study of intraspecific diversity. Appl. Environ. Microbiol. *74*, 3058–3068.

Legault, B., Lopez-Lopez, A., Alba-Casado, J., Doolittle, W.F., Bolhuis, H., Rodriguez-Valera, F., and Papke, R.T. (2006). Environmental genomics of "*Haloquadratum walsbyi*" in a saltern crystallizer indicates a large pool of accessory genes in an otherwise coherent species. BMC Genomics. *7*, 171.

Makhdoumi-Kakhki, A., Amoozegar, M.A., and Ventosa, A. (2012). *Salinibacter iranicus* sp. nov. and *Salinibacter luteus* sp. nov., isolated from a salt lake, and emended descriptions of the genus *Salinibacter* and of *Salinibacter ruber*. Int. J. Syst. Evol. Microbiol. *62*, 1521–1527.

Marshall, A.G. (2004). Accurate mass measurement: taking full advantage of nature's isotopic complexity. Physica B *346–347*, 503–508.

Maturrano, L., Santos, F., Rosselló-Mora, R., and Antón, J. (2006a). Microbial diversity in Maras salterns, a hypersaline environment in the Peruvian Andes. Appl. Environ. Microbiol. *72*, 3887–3895.

Maturrano, L., Valens-Vadell, M., Rosselló-Mora, R., and Antón, J. (2006b). *Salicola marasensis* gen. nov., sp. nov., an extremely halophilic bacterium isolated from the Maras solar salterns in Peru. Int. J. Syst. Evol. Microbiol. *56*, 1685–1691.

Mongodin, E.F., Nelson, K.E., Daugherty, S., DeBoy, R.T., Wister, J., Khouri, H., Weidman, J., Walsh, D.A., Papke, R.T., Sanchez Perez, G., *et al.* (2005). The genome of *Salinibacter ruber*. Convergence and gene exchange among hyperhalophilic Bacteria and Archaea. Proc. Natl. Acad. Sci. U.S.A. *102*, 18147–18152.

Mutlu, M.B., Martínez-García, M., Santos, F., Peña, A., Guven, K., and Antón, J. (2008). Prokaryotic diversity in Tuz Lake, a hypersaline environment in inland Turkey. FEMS Microbiol. Ecol. *65*, 474–483.

Narasingarao, P., Podell, S., Ugalde, J.A., Brochier-Armanet, C., Emerson, J.B., Brocks, J.J., Heidelberg, K.B., Banfield, J.F., and Allen, E.E. (2011). De novo metagenomic assembly reveals abundant novel major lineage of Archaea in hypersaline microbial communities. ISME J. *6*, 81–93.

Oren, A. (2013). *Salinibacter*: an extremely halophilic bacterium with archaeal properties. FEMS Microbiol. Lett. *342*, 1–9.

Oren, A., and Rodríguez-Valera, F. (2001). The contribution of halophilic Bacteria to the red coloration of saltern crystallizer ponds. FEMS Microbiol. Ecol. *36*, 123–130.

Papke, R.T., and Gogarten, J.P. (2012). How bacterial lineages emerge. Science *336*, 45–46.

Papke, R.T., Koenig, J.E., Rodríguez-Valera, F., and Doolittle, W.F. (2004). Frequent recombination in a saltern population of *Halorubrum*. Science *306*, 1928–1929.

Papke, R.T., Zhaxybayeva, O., Feil, E.J., Sommerfeld, K., Muise, D., and Doolittle, W.F. (2007). Searching for species in haloarchaea. Proc. Natl. Acad. Sci. U.S.A. *104*, 14092–14097.

Pašić, L., Bartual, S.G., Ulrih, N.P., Grabnar, M., and Velikonja, B.H. (2005). Diversity of halophilic Archaea in the crystallizers of an Adriatic solar saltern. FEMS Microbiol. Ecol. *54*, 491–498.

Peña, A., Valens, M., Santos, F., Buczolits, S., Antón, J., Kämpfer, P., Busse, H.J., Amann, R., and Rosselló-Mora, R. (2005). Intraspecific comparative analysis of the species *Salinibacter ruber*. Extremophiles. *9*, 151–161.

Peña, A., Teeling, H., Huerta-Cepas, J., Santos, F., Yarza, P., Brito-Echeverria, J., Lucio, M., Schmitt-Kopplin, P., Meseguer, I., Schenowitz, C., *et al.* (2010). Fine-scale evolution: genomic, phenotypic and ecological differentiation in two coexisting *Salinibacter ruber* strains. ISME J. *4*, 882–895.

Peña, A., Teeling, H., Huerta-Cepas, J., Santos, F., Meseguer, I., Lucio, M., Schmitt-Kopplin, P., Dopazo, J., Rosselló-Mora, R., Schüler, M., *et al.* (2011). From genomics to microevolution and ecology: the case of *Salinibacter ruber*. In Halophiles and Hypersaline Environments, Ventosa, A., Oren, A., and Ma, Y., eds. (Springer-Verlag, Berlin), pp. 109–122.

Ramette, A. (2007). Multivariate analyses in microbial ecology. FEMS Microbiol. Ecol. *62*, 142–160.

Rodriguez-Brito, B., Li, L.-L., Wegley, L., Furlan, M., Angly, F., Breitbart, M., Buchanan, J., Desnues, C., Dinsdale, E., Edwards, R., *et al.* (2010). Viral and microbial community dynamics in four aquatic environments. ISME J. *4*, 739–751.

Rodríguez-Valera, F., Acinas, S.G., and Antón, J. (1999). Contribution of molecular techniques to the study of microbial diversity in hypersaline environments. In Microbiology and Biogeochemistry of Hypersaline Environments, Oren, A., ed. (CRC Press, Boca Raton), pp. 27–38.

Rosselló-Mora, R., Lucio, M., Peña, A., Brito-Echeverria, J., López-López, A., Valens-Vadell, M., Frommberger, M., Antón, J., and Schmitt-Kopplin, P. (2008). Metabolic evidence for biogeographic isolation of the extremophilic bacterium *Salinibacter ruber*. ISME J. *2*, 242–253.

Santos, F., Meyerdierks, A., Peña, A., Rosselló-Mora, R., Amann, R., and Antón, J. (2007). Metagenomic approach to the study of halophages: the environmental halophage 1. Environ. Microbiol. *9*, 1711–1723.

Santos, F., Yarza, P., Parro, V., Briones, C., and Antón, J. (2010). The metavirome of a hypersaline environment. Environ. Microbiol. *12*, 2965–2976.

Santos, F., Moreno-Paz, M., Meseguer, I., López, C., Rosselló-Mora, R., Parro, V., and Antón, J. (2011). Metatranscriptomic analysis of extremely halophilic viral communities. ISME J. *5*, 1621–1633.

Santos, F., Yarza, P., Parro, V., Meseguer, I., Rosselló-Mora, R., and Antón, J. (2012). Culture-independent approaches for studying viruses from hypersaline environments. Appl. Environ. Microbiol. *78*, 1635–1643.

Sharma, A., Walsh, D., Bapteste, E., Rodriguez-Valera, F., Doolittle, W.F., and Papke, R.T. (2007). Evolution of rhodopsin ion pumps in haloarchaea. BMC Evol. Biol. *7*, 79.

Soria-Carrasco, V., Valens-Vadell, M., Peña, A., Antón, J., Amann, R., Castresana, J., and Rosselló-Mora, R. (2007). Phylogenetic position of *Salinibacter ruber* based on concatenated protein alignments. Syst. Appl. Microbiol. *30*, 171–179.

Traxler, M.F., and Kolter, R. (2012). A massively spectacular view of the chemical lives of microbes. Proc. Natl. Acad. Sci. U.S.A. *109*, 10128–10129.

Vaisman, N., and Oren, A. (2009). *Salisaeta longa* gen. nov., sp. nov., a red, halophilic member of the Bacteroidetes. Int. J. Syst. Evol. Microbiol. *59*, 2571–2574.

Want, E.J., Nordström, A., Morita, H., and Siuzdak, G. (2007). From exogenous to endogenous: the inevitable imprint of mass spectrometry in metabolomics. J. Proteome Res. *6*, 459–468.

Watrous, J., Roach, P., Alexandrov, T., Heath, B.S., Yang, J.Y., Kersten, R.D., van der Voort, M., Pogliano, K., Gross, H., Raaijmakers, J.M., *et al.* (2012). Mass spectral molecular networking of living microbial colonies. Proc. Natl. Acad. Sci. U.S.A. *109*, E1743–E1752.

Whitaker, R.J., Grogan, D.W., and Taylor, J.W. (2003). Geographic barriers isolate endemic populations of hyperthermophilic Archaea. Science *301*, 976–978.

Zhaxybayeva, O., Stepanauskas, R., Mohan, N.R., and Papke, R.T. (2013). Cell sorting analysis of geographically separated hypersaline environments. Extremophiles *17*, 265–275.

Horizontal Gene Transfer in Halobacteria

3

Matthew S. Fullmer, J. Peter Gogarten and R. Thane Papke

Abstract

The Halobacteria are a class of Archaea that have been fundamentally shaped by Horizontal Gene Transfer (HGT). The mechanisms for HGT are not well understood, or are unreported. A noteworthy exception exists for the genus *Haloferax*, where a novel mating system exists that includes the fusion of cytoplasm between two cells. Despite shallow insight into mechanisms evidence from phylogenetics and population genetics studies demonstrate that these organisms have been able to exchange genes since their distant origins and continue to actively do so today. Single gene studies have uncovered transfer of Halobacterial rhodopsins into diverse lineages such as the fungi and multiple bacterial taxa, construction of novel biosynthetic pathways, homologous recombination of parts or whole ribosomal proteins and RNAs, as well as divergent tRNA synthetases being exchanged between distant lineages. Furthermore, the very origin of the Halobacteria appears to have resulted from an influx of genes from the bacterial domain, which reshaped the fundamental metabolism from an anaerobic chemoautolithotrophic methanogen into a facultative aerobic heterotroph. Population genetics analysis demonstrated that gene flow with phylogenetically defined populations is so frequent that allele distributions resemble that of sexually reproducing eukaryotes, and acts as both a homogenizing and diversifying evolutionary force. Given all of the evidence for abundant recombination into, out of and between these lineages, how then do new, distinct, lineages such as these stably emerge? The answer appears to lie in a balance between recombination as a cohesive force holding populations together as entities recognizable as taxonomic units, and barriers to that transfer for promoting diversification. A primary candidate appears to be geographic barriers that reduce gene transfer between populations sufficiently to allow regional signatures to emerge.

Introduction

The Halobacteria (often referred to colloquially as the Haloarchaea) are a highly recombinant class of salt-loving organisms in the archaeal phylum Euryarchaeota. These remarkable organisms thrive in brines around the globe that have salt concentrations as high as 35% (sea water is ~3.5%). Most strains cannot tolerate concentrations below 10%. Their metabolism is primarily aerobic heterotroph and many garner extra energy by harvesting light via rhodopsin pigments in their membranes. Perhaps most profound, the Halobacteria have been directly constructed or inextricably molded by the exchange of genes between organisms in a non-Mendelian manner.

Mechanisms of horizontal gene transfer

The classic description of HGT includes three mechanisms for transfer of genetic material (Thomas and Nielsen, 2005). These mechanisms are conjugation, transduction and natural transformation. In recent years a fourth process known as gene transfer agents (GTAs) has been elucidated (Lang *et al.*, 2012). Conjugation and transduction are considered mechanisms for the propagation of selfish DNA elements (plasmids and viruses/phage respectively) that only coincidentally transfer their infected host's genes to another cell, while natural transformation most likely evolved to import environmental DNA specifically to benefit the individual cell: it requires many genes and is typically a highly regulated process (Redfield, 2001). GTAs though seemingly more like transduction because DNA is packaged in a virus-like protein coat it is also similar to transformation because it is also a tightly regulated process that specifically moves host DNA from cell to cell. In each of these cases DNA is transferred unidirectionally from a donor to a recipient.

The protein machinery for conjugation is typically encoded by *tra* genes located on conjugative plasmids. These gene products form a specialized pilus which links the cell containing the plasmid (i.e. donor: F′) to a cell that does not (i.e. recipient: F^-). The conjugative plasmids are copied by rolling circle replication, which requires one strand of DNA to be nicked and peeled off while the intact strand is used as a template for DNA synthesis. The non-template strand is then transferred through the pilus into the recipient cell. In the recipient cell the ssDNA is re-circularized and DNA polymerase creates a complementary strand to result in a double-stranded DNA molecule. Sometimes the plasmid will integrate into the host chromosome, or a fraction of the host chromosome will integrate into the plasmid and when the plasmid is transferred it will also bring host genes to the recipient. In these cases, the host chromosome is transferred inadvertently.

Conjugation is a notable unidirectional method of transfer; only cells with the plasmid can act as a donor, and only cells without the plasmid can be recipients. Cells already containing a plasmid cannot receive an additional copy from another donor. Transfer of plasmids can be picky, and promiscuous: sometimes they cannot be transferred to potential donor cells from the same species, yet can infect multiple species.

Transduction is the transfer of genetic material from one cell to another via a viral intermediate. Viruses can pursue two strategies once they have penetrated a host's external defences. The simplest strategy is to enter the lytic cycle. In this process the virus co-opts the host's nucleic acid and protein synthesis machinery for its own purposes while suppressing the host's usage. The virus creates many copies of its genome while also creating new proteinaceous shells (capsids) to transport these genomes to a new host. Lysogenic, or temperate phages as they are also called, do not immediately co-opt the host's molecular machinery for reproduction. Instead, they integrate their genetic material into their host's chromosome. While inserted, they replicate when the host reproduces. Thus, these viruses are able to expand their population through vertical inheritance. Once they detect a signal from the host (usually a stress response) they excise their DNA and begin the lytic phase. When excising their genome and packaging copies into capsids the phage are sometimes imprecise. Some virions are packaged with host DNA. Yet, the virions themselves are fully intact and infectious. Thus, the virions can still transfer their DNA into a new host. Once in the new host the DNA may be recombined into the chromosome through homologous replacement or as part of a non-functional genomic island (the phage 'genome'). The net result is transfer of genetic material from one organism to another in a non-Mendelian fashion.

Transformation is the uptake of DNA from the environment and its integration into the chromosome by a cell (Chen *et al.*, 2005). Natural transformation is intrinsically linked to the process of competence, which is a state that enables cells to transport DNA from the environment across their membrane and into the cell as a high molecular weight molecule. Some bacteria are competent and may use DNA for nutritional purposes but not recombine it into their chromosomes (i.e. not undergo transformation; (Finkel and Kolter, 2001; Redfield, 2001). Movement of DNA from the environment into the cell is complex and requires a set of numerous proteins acting in concert. Many proteins are homologous and conserved between Gram-positive and -negative bacteria yet some of the molecular machinery is more specialized to accommodate the specificities of the cell wall and membrane characteristics. Only few instances of natural competence or transformation have been described for Archaea (Bertani and Baresi, 1987; Johnsborg *et al.*, 2007; Lipscomb *et al.*, 2011; Sato *et al.*, 2003; Worrell *et al.*, 1988).

Gram-negative and -positive model organisms frequently employ mechanisms for biasing the DNA that they import. Pathogenic Gram-negative organisms such as *Neisseria gonorrhoeae* and *Haemophilus influenzae* (among many others) use DNA uptake sequences (DUS) (Smith *et al.*, 1995), which are ~10mer DNA repeats present in high copy number (>1000×) and on both strands of the chromosome, to delineate DNA for import: cell machinery must first recognize a DUS before the DNA can be taken in by the cell. DUSs are typically thought to be species specific, though close relatives can sometimes recognize the same signal. Organisms that utilize DUSs typically constitutively produce competence proteins and can import DNA at any time (Hamilton and Dillard, 2006). This is in contrast to model Gram-positive organisms like *Bacillus subtilis* and *Streptococcus pneumoniae* (Havarstein and Morrison, 1999), which can import any kind of DNA but regulate when they take it up through quorum sensing. This population size-associated timing of protein expression and thus when DNA is imported ensures a large number of their own species are available to donate DNA. Further, the biofilms in which these organisms typically live, diffusional processes like quorum sensing up-regulate competence in only a fraction of the population (Steinmoen *et al.*, 2002). This is relevant because in some instance fratricide is an important aspect of competence: in addition to up-regulating competence, the quorum sensing regulation simultaneously signals for the production of a toxin/antitoxin system (Claverys and Havarstein, 2007). Competent cells produce the toxin and antitoxin while non-competent cells do not. Therefore non-competent cells are lysed and spill out their DNA for the competent cells to utilize.

HGT via gene transfer agents can be broadly thought of as similar to transduction. More specifically, it can be thought of as decayed prophages incapable of packaging their own DNA: the GTA randomly packages host DNA for transfer of DNA to a new host. Furthermore, production of GTAs has been observed to be under the control of a quorum sensing process. While GTA similarity to phage is clear, it is also helpful to think of the process as a 'protected' natural transformation.

Mechanisms of horizontal gene transfer in Halobacteria

The single largest unifying feature of the above mechanisms is the polarity of their transfer. Each of them has a donor from where the genetic material leaves and a recipient to where the material arrives. Thus, a single HGT event can only alter one of the two parties. In contrast,

Haloferax, a genus from the Class Halobacteria (Phylum Euryarchaeota) utilizes a novel mechanism of HGT in which a physical connection is required for transfer, but the donor–recipient relationship is non-existent: all cells are donors and recipients. The evidence for this non-canonical transfer mechanism called mating is multifaceted:

- Shaking cultures rarely if ever produce recombinants. Recombinants are formed readily, however, after cells are pelleted from centrifugation and when cells are incubated on a substratum, indicating the need for cell–cell contact (Mevarech and Werczberger, 1985).
- Nuclease treatment of cells still results in a high frequency of recombinants, indicating natural competence is not involved (Mevarech and Werczberger, 1985).
- Medium filtrate (0.22 μm) from one auxotroph when applied to another does not form recombinants, indicating the process is not virus nor GTA mediated (Mevarech and Werczberger, 1985).
- Auxotrophs form prototrophs in every strain combination ever tried (Mevarech and Werczberger, 1985), whereas in standard conjugation experiments transfer only occurs between specific cell types, thus indicating no donor–recipient relationship for *Haloferax.*
- Plasmid and chromosome-based traits are transferred at the same rate (Naor *et al.*, 2012). In conjugation, chromosomes are much less frequently transferred.
- Electron microscopy reveals the presence of multiple intercellular 'bridges' between cells (Rosenshine *et al.*, 1989), and multiple cells are connected like beads on strings (Mullakhanbhai and Larsen, 1975).
- Cell walls and membranes appear contiguous between cells (Mullakhanbhai and Larsen, 1975). In contrast, conjugation requires the formation of a protein pilus.
- A low concentration of Mg^{2+}, which interferes with the stability of cell membranes, also prevents the formation of bridges (Rosenshine *et al.*, 1989).
- Unusually large segments of chromosomal DNA (10–17% of the total chromosome) are transferred (Naor *et al.*, 2012).
- Plasmids from cells containing multiple plasmids (e.g. *Haloferax volcanii* strain DS2) are transferred in an apparently random process (Naor *et al.*, 2012).
- Mating is a pilin-independent process (Tripepi *et al.*, 2010).

A four-step model has been hypothesized to explain the data (Ortenberg *et al.*, 1999). In step 1, 'haploid' cells grow bridges (pseudopodia-like appendages are seen extending from cells) until they fuse with nearby cells. During step 2, bridges widen to accommodate the flow of cytosolic contents, including chromosomes and plasmids producing 'diploid' or heterochromosomal cells. Step 3 is when recombination of chromosomes occurs, and step 4 is the segregation of chromosomes and plasmids into separate cells. [Quotations around haploid and diploid are used to indicate the inaccuracy of the terms. *Haloferax* spp. and other Halobacteria have recently been determined to be polyploids, containing tens of chromosomes even during stationary phase (Breuert *et al.*, 2006).] This process is remarkably analogous to the haploid/diploid life cycle of sexually reproducing eukaryotes, with the caveat that no reproduction is occurring during the mating activity of Halobacteria. As intriguing as this process is, how this process works and its uniqueness in the biological world remain open questions.

Very little is known about transduction and natural competence in Halobacteria. For instance, viruses are known to infect Halobacterial hosts (e.g. Atanasova *et al.*, 2012; Dyall-Smith *et al.*, 2003), there is evidence for viruses embedded in Halobacterial chromosomes (e.g. Cuadros-Orellana *et al.*, 2007) and virus DNA from hypersaline environments has been sequenced (Rodriguez-Brito *et al.*, 2010), but no virus mediated genetic system has been developed and transduction as a process has not been reported (Allers and Mevarech, 2005) (see also Chapter 4). The genetic systems developed for Halobacteria are based on the formation of spheroplasts for chemical-induced competence (e.g. Charlebois *et al.*, 1987; Cline and Doolittle, 1987), rather than natural competence. Evidence for natural transformation has not been reported.

Evidence for horizontal gene transfer in Halobacteria from single gene studies

Rhodopsins

Even though we do not have a deep understanding for the mechanisms of horizontal gene transfer in Halobacteria, we know it occurs because their chromosomes record ample evidence for rampant HGT. As early as the first sequenced Halobacterial genome, *Halobacterium* sp. strain NRC-1, the respiratory apparatus was identified as having originated in Bacteria (Ng *et al.*, 2000). Subsequent work has reaffirmed the bacterial origin for this system (Nelson-Sathi *et al.*, 2012). The movement of respiration genes into a methanogen-like Halobacterial ancestor probably played a crucial role in the origins of the Halobacteria, and will be discussed in more depth in a following section.

A signature trait of many Halobacteria is the presence of light-harvesting rhodopsins that have three major functions: bacteriorhodopsins (bR), which pump protons out of the cell thus creating an ATP generating proton motive force (Lozier *et al.*, 1975); halorhodopsin (hR), an anion transporter (primarily chloride) for maintaining iso-osmotic balance (Kolbe *et al.*, 2000); and sensory rhodopsin (I and II) to regulate phototaxis: cells move towards useful wavelengths utilized by bR and hR and away from harmful ultraviolet range wavelengths (Spudich and Bogomolni, 1988). Both the Chloride and Sensory paralogues have been reported in HGT events across the domain level. *Salinibacter ruber* contains four rhodopsin homologues. Two of these are sensory in nature and the third is a chloride transporter. Phylogenetic analysis places the *Salinibacter* genes squarely within the Halobacterial clade, suggesting they were recent acquisitions, across domain boundaries, by the bacteria, and not the other way round. Interestingly, a fourth rhodopsin, which is a H^+ pump of no apparent relation to the archaeal version, appears to have been acquired from a divergent bacterial phylum in another HGT event (Sharma *et al.*, 2006).

Fungal rhodopsins also share ancestry with halorhodopsins. Once rapidly evolving sequences that have changed function are screened out of the fungal rhodopsins, it is clear they group with the hR chloride transporters (Sharma *et al.*, 2006). In contrast with the *Salinibacter* example, this appears to be a single transfer event into the ancestor of Fungi as fungal sequences across the entire kingdom group together inside the Halobacteria (Sharma *et al.*, 2006).

Rubrobacter xylanophilus is a member of the Actinobacteria. It is noted for its radiation resistance, and the presence of a rhodopsin (Carreto *et al.*, 1996). Another actinobacterium,

Kinecoccus radiotolerans, has a rhodopsin that is bacterial in origin (Sharma *et al.*, 2006). However, the rhodopsin from *R. xylanophilus* groups within the Halobacterial clade, albeit not clearly within any of the major functional clusters. Both of those rhodopsin proteins fit the primary sequence and secondary structure profiles of proton transporters. Since *R. xylanophilus* is a member of the most basal lineage of the Actinobacteria it was inferred that this transfer occurred after its divergence from the rest of the phylum.

Unsurprisingly, the HGT of rhodopsins is not solely across vast phylogenetic distances. Within the Halobacteria HGT appears to have been a regular occurrence. Rhodopsin genes can be found across the Halobacterial class. However, their distribution is patchy. This has led to suggestions of all four functionalities having existed in the most-recent Halobacterial common ancestor (Ihara *et al.*, 1999). An examination of the specific rhodopsin genes against a species tree finds phylogenetic disagreement (Sharma *et al.*, 2007). Using the rpoB′ protein as a marker for vertical descent both the bacteriorhodopsin and halorhodopsin gene trees displayed discordance. Overall, the bacteriorhodopsin gene tree agreed with the 16S rRNA sequence-based species tree but for two instances. These two cases each saw the rhodopsin linked to two adjacent genes, evidentially transferred as a unit. It is not clear if these were two separate events or a single ancestral transfer event. The halorhodopsins appear to have at least two HGT events causing their gene tree to diverge from their bacteriorhodopsin homologues (Sharma *et al.*, 2007).

The linkage of the bacteriorhodopsins with two adjacent genes is worthy of extra note. Sharma *et al.* (2007) found these genes (*bac* and *bap*) to be more frequently linked to bacteriorhodopsins than other genes previously shown to be important in the proton pump function. It is also interesting that in the absence of a bacteriorhodopsin these linked genes are also absent. When this is combined with other observations such as that the *Haloferax volcanii* strain DS2 does not harbour rhodopsins, (it was cultivated from Dead Sea mud, which is not transparent to sunlight, and it utilizes nitrate reduction in conditions unfavourable to rhodopsin use) and the diversification of function in the rhodopsins, one can view rhodopsins as niche-specific functional units that can be lost and gained as the ecology demands (Sharma *et al.*, 2007).

New biochemical pathways assembled through horizontal gene transfer

Halobacteria have built via HGT a new pathway tailored to their environments. In salt lakes, for example, blooms of microorganisms are often rare and short-lived. Halobacteria store carbon in the form of polyhydroxyalkanoate. As a result, pathways for assimilation of carbon may be advantageous. *Haloarcula marismortui* has assembled a novel acetate-assimilating pathway from glutamate fermentation, acetate fermentation and propionate assimilation genes all of which originated from the Bacteria (Khomyakova *et al.*, 2011). The resulting pathway, known as the methylaspartate pathway, allows these halophiles to assimilate acetyl-CoA in microaerobic environments where the canonical glyoxylate pathway cannot function (Khomyakova *et al.*, 2011). The discoverers suggest that this pathway could serve as a mechanism to cope with glutamate overloads.

Ribosomal RNA and proteins

16S rRNA sequences are typically considered the 'gold standard' for phylogenetics, primarily for historical reasons. Early prokaryotic phylogenies that described relationships between taxa used 16S rRNA gene sequence data (Woese and Fox, 1977; Woese *et al.*, 1990) and the marker became the standard for describing a newly discovered strain's place in the tree of life (Oren *et al.*, 1997; Stackebrandt and Goebel, 1994). With the advent of PCR, environmental 16S rRNA gene sequencing for understanding community composition independently of cultivation vastly expanded the database. After the discoveries of frequent HGT, it was argued that the 16S rRNA gene, and other informational genes, such as those involved in DNA transcription and translation, are resistant to horizontal transfer (Complexity Hypothesis: (Jain *et al.*, 1999)). Transferred genes whose products participated in many macromolecular interactions would be poorly optimized to interact efficiently as part of a complex in their new host. This contrasts with so-called accessory or operational genes that often are involved in few interactions (Jain *et al.*, 1999) and which are frequently gained and lost from lineages (Lawrence and Roth, 1996). For example, a cytosolic catalytic enzyme might not need to interact with any other gene-product to fold, become active and metabolize its substrate. The result is that many accessory genes may be able to function at normal efficiency in foreign environments whereas an informational gene might cause a substantial penalty, and decrease the fitness of the organism it arrived in.

Though the complexity hypothesis is probably enforced most of the time, the examples of apparent ribosomal RNA and ribosomal-associated protein gene transfers have been growing (e.g. Badger *et al.*, 2005; Boucher *et al.*, 2004a; Gogarten *et al.*, 2002; Gupta *et al.*, 2003; Wellner *et al.*, 2007; Williams *et al.*, 2012; Zhaxybayeva *et al.*, 2006, 2009) indicating it is not always deleterious, and perhaps sometimes useful to have a foreign-derived informational gene. Specific examples can be found in the Halobacteria. *H. marismortui* possesses two rRNA operons. The SSU genes are 5.0% divergent from each other while the large subunit (LSU) genes are 1.3% dissimilar (Mylvaganam and Dennis, 1992). *Halosimplex carlsbadense* possesses SSU genes 6.8% divergent and LSUs 2.7% divergent; *Natrinema* sp. strain XA3-1 exhibits four operons wherein three SSU genes are ~0.1% distant and the fourth is ~5.0%; all four LSUs are ~1–2% distant; in both *Har. marismortui* and *Hsx. carlsbadense* the rRNA intergenic spacers (ITS), are 24.6% and 49.1% divergent, respectively; the *Natrinema* strain displays zero divergence in its ITS region (Boucher *et al.*, 2004a).

Each *Har. marismortui, Hsx. carlsbadense* and *Natrinema* SSU gene clearly does not share a common recent history with its intragenomic relatives. In *Natrinema* the three nearly identical copies fit expectations but the fourth with ~5.0% divergence did not display strong affinity with any other known species at the time. In the other two strains, the very high level of divergence between the two rRNA operons suggest that at least one version likely evolved in a different lineage and was horizontally transferred into its current genome.

The *Natrinema* sequences are particularly interesting. The ITS regions are among the most hyper-variable and should diverge much faster than the rRNA genes themselves, which have large amounts of their sequence under strong purifying selection. However, all four are identical. It is possible that the ITS locus is under strong purifying selection though no function is known; more likely recombination between the rRNA operons coexisting in the same genome recently purged the diversity in this region, as was discussed for the rRNA operons in *Thermomonospora chromagena* (Gogarten *et al.*, 2002; Yap *et al.*, 1999). In contrast, the LSU genes themselves have all accrued ~1.0% separation from each other.

A more detailed examination of the evolutionary signals within the divergent genes identified an apparent transfer between *Natrinema* sp. XA3–1 strain and *Natrialba magadii*. A 100-bp region from one *Natrinema* copy showed strong similarity to the corresponding region in *N. magadii*'s LSU genes (Boucher *et al.*, 2004a). This indicates that the region was transferred into *Natrinema*, a homologous recombination event occurring across a distance of ~5%, while the remainder of the LSU was not.

There are many problems for microbial phylogeneticists, ecologists and taxonomists associated with highly divergent intragenomic rRNA heterogeneity. First, it is greater than that typically seen within species, which usually is below ~2.0%. This creates difficulties for the suitability of rRNA genes as a marker for classification in strains featuring such heterogeneity: which SSU copy is to be used for determining relationships? Further, with divergent copies in a single chromosome, the difficulties of obtaining a reliable sequence are immense. Multiple divergent copies will cause a PCR product that is directly sequenced (i.e. without cloning first) to have dozens of sites across the alignment be unresolved, with peaks in sequencing chromatograms for two different nucleotides at the same position. If the PCR product is cloned and then sequenced, chimera artefacts will likely occur leading to overestimating the number of copies, and sequence variation per copy. For example, when *Halosimplex carlsbadense* was first described, it reportedly contained three distinct 16S rRNA genes, with A and B being 97.7% similar and C being 93.8 and 92.2% similar to A and B respectively (Vreeland *et al.*, 2002). Later, using PCR independent techniques, it was demonstrated that *Hsx. carlsbadense* has only two distinct copies that are 6.8% divergent (Boucher *et al.*, 2004a). Multiple divergent operons from single cells will also over estimate community species richness and abundance (in addition to forming PCR induced chimeric sequences) when applying standard PCR and cloning techniques to analysis of environmental DNA from saturated brines. Finally, fractions of rRNA genes can be transferred independently of the entire gene or operon (Boucher *et al.*, 2004a). Thus, a single recombination event might sufficiently obscure the 'true' evolutionary signal. A single transfer of several hundred nucleotides might be spotted, if multiple sequence alignments are analysed; however, mosaics formed as the result of multiple gene conversion events that ultimately lead to homogenization of the initially divergent copies coexisting in a genome likely remain undetectable in most instances.

Gene conversion, the effect of homologous recombination between two copies of the same gene on the same chromosome, is known to occur in Halobacteria (Lange *et al.*, 2011) and would prevent divergence of loci (Liao, 1999) unless subfunctionalization occurs and variation is selected for. Without subfunctionalization and selection, there would likely be no bias towards maintaining the original copy and both copies are expected to contribute to the homogenized mosaic resulting from many gene conversion events. In many Halobacteria, rRNA heterogeneity appears stably maintained and may be quite common within the class (Dennis *et al.*, 1998; Grant *et al.*, 1998). Evidence from analysis of rRNA operons in *Haloarcula* and *Halomicrobium* species has demonstrated that the divergent copies are expressed under different laboratory conditions, e.g. temperature and salinity, and probably confer adaptive advantages when the organism finds itself in fluctuating environmental circumstances (Cui *et al.*, 2009; Lopez-Lopez *et al.*, 2007).

tRNA synthases

The Halobacteria have some tRNA synthases with unusual evolutionary histories. The class contains two versions of the Leucine tRNA synthase (Andam *et al.*, 2012). The first type is the archaeal version that falls within the Euryarchaeota (LeuRS-A). The other type, more common among the Halobacteria, is more similar to bacterial homologues (LeuRS-B). In phylogenetic reconstruction the B-type does not actually group inside the cluster of bacterial homologues. Rather, it branches deeper than the last common bacterial ancestor. Thus, it appears that the Halobacteria received the LeuRS-B genes from an unsampled, or possibly extinct, lineage that diverged from the Bacteria before that domain's common ancestor appeared.

The Halobacterial LeuRS gene distribution is the product of HGT. When the presence of LeuRS-A and B are mapped onto an MLSA tree, created using concatenated housekeeping genes, an unusual pattern emerges. All of the strains that possess an A-version lack a B-version. Likewise, all of the strains with a B-version do not have an A-version. What is striking is that the A and B gene type phylogenetic trees and the MLSA derived versions are conflicted, and carriers do not form monophyletic groups. Apparently, several lineages have replaced their own LeuRS genes with extremely divergent out-paralogues (xenologues). The LeuRS A and B loci are not in syntenic gene neighbourhoods suggesting that lineages gained the xenologous copy through non-homologous recombination and that after co-existing for some time one or the other LeuRS version was lost.

The possibility that lineages maintained the two divergent LeuRSes is possibly supported by the evolutionary dynamics of the B-version. The B-version has a split inside its own group, which divides it into two subgroups, designated B′ and B″. Every strain with a B″ copy is also a carrier of B′. However, B′ is often a solitary copy. There are two possible explanations for this arrangement. The first is that in taxa with both versions the protein has evolved to function as a hetero-dimer. Second, this could be evidence that multiple versions can be maintained for evolutionarily meaningful time before differential loss purges one or the other. Presumably, if this is the case, both copies confer some form of selective advantage under different conditions to enforce their maintenance (Andam *et al.*, 2012).

The origins of the Halobacteria may be rooted in horizontal gene transfer

The different studies above describing the number of gene families that have been transferred into or among the Halobacteria suggests that this class of organisms have been heavily impacted by participation in HGT events throughout their evolutionary history. New evidence is now suggesting that the very origins of the Halobacteria are founded on the influx of genes from elsewhere, especially from the Bacteria.

Phylogenetic analysis, including that of the 16S rRNA gene and ribosomal proteins, place the Halobacteria as sister to *Methanosarcina*, and unrelated to *Methanococcus* and *Methanobacterium* (Matte-Tailliez *et al.*, 2002). In contrast, phylogenies drawn completely from the presence and absence of genes across the entire Archaea place the Halobacteria near the root of the domain (Korbel *et al.*, 2002). The methanogen placement has gained overall acceptance. The supposition that the ancestor to all Halobacteria was a methanogen is rather interesting, as they are dissimilar in almost every metabolic pathway. Methanogens are obligate anaerobes and chemoautolithotrophs. They create methane from H_2 and CO_2 to fuel synthesis of ATP. They also use the acetyl-CoA pathway to convert CO_2 into cellular material. Halobacteria are obligate heterotrophs. They are also either obligate or facultative aerobes. Many supplement energy production by generating a proton gradient via their light driven proton pump (bacteriorhodopsins). However, this is not an essential component to their growth and live exclusively from oxidizing organic carbon substrates (Oren, 2008).

Both the Halobacteria and their Methanosarcinales neighbours show large bacterial signatures in their genomes (Nelson-Sathi *et al.*, 2012). In the case of the Halobacteria, as many as 1089 genes (of 1479 common to at least two Halobacterial genomes from a dataset of 10) place the Halobacterial version as branching within or next to the Bacteria. Among these genes are those required to transform Halobacteria into the oxidative heterotrophs they are today. Critical are the genes encoding the electron transport chain. In total, almost half (482, 44%) of the putative transfers from bacteria are related to metabolism (Nelson-Sathi *et al.*, 2012). Several genes in the isoprenoid biosynthesis pathway are also among these. IDI1, a type 1 isopentenyl diphosphate isomerase is an analogue of the ubiquitous Archaeal version but was acquired from the Bacteria and the Halobacterial glycerol dehydrogenase proteins group within the bacterial clade (Boucher *et al.*, 2004b).

The question then arises whether or not these genes arrived largely simultaneously or as piecemeal acquisitions. Genes present in all ten genomes (473 total) displayed similar topologies within their phylogenetic trees that match the topology of a tree generated from 56 universally distributed archaeal genes, as confirmed by a goodness of fit test. This agreement of topology suggests that these genes share a common vertically descended history. Thus, a considerable fraction of the bacterially acquired genes appear to have arrived at the root of the class, prior to the genetic radiation of the modern Halobacteria (Nelson-Sathi *et al.*, 2012). Nelson-Sathi *et al.* (2012) then boldly suggest that the transferred genes came from a single bacterial donor, similar in concept to the acquisition of mitochondria and thus the origin of eukaryotes popularized by the Hydrogen Hypothesis (Martin and Muller, 1998). However, in the case of the Halobacteria, transfer was ultimately a failed attempt at eukaryogenesis, resulting in 'just' aerobic heterotrophic Archaea. While certainly daring and thought provoking, the hypothesis of a single bacterial donor is not supported by the data.

Homologous recombination within and between Halobacterial lineages

As the studies above demonstrate, recombination is an important force shaping the Halobacteria. This force profoundly affects how their taxonomy and species must be thought of. It has been known for more than 15 years that prokaryotes do not recombine at the same rate with every other lineage of prokaryotes. More specifically, a log-linear distance–decay relationship has been observed between the recombination rate and the sequence divergence between the two organisms (Fraser *et al.*, 2007; Vulic *et al.*, 1997; Williams *et al.*, 2012). Two members of the same species might recombine at a very high rate whereas individuals from different genera would do so at a much lower rate. The result being that transfers across phyla or domains are exceedingly uncommon. The Halobacteria have been demonstrated to have a log-linear distance–decay relationship for homologous recombination, but it appears to be less of a barrier than that seen in bacterial model organisms (Naor *et al.*, 2012; Williams *et al.*, 2012).

HGT has been proposed as a homogenizing force upon populations of related cells (Andam *et al.*, 2010a; Doolittle and Zhaxybayeva, 2009; Feil *et al.*, 2000; Gogarten *et al.*, 2002; Lawrence, 2002; Papke, 2009; Smith *et al.*, 2000; Whitaker *et al.*, 2005). Contrary arguments have been put forth that the level of HGT required to homogenize a population is much higher than that observed in nature (Cohan, 2002, 2006). The question then centres on how much recombination is required to create the phylogenetic clustering or 'clumpiness' observed in nature. Fraser *et al.* (2007) explored some of the impacts of varying recombination rates via a neutral (i.e. selectionless) model. When recombination rates are far below the mutation rate clusters of related organisms form. These clusters are composed of individual cells that can be thought to have their own reproductive fate. They behave in a manner similar to that posited by ecotype-style hypotheses (Cohan, 2006). When recombination is much higher than the mutation rate a different scenario emerges. The diversity of alleles remains the same. However, the number of unique genotypes is higher than in the low recombination situation. The clustering seen under the high recombination to mutation regime is more pronounced but mostly transient. However, when the recombination ratio is between 0.25 and four times that of the mutation rate coherent populations are easily discerned and stable. Two extraordinary conclusions, which are in direct conflict with the ecotype hypothesis, are drawn from this analysis: (1) recombination can act as a cohesive force at biologically observed recombination rates and (2) selection is not required to create the appearance of unique genetic clusters (i.e. species): random birth/death processes are sufficient (Fraser *et al.*, 2007).

The first consequence of recombination occurring at the observed rates is that, as mentioned above, it can act as a cohesive force for populations of organisms (Andam and Gogarten, 2011; Andam *et al.*, 2010b; Gogarten *et al.*, 2002). These clustered populations can be thought of as species, although that definition is still quite inadequate, as will be discussed below. The recombination-dictated clustering, combined with the distance-decay frequencies for recombination, suggest that many gene phylogenies within a 'species' will be conflicted while those outside the species will not be so – although their own intra-clustering trees will be conflicted, too (Dykhuizen and Green, 1991; Lawrence, 2002). The second consequence is a lack of clonality within a species, despite asexual reproduction. As genes are transferred between organisms new combinations of alleles will be created that may have been impossible, or at least statistically impossible, without recombination. This has already been observed via computer modelling (Fraser *et al.*, 2007), and is similar to the 'shuffling of a deck of cards' metaphor used by Dawkins (1976) in his book *The Selfish Gene* to describe sexual recombination. If one were playing a game of cards where each player had a hand and could only change their hand one card at a time, in gradual intervals, from a draw pile it might be thought of as analogous to the situation predicted by an ecotype hypothesis. Here, players would be largely stuck with the hand they had been dealt and when a player is forced to fold the diversity (cards) they possess is lost to the population (card players). Each card in a player's hand would be linked to the fate of the hand as a whole, no matter how advantageous that card may be. However, if the game allowed players to swap cards with each other, either through a communal pool or by directly taking cards from another player's hand, players could radically alter their hands in short order and sample new card combinations which may prove fruitful. Each card would also find itself less linked to the fate of the hand (player) as a whole. Even if a player were to fold, a strong card from their hand may have already found new life in another player's hand.

How do the Halobacteria fare compared to these predictions and consequences developed for bacterial models? Multilocus sequence analysis of over 150 *Halorubrum* strains cultivated from three sites (two different salinity ponds from a saltern in Santa Pola, Spain, and a salt lake in Algeria) demonstrated that recombination was so frequent within phylogenetically defined clusters of closely related strains (< 1% DNA divergence; called phylogroups) that alleles at loci were unlinked to the extent that there was no association of alleles and loci (i.e. there was no evidence for clonality) (Papke *et al.*, 2004, 2007). Measurements of recombination to mutation demonstrated that loci change twice as frequently by recombination than mutation, which is right in the middle of the range identified by Fraser *et al.* (2007) – see above. The two above observations for random association of alleles and loci, and a rate of homologous recombination in agreement with computer model studies, strongly supports that *Halorubrum* clusters are homogenized by homologous recombination rather than by a selective sweeps model proposed for Cohan's ecotype hypothesis.

Because genes were unlinked through recombination, if selection were applied to a single allele, in theory it could rise in frequency independently of the genes in the rest of the chromosome. Evidence for the fixation of a single allele was seen in the bacteriorhodopsin locus for two *Halorubrum* phylogroups. The *bop* showed minimal nucleotide diversity; approximately 90% of the strains had the same allele in their respective phylogroup, and the only mutations seen were in two strains where A and T mutations were observed in neutral third codon positions. These are likely transient mutations as Halobacteria are well know for their high G + C genomic content, which is especially pronounced at 3rd codon positions, reaching approximately 90%. The other loci examined in that study all demonstrated more polymorphisms in 3rd codon positions, in some cases nearly 10 times more, and for each of the additional loci, no allele was found in more than 50% of the strains. The lack of diversity at the third codon positions, which due to the redundancy in the genetic code can be assumed to be neutrally evolving at least within phylogenetic clusters, while maintaining diversity at other loci in the third codon position is indicative of a recent fixation of the *bop* allele only. If frequent recombination did not occur and genes remained linked, then selection on one adaptive allele that led to its fixation would have also fixed all the other loci in the population, and all diversity within the *Halorubrum* phylogroups would have been purged. Because diversity was purged in only one locus, the only possible explanation is that there is frequent recombination that enables the fixation of single alleles.

Another discussion point to be made from that study is the appearance of well-supported clusters. Clustering is often considered evidence for species, which of course has a lot of intellectual baggage, one of which would be from the Darwinian model that two sister species have a common ancestor. However, that outcome was not necessarily observed. For instance, despite the fact that individuals typically generated the same clusters, irrespective of the locus examined, each gene supported a different sister lineage, indicating the lack of a common ancestor for those 'species'. The only reasonable explanation is that genes originating from outside the population are continuously homologously recombined and fixed in the population, in the same way as described above for the *bop* genes. Furthermore, the 16S rRNA gene tree was completely incongruent with the concatenated gene tree, indicating it is probably the most frequently recombined gene among closely related phylogenetic clusters, an outcome of its extreme sequence conservation and the log-linear frequency of recombination and genetic distance relationship. Interesting to point out is that it is not the strain relatedness but the gene relatedness that is important for frequency of recombination, indicating that numerous recombination events could be ongoing at multiple loci between multiple 'lineages', simultaneously (Papke *et al.*, 2007).

Given that recombination is the diving force behind diversification (and by extension selection) in the Halobacteria, it is logical to attempt to quantify this level of influence. Recent studies have confirmed that the class as a whole is highly recombinogenic (Naor *et al.*, 2012; Williams *et al.*, 2012). Both Williams *et al.* (2012) and Naor *et al.* (2012) recover a distance–decay relationship between relatedness and frequency of recombination. In Williams' case the frequency of recombination and genetic distance is measured using ribosomal protein tree as a query tree to discover recombined genes on the chromosomes and ribosomal protein divergence to estimate how related the donor and recipient are. Naor measured directly the frequency of recombination using genetically manipulatable strains with sequenced genomes: *Hfx. volcanii*, and *Hfx. mediterranei*. Williams found the log-linear relationship to hold throughout the entire class (using 21 genomes, representing all of the major Halobacterial groups). Examining the relaxed core (defined as genes found in 15 of the 21 Halobacterial genomes) it was estimated that 11–20% of genes evolved in other taxa. Naor's work focused on putative barriers to mating (cell fusion) and recombination within and between species that display approximately 14% nucleotide divergence. As mentioned above, mating is a multistep process that first involves cell membrane fusion, and then recombination. The rate of cell fusion (a 'premating' barrier) was measured by taking advantage of the fact that mated cells remain in a heterochromosomal (diploid) state that preserves the presence of molecular markers from both of the mated strains: PCR amplification was used to assess the presence of each molecular marker. The numbers of between species cell fusion events was less than an order of magnitude smaller compared to within species cell fusions. After measuring the successful fusion events, selection pressure was added to heterochromosomal colonies to identify any hybrid strains that underwent recombination, and an estimate for recombination (a 'postmating' barrier) frequency was made. Surprisingly, the difference within to between species recombination was less than an order of magnitude different. The results recapitulated the expected distance–decay relationship. However, the slope was drastically reduced compared to expectations (e.g. see Vulic *et al.*, 1997; Zawadzki *et al.*, 1995). Instead of the multifold orders of magnitude drop-off in recombination seen in bacterial data (and the basis for computer simulations of Fraser *et al.*), less than an order of magnitude decrease was observed. However, these data still support a 'clumpiness' to organisms in nature defined by groups or clusters that engage in more recombination amongst each other than between groups, something we refer to as preferred trading partners (Papke and Gogarten, 2012).

Geographic isolation and barriers to recombination

From the evidence, it becomes fairly easy to conclude that homologous recombination is an overwhelming homogenizing force for maintaining genetic cohesion within populations of gene trading partners. New alleles that provide a selective advantage arising from mutation can sweep through populations without affecting even nearby loci on a chromosome. In comparison, if loci were linked on chromosomes, the selection process would affect all loci equally bringing them all to zero diversity. Most data indicate that loci have varying amounts of sequence variation (especially at 3rd codon positions), the best explanation for this observation is that genes are unlinked due to recombination. Alleles that originate from outside a recombining population and provide an adaptive advantage will have the same fate as one that arose by mutation alone. However, it will leave a 'foreign' signature because the phylogenetic reconstruction will show a different relationship among relatives for the same gene. When new alleles are invading populations on a regular basis, the evolutionary history of the group will acquire a mosaic appearance, in comparison to any relatives. In a recent publication on marine *Vibrio* populations, comparative genome analysis demonstrated that a very small fraction (approximately only 100 loci out of thousands shared) had the same phylogenetic topology, indicating that genes were invading and being fixed in populations on a regular basis (Shapiro *et al.*, 2012). Within Halobacterial populations, gene flow is fast enough to unlink alleles and loci to the point of random association, to fix alleles in populations and to generate every possible evolutionary relationship for individual genes on a chromosome (Papke *et al.*, 2007). The important point to emphasize here is that there is an additional layer of complexity between diverged strains: the rate of homologous recombination though slower between than within phylogenetic clusters, can still be quite high, dictating that once diverged, two populations could re-merge given an increase in trading frequency (for interesting possibilities see Sheppard *et al.*, 2008; Zhaxybayeva *et al.*, 2009). The conclusion drawn is that preventing recombination between closely related cells (e.g. sister clones) is very difficult. Yet divergence occurs nonetheless! The question, then, is 'How?' We suggest a role for geographic isolation.

The geographic distribution of the Halobacteria may play a major role in the dynamic interplay between homogenizing and diverging forces. A fundamental fact is that the same genera, and some might argue species within the Halobacteria have a global distribution (e.g. *Haloquadratum walsbyi* (Dyall-Smith *et al.*, 2011; Oh *et al.*, 2010)). Contrasted against this is the reality that not every halophilic environment is the same (e.g. thalassohaline vs. athalassohaline; basic vs. neutral pH). Examination of community composition for solar saltern saturated brines (~35%NaCl) located in Alicante, Spain and Chula Vista, California, revealed dramatically different species richness and abundance, despite both sites being sampled at the same time, being derived from seawater, and existing at the same latitude (Zhaxybayeva *et al.*, 2013). Alicante hosted almost completely Halobacterial inhabitants, and primarily the genus *Haloquadratum*, while Chula Vista possessed an approximately 50:50 split between archaeal and bacterial communities. Each site was uniquely constructed, and had limited shared operational taxonomic units even when applying 95% identity definitions, suggesting enormous difficulty for most halophilic prokaryotes to freely disperse across great distances. Interestingly, saturated brines from Australia that have similar chemical composition also displayed clear difference in species richness and abundance between locations, and very few shared OTUs in comparison to site-specific ones (Oh *et al.*, 2010), indicating even within continents it is difficult to disperse between sites. In the case for strains from the same species, evidence for geographic affect is also found. Comparison of two *Haloquadratum walsbyi* genomes from Spain and Australia demonstrated approximately 1.4% sequence divergence across roughly 84% of their shared genomic regions (Dyall-Smith *et al.*, 2011). Given the known high recombination rates of Halobacteria, which act as a homogenizing force, it is unlikely these two isolates have lived in the same location for some time.

Despite finding a few recognizable Halobacterial OTUs across the globe, the evidence indicates that each halophilic site is unique in composition. This on the surface seems paradoxical, as with the ability to freely disperse, one would expect that each site should be composed similarly. There could be at least two explanations for the observation. Sampling could be a big issue: how deeply sites are sampled could affect the observations and conclusions made. If methods cannot obtain a statistically significant representation of the true diversity and its abundance than we are in deep trouble as ecologists and evolutionists. Molecular techniques though having limitations and biases seem to be very sensitive for estimating diversity and abundance, however. Underlying some of the objection to the observation of differentially composed communities, and the push for the 'poor sampling' hypothesis is the culturally ingrained and ancient microbiological dictum 'everything is everywhere', which is taken to mean that if two distant locations have similar abiotic conditions they should be composed of the same microbiota (i.e. no dispersal or invasiveness limitations). This idea came out of the Darwinian revolution (i.e. only fitness affects speciation) and before ideas of geographic endemism were accepted for macroflora and fauna (O'Malley, 2007). Add to this layer a dearth of technological knowhow for most of the twentieth century for identifying prokaryotic diversity, and a real possibility that we may never know what bacterial species are, it is no wonder that we can see the 'same thing' everywhere. In the world of macrofauna and flora biologists, geographic isolation is the null hypothesis for speciation, and proving deviations from the null model is required to propose sympatric speciation. Though sympatric speciation occurs in animals and plants, only a few rare conditions exist that can promote it in freely recombining populations (Hey and Pinho, 2010). We microbiologists do not have a null model for speciation, but we seem to accept sympatric speciation frequently without examining the global distribution. Data are accruing in favour of geographic endemism for halophiles and Halobacteria. The same types of observation are frequently made: similar species and genera are found in many places but often there are radiations of microdiversity in specific locations, and that similar sites are composed of different diversity. We think the data support a hypothesis that periods of endemism occur in sites all over the world, but that it happens over short geological time. Short durations of endemism are followed by dispersal events. If recombination prevents divergence in populations, and there is excellent evidence indicating that this is the rule, not the exception, but even if it is not and the homogenizing force is selective sweeps (e.g. see Papke and Ward, 2004), then migration to a new location will counteract both of those forces and divergence can ensue. This means that speciation will be easily facilitated if the rate of mutation and recombination between 'species' is faster as an evolutionary force at generating diversity than the rate of dispersal is at finding places already occupied by the same 'species'.

References

Allers, T., and Mevarech, M. (2005). Archaeal genetics – the third way. Nat. Rev. Genet. *6*, 58–73.

Andam, C.P., and Gogarten, J.P. (2011). Biased gene transfer and its implications for the concept of lineage. Biol. Direct 6, 47.

Andam, C.P., Harlow, T.J., Papke, R.T., and Gogarten, J.P. (2012). Ancient origin of the divergent forms of leucyl-tRNA synthetases in *Halobacteriales*. BMC Evol. Biol. *12*, 85.

Andam, C.P., Williams, D., and Gogarten, J.P. (2010a). Biased gene transfer mimics patterns created through shared ancestry. Proc. Natl. Acad. Sci. U.S.A. *107*, 10679–10684.

Andam, C.P., Williams, D., and Gogarten, J.P. (2010b). Natural taxonomy in light of horizontal gene transfer. Biol. Philos. *25*, 589–602.

Atanasova, N.S., Roine, E., Oren, A., Bamford, D.H., and Oksanen, H.M. (2012). Global network of specific virus–host interactions in hypersaline environments. Environ. Microbiol. *14*, 426–440.

Badger, J.H., Eisen, J.A., and Ward, N.L. (2005). Genomic analysis of *Hyphomonas neptunium* contradicts 16S rRNA gene-based phylogenetic analysis: implications for the taxonomy of the orders '*Rhodobacterales*' and *Caulobacterales*. Int. J. Syst. Evol. Microbiol. *55*, 1021–1026.

Bertani, G., and Baresi, L. (1987). Genetic transformation in the methanogen *Methanococcus voltae* PS. J. Bacteriol. *169*, 2730–2738.

Boucher, Y., Douady, C.J., Sharma, A.K., Kamekura, M., and Doolittle, W.F. (2004a). Intragenomic heterogeneity and intergenomic recombination among haloarchaeal rRNA genes. J. Bacteriol. *186*, 3980–3990.

Boucher, Y., Kamekura, M., and Doolittle, W.F. (2004b). Origins and evolution of isoprenoid lipid biosynthesis in archaea. Mol. Microbiol. *52*, 515–527.

Breuert, S., Allers, T., Spohn, G., and Soppa, J. (2006). Regulated polyploidy in halophilic archaea. PLoS One *1*, e92.

Carreto, L., Moore, E., Fernanda-Nobre, M., Wait, R., Riley, P., Sharp, R.J., and Da Costa, M.S. (1996). *Rubrobacter xylanophilus* sp. nov., a new thermophilic species isolated from a thermally polluted effluent. IJSEM *46*, 460–465.

Charlebois, R.L., Lam, W.L., Cline, S.W., and Doolittle, W.F. (1987). Characterization of pHV2 from *Halobacterium volcanii* and its use in demonstrating transformation of an archaebacterium. Proc. Natl. Acad. Sci. U.S.A. *84*, 8530–8534.

Chen, I., Christie, P.J., and Dubnau, D. (2005). The ins and outs of DNA transfer in bacteria. Science *310*, 1456–1460.

Claverys, J.P., and Havarstein, L.S. (2007). Cannibalism and fratricide: mechanisms and *raisons d'etre*. Nat. Rev. Microbiol. *5*, 219–229.

Cline, S.W., and Doolittle, W.F. (1987). Efficient transfection of the archaebacterium *Halobacterium halobium*. J. Bacteriol. *169*, 1341–1344.

Cohan, F.M. (2002). Sexual isolation and speciation in bacteria. Genetica *116*, 359–370.

Cohan, F.M. (2006). Towards a conceptual and operational union of bacterial systematics, ecology, and evolution. Philos. Trans. R. Soc. Lond. B Biol. Sci. *361*, 1985–1996.

Cuadros-Orellana, S., Martin-Cuadrado, A.B., Legault, B., D'Auria, G., Zhaxybayeva, O., Papke, R.T., and Rodriguez-Valera, F. (2007). Genomic plasticity in prokaryotes: the case of the square haloarchaeon. ISME J. *1*, 235–245.

Cui, H.L., Zhou, P.J., Oren, A., and Liu, S.J. (2009). Intraspecific polymorphism of 16S rRNA genes in two halophilic archaeal genera, *Haloarcula* and *Halomicrobium*. Extremophiles *13*, 31–37.

Dawkins, R. (1976). The Selfish Gene (Oxford University Press, Oxford, UK).

Dennis, P.P., Ziesche, S., and Mylvaganam, S. (1998). Transcription analysis of two disparate rRNA operons in the halophilic archaeon *Haloarcula marismortui*. J. Bacteriol. *180*, 4804–4813.

Doolittle, W.F., and Zhaxybayeva, O. (2009). On the origin of prokaryotic species. Genome Res. *19*, 744–756.

Dyall-Smith, M., Tang, S.L., and Bath, C. (2003). Haloarchaeal viruses: how diverse are they? Res. Microbiol. *154*, 309–313.

Dyall-Smith, M.L., Pfeiffer, F., Klee, K., Palm, P., Gross, K., Schuster, S.C., Rampp, M., and Oesterhelt, D. (2011). *Haloquadratum walsbyi*: limited diversity in a global pond. PLoS One *6*, e20968.

Dykhuizen, D.E., and Green, L. (1991). Recombination in *Escherichia coli* and the definition of biological species. J. Bacteriol. *173*, 7257–7268.

Feil, E.J., Smith, J.M., Enright, M.C., and Spratt, B.G. (2000). Estimating recombinational parameters in *Streptococcus pneumoniae* from multilocus sequence typing data. Genetics *154*, 1439–1450.

Finkel, S.E., and Kolter, R. (2001). DNA as a nutrient: novel role for bacterial competence gene homologs. J. Bacteriol. *183*, 6288–6293.

Fraser, C., Hanage, W.P., and Spratt, B.G. (2007). Recombination and the nature of bacterial speciation. Science *315*, 476–480.

Gogarten, J.P., Doolittle, W.F., and Lawrence, J.G. (2002). Prokaryotic evolution in light of gene transfer. Mol. Biol. Evol. *19*, 2226–2238.

Grant, W.D., Gemmell, R.T., and McGenity, T.J. (1998). Halobacteria: the evidence for longevity. Extremophiles *2*, 279–287.

Gupta, R.S., Pereira, M., Chandrasekera, C., and Johari, V. (2003). Molecular signatures in protein sequences that are characteristic of Cyanobacteria and plastid homologues. Int. J. Syst. Evol. Microbiol. *53*, 1833–1842.

Hamilton, H.L., and Dillard, J.P. (2006). Natural transformation of *Neisseria gonorrhoeae*: from DNA donation to homologous recombination. Mol. Microbiol. *59*, 376–385.

Havarstein, L.S., and Morrison, D.A. (1999). Quorum sensing and peptide pheromones in streptococcal competence for genetic transfromation. In Cell–Cell Signaling in Bacteria, Dunny, G.M., and Winans, S.C., eds. (ASM Press, Washington DC).

Hey, J., and Pinho, C. (2010). Divergence with gene flow: models and data. Annu. Rev. Ecol. Evol. Syst. *41*, 215–230.

Ihara, K., Umemura, T., Katagiri, I., Kitajima-Ihara, T., Sugiyama, Y., Kimura, Y., and Mukohata, Y. (1999). Evolution of the archaeal rhodopsins: evolution rate changes by gene duplication and functional differentiation. J. Mol. Biol. *285*, 163–174.

Jain, R., Rivera, M.C., and Lake, J.A. (1999). Horizontal gene transfer among genomes: the complexity hypothesis. Proc. Natl. Acad. Sci. U.S.A. *96*, 3801–3806.

Johnsborg, O., Eldholm, V., and Havarstein, L.S. (2007). Natural genetic transformation: prevalence, mechanisms and function. Res. Microbiol. *158*, 767–778.
Khomyakova, M., Bukmez, O., Thomas, L.K., Erb, T.J., and Berg, I.A. (2011). A methylaspartate cycle in haloarchaea. Science *331*, 334–337.
Kolbe, M., Besir, H., Essen, L.O., and Oesterhelt, D. (2000). Structure of the light-driven chloride pump halorhodopsin at 1.8 A resolution. Science *288*, 1390–1396.
Korbel, J.O., Snel, B., Huynen, M.A., and Bork, P. (2002). SHOT: a web server for the construction of genome phylogenies. Trends Genet. *18*, 158–162.
Lang, A.S., Zhaxybayeva, O., and Beatty, J.T. (2012). Gene transfer agents: phage-like elements of genetic exchange. Nat. Rev. Microbiol. *10*, 472–482.
Lange, C., Zerulla, K., Breuert, S., and Soppa, J. (2011). Gene conversion results in the equalization of genome copies in the polyploid haloarchaeon *Haloferax volcanii*. Mol. Microbiol. *80*, 666–677.
Lawrence, J.G. (2002). Gene transfer in bacteria: speciation without species? Theor. Pop. Biol. *61*, 449–460.
Lawrence, J.G., and Roth, J.R. (1996). Selfish operons: horizontal transfer may drive the evolution of gene clusters. Genetics *143*, 1843–1860.
Liao, D. (1999). Concerted evolution: molecular mechanism and biological implications. Am. J. Hum. Genet. *64*, 24–30.
Lipscomb, G.L., Stirrett, K., Schut, G.J., Yang, F., Jenney, F.E. Jr., Scott, R.A., Adams, M.W., and Westpheling, J. (2011). Natural competence in the hyperthermophilic archaeon *Pyrococcus furiosus* facilitates genetic manipulation: construction of markerless deletions of genes encoding the two cytoplasmic hydrogenases. Appl. Environ. Microbiol. *77*, 2232–2238.
Lopez-Lopez, A., Benlloch, S., Bonfa, M., Rodriguez-Valera, F., and Mira, A. (2007). Intragenomic 16S rDNA divergence in *Haloarcula marismortui* is an adaptation to different temperatures. J. Mol. Evol. *65*, 687–696.
Lozier, R.H., Bogomolni, R.A., and Stoeckenius, W. (1975). Bacteriorhodopsin: a light-driven proton pump in *Halobacterium halobium*. Biophys. J. *15*, 955–962.
Martin, W., and Muller, M. (1998). The hydrogen hypothesis for the first eukaryote. Nature *392*, 37–41.
Matte-Tailliez, O., Brochier, C., Forterre, P., and Philippe, H. (2002). Archaeal phylogeny based on ribosomal proteins. Mol. Biol. Evol. *19*, 631–639.
Mevarech, M., and Werczberger, R. (1985). Genetic transfer in *Halobacterium volcanii*. J. Bacteriol. *162*, 461–462.
Mullakhanbhai, M.F., and Larsen, H. (1975). *Halobacterium volcanii* spec. nov., a Dead Sea halobacterium with a moderate salt requirement. Arch. Microbiol. *104*, 207–214.
Mylvaganam, S., and Dennis, P.P. (1992). Sequence heterogeneity between the two genes encoding 16S rRNA from the halophilic archaebacterium *Haloarcula marismortui*. Genetics *130*, 399–410.
Naor, A., Lapierre, P., Mevarech, M., Papke, R.T., and Gophna, U. (2012). Low species barriers in halophilic archaea and the formation of recombinant hybrids. Curr. Biol. *22*, 1444–1448.
Nelson-Sathi, S., Dagan, T., Landan, G., Janssen, A., Steel, M., McInerney, J.O., Deppenmeier, U., and Martin, W.F. (2012). Acquisition of 1,000 eubacterial genes physiologically transformed a methanogen at the origin of Haloarchaea. Proc. Natl. Acad. Sci. U.S.A. *109*, 20537–20542.
Ng, W.V., Kennedy, S.P., Mahairas, G.G., Berquist, B., Pan, M., Shukla, H.D., Lasky, S.R., Baliga, N.S., Thorsson, V., Sbrogna, J., *et al.* (2000). Genome sequence of *Halobacterium* species NRC-1. Proc. Natl. Acad. Sci. U.S.A. *97*, 12176–12181.
O'Malley, M.A. (2007). The nineteenth century roots of 'everything is everywhere'. Nat. Rev. Microbiol. *5*, 647–651.
Oh, D., Porter, K., Russ, B., Burns, D., and Dyall-Smith, M. (2010). Diversity of *Haloquadratum* and other haloarchaea in three, geographically distant, Australian saltern crystallizer ponds. Extremophiles *14*, 161–169.
Oren, A. (2008). Microbial life at high salt concentrations: phylogenetic and metabolic diversity. Saline Syst. *4*, 2.
Oren, A., Ventosa, A., and Grant, W.D. (1997). Proposed minimal standards for description of new taxa in the order Halobacteriales. Int. J. Syst. Bacteriol. *47*, 233–238.
Ortenberg, R., Tchelet, R., and Mevarech, M. (1999). A model for the genetic exchange system of the extremely halophilic archaeon *Haloferax volcanii*. In Microbiology and Biogenochemistry of Hypersaline Environments, Oren, A., ed. (CRC Press, Boca Raton), pp. 331–338.
Papke, R.T. (2009). A critique of prokaryotic species concepts. Methods Mol. Biol. *532*, 379–395.
Papke, R.T., and Gogarten, J.P. (2012). How bacterial lineages emerge. Science *336*, 45–46.

Papke, R.T., and Ward, D.M. (2004). The importance of physical isolation to microbial diversification. FEMS Microbiol. Ecol. *48*, 293–303.
Papke, R.T., Koenig, J.E., Rodriguez-Valera, F., and Doolittle, W.F. (2004). Frequent recombination in a saltern population of *Halorubrum*. Science *306*, 1928–1929.
Papke, R.T., Zhaxybayeva, O., Feil, E.J., Sommerfeld, K., Muise, D., and Doolittle, W.F. (2007). Searching for species in haloarchaea. Proc. Natl. Acad. Sci. U.S.A. *104*, 14092–14097.
Redfield, R.J. (2001). Do bacteria have sex? Nat. Rev. Genet. *2*, 634–639.
Rodriguez-Brito, B., Li, L., Wegley, L., Furlan, M., Angly, F., Breitbart, M., Buchanan, J., Desnues, C., Dinsdale, E., Edwards, R., *et al.* (2010). Viral and microbial community dynamics in four aquatic environments. ISME J. *4*, 739–751.
Rosenshine, I., Tchelet, R., and Mevarech, M. (1989). The mechanism of DNA transfer in the mating system of an archaebacterium. Science *245*, 1387–1389.
Sato, T., Fukui, T., Atomi, H., and Imanaka, T. (2003). Targeted gene disruption by homologous recombination in the hyperthermophilic archaeon *Thermococcus kodakaraensis* KOD1. J. Bacteriol. *185*, 210–220.
Shapiro, B.J., Friedman, J., Cordero, O.X., Preheim, S.P., Timberlake, S.C., Szabo, G., Polz, M.F., and Alm, E.J. (2012). Population genomics of early events in the ecological differentiation of bacteria. Science *336*, 48–51.
Sharma, A.K., Spudich, J.L., and Doolittle, W.F. (2006). Microbial rhodopsins: functional versatility and genetic mobility. Trends Microbiol. *14*, 463–469.
Sharma, A., Walsh, D., Bapteste, E., Rodriguez-Valera, F., Ford Doolittle, W., and Papke, R.T. (2007). Evolution of rhodopsin ion pumps in haloarchaea. BMC Evol. Biol. *7*, 79.
Sheppard, S.K., McCarthy, N.D., Falush, D., and Maiden, M.C. (2008). Convergence of *Campylobacter* species: implications for bacterial evolution. Science *320*, 237–239.
Smith, H.O., Tomb, J.F., Dougherty, B.A., Fleischmann, R.D., and Venter, J.C. (1995). Frequency and distribution of DNA uptake signal sequences in the *Haemophilus influenzae* Rd genome. Science *269*, 538–540.
Smith, J.M., Feil, E.J., and Smith, N.H. (2000). Population structure and evolutionary dynamics of pathogenic bacteria. Bioessays *22*, 1115–1122.
Spudich, J.L., and Bogomolni, R.A. (1988). Sensory rhodopsins of Halobacteria. Annu. Rev. Biophys. Biophys. Chem. *17*, 193–215.
Stackebrandt, E., and Goebel, B.M. (1994). Taxonomic note: a place for DNA–DNA reassociation and 16S rRNA sequence analysis in the present species definition in bacteriology. Int. J. Syst. Evol. Microbiol. *44*, 846–849.
Steinmoen, H., Knutsen, E., and Havarstein, L.S. (2002). Induction of natural competence in *Streptococcus pneumoniae* triggers lysis and DNA release from a subfraction of the cell population. Proc. Natl. Acad. Sci. U.S.A. *99*, 7681–7686.
Thomas, C.M., and Nielsen, K.M. (2005). Mechanisms of, and barriers to, horizontal gene transfer between bacteria. Nat. Rev. Microbiol. 3, 711–721.
Tripepi, M., Imam, S., and Pohlschroder, M. (2010). *Haloferax volcanii* flagella are required for motility but are not involved in PibD-dependent surface adhesion. J. Bacteriol. *192*, 3093–3102.
Vreeland, R.H., Straight, S., Krammes, J., Dougherty, K., Rosenzweig, W.D., and Kamekura, M. (2002). *Halosimplex carlsbadense* gen. nov., sp. nov., a unique halophilic archaeon, with three 16S rRNA genes, that grows only in defined medium with glycerol and acetate or pyruvate. Extremophiles *6*, 445–452.
Vulic, M., Dionisio, F., Taddei, F., and Radman, M. (1997). Molecular keys to speciation: DNA polymorphism and the control of genetic exchange in enterobacteria. Proc. Natl. Acad. Sci. U.S.A. *94*, 9763–9767.
Wellner, A., Lurie, M.N., and Gophna, U. (2007). Complexity, connectivity, and duplicability as barriers to lateral gene transfer. Genome Biol. *8*, R156.
Whitaker, R.J., Grogan, D.W., and Taylor, J.W. (2005). Recombination shapes the natural population structure of the hyperthermophilic archaeon *Sulfolobus islandicus*. Mol. Biol. Evol. *22*, 2354–2361.
Williams, D., Gogarten, J.P., and Papke, R.T. (2012). Quantifying homologous replacement of loci between haloarchaeal species. Genome Biol. Evol. *4*, 1223–1244.
Woese, C.R., and Fox, G.E. (1977). Phylogenetic structure of the prokaryotic domain: the primary kingdoms. Proc. Natl. Acad. Sci. U.S.A. *74*, 5088–5090.
Woese, C.R., Kandler, O., and Wheelis, M.L. (1990). Towards a natural system of organisms: proposal for the domains Archaea, Bacteria, and Eucarya. Proc. Natl. Acad. Sci. U.S.A. *87*, 4576–4579.
Worrell, V.E., Nagle, D.P. Jr., McCarthy, D., and Eisenbraun, A. (1988). Genetic transformation system in the archaebacterium *Methanobacterium thermoautotrophicum* Marburg. J. Bacteriol. *170*, 653–656.

Yap, W.H., Zhang, Z., and Wang, Y. (1999). Distinct types of rRNA operons exist in the genome of the actinomycete *Thermomonospora chromogena* and evidence for horizontal transfer of an entire rRNA operon. J. Bacteriol. *181*, 5201–5209.

Zawadzki, P., Roberts, M.S., and Cohan, F.M. (1995). The log–linear relationship between sexual isolation and sequence divergence in *Bacillus* transformation is robust. Genetics *140*, 917–932.

Zhaxybayeva, O., Gogarten, J.P., Charlebois, R.L., Doolittle, W.F., and Papke, R.T. (2006). Phylogenetic analyses of cyanobacterial genomes: quantification of horizontal gene transfer events. Genome Res. *16*, 1099–1108.

Zhaxybayeva, O., Doolittle, W.F., Papke, R.T., and Gogarten, J.P. (2009). Intertwined evolutionary histories of marine *Synechococcus* and *Prochloroccus marinus*. Genome Biol. Evol. *1*, 325–339.

Zhaxybayeva, O., Stepanauskas, R., Mohan, N.R., and Papke, R.T. (2013). Cell sorting analysis of geographically separated hypersaline environments. Extremophiles *17*, 265–275.

Comparative Genomics of Haloarchaeal Viruses

4

Elina Roine

Abstract

Haloarchaeal viruses have been among the least studied group with prokaryotic (i.e. bacteria and Archaea) hosts. However, recent efforts have brought the number of isolated haloarchaeal viruses from 20 to almost 70, environmental phage genomes from 1 to more than 40 and increased the amount of environmental sequences of halophilic viral fraction with several megabases. Also, new research has further reinforced the observation that haloarchaeal viruses are more reminiscent of bacteriophages (i.e. bacteria-infecting viruses) than the morphologically varying crenarchaeal viruses. This review will update the latest advances in genomics of haloarchaeal viruses, its impact on taxonomy, and suggest some additional aspects of the bacteriophage genomics that may prove to be of significance also in haloarchaeal viral genomics.

Introduction

Viruses at their simplest are nucleoprotein complexes parasitizing cellular life. Suttle (2005) postulated that where ever there is cellular life there will also be viruses. Because of the obligate parasitic nature of viruses, the interplay with their hosts is bound to be very close. This applies especially to the aspects of genome replication for which many cellular enzymes are often utilized as well as to transcription and translation of viral components.

Hypersaline environments have been considered to be restricted in organismic diversity. The major players among micro-organisms have traditionally included halophilic prokaryotes from both bacterial and archaeal domains as well as the halophilic alga *Dunaliella salina* (Oren, 2008). There are, however, also a range of other eukaryotic organisms such as extremely halotolerant black yeasts or melanized yeast-like fungi (Oren, 2008; Plemenitaš and Gunde-Cimerman, 2011). Environmental studies have suggested that in the hypersaline environment the amount of viruses increases along the salt gradient (Pedrós-Alió *et al.*, 2000) and exceeds the amount of their hosts at least 10-fold (Santos *et al.*, 2012, and references therein). Although viruses were found to be the only parasites in some hypersaline environments (Guixa-Boixareu *et al.*, 1996), actively grazing flagellates have also been found in other hypersaline systems (Park *et al.*, 2003). When studying the direct impact of haloarchaeal viruses on their hosts, it is important to make a distinction between a functional virus and genetic element that seems to originate from a virus with no information about its ability to infect an organism. Since real viruses can be identified on the basis of their ability to produce infective viral particles, virions, with few exceptions this review will mainly concentrate on discussing those genomes that are shown to originate from functional viruses.

The goal of this review is to synthesize the results from the many fresh reports of viral genomic sequences obtained from the hypersaline environments. I will concentrate on viruses of haloarchaea, but by introducing first the general concepts prevailing in genomics of prokaryotic viruses. While there are plenty of examples of the more indirect influence of prokaryotic virus infections on their hosts (Brüssow *et al.*, 2004; Casjens, 2003; Hendrix *et al.*, 2000; Suttle, 2005, 2007), we can only predict similar phenomena to be found also in the interactions between haloarchaeal viruses and their hosts. Information presented in many extensive previous reviews of haloarchaeal viruses will for the most part not be duplicated here. (e.g. Dyall-Smith *et al.*, 2003; Porter and Dyall-Smith, 2006; Porter *et al.*, 2007, 2008; Reiter *et al.*, 1988; Roine and Oksanen, 2011).

Haloarchaeal versus prokaryotic viruses

In 2007 there were approximately 5500 prokaryotic viruses described by electron microscopy (Ackermann, 2007). Of these 96% were tailed viruses belonging to the order *Caudovirales*. Most of the tailed viruses (61%) belong to the family *Siphoviridae*, tailed viruses with flexible, but non-contractile tail (Fig. 4.1A; Ackermann, 2007). Family *Myoviridae*, tailed viruses with contractile tail contains 25% of the order and the rest (14%) belong to the family *Podoviridae*. The remaining 4% of all known prokaryotic viruses consists of viruses with various morphotypes such as simple icosahedral viruses or icosahedral viruses with inner membrane, filamentous viruses and viruses containing an outer membrane (an envelope; Fig. 4.1A). Crenarchaeal viruses exhibit wider diversity in the virion morphotypes, whereas the characterized haloarchaeal viruses are more similar to bacteriophage morphotypes (Dyall-Smith *et al.*, 2003; Pina *et al.*, 2011). The first haloarchaeal virus isolated and characterized was a tailed virus (Torsvik and Dundas, 1974). This discovery was followed by almost a dozen more tailed viruses (Porter *et al.*, 2008). Still to date, tailed morphotypes represent by far the majority among haloarchaeal viruses (Fig. 4.1B and Table 4.1). The first non-tailed haloarchaeal virus isolated was His1, a spindle shaped virus (Bath and Dyall-Smith, 1998) and SH1, an icosahedral virus infecting *Haloarcula hispanica*, was the first icosahedral membrane containing virus among haloarchaeal viruses (Porter *et al.*, 2005). In a recent study (Atanasova *et al.*, 2012) we reported the results of screening for haloarchaeal viruses world-wide using the culture based method. We isolated 28 new tailed viruses and among them the first member of haloarchaeal podovirus '*Haloarcula sinaiiensis*' tailed virus 1, HSTV-1. In addition to the pleomorphic viruses HRPV-1 (Pietilä *et al.*, 2009) and HHPV-1 (Roine *et al.*, 2010), three new pleomorphic viruses were reported. Also, one new icosahedral virus *Haloarcula hispanica* icosahedral virus 2 (HHIV-2) that is closely related to SH1 was isolated. In conclusion, the morphotypes discovered in the study of Atanasova *et al.* (2012) reflect well the proportions of morphotypes found in bacteriophages except for the filamentous viruses that have not (yet) been found among haloarchaeal viruses. Also, the majority of isolated haloarchaeal tailed viruses are myoviruses instead of siphoviruses (Table 4.1). The viral genome home page at NCBI (http://www.ncbi.nlm.nih.gov/genomes/GenomesGroup.cgi?taxid=10239&opt=Virus&sort=genome) currently lists almost 700 complete genome sequences of bacteriophages (August 2012). There are approximately 40 complete genome sequences of archaeal viruses listed at the same site

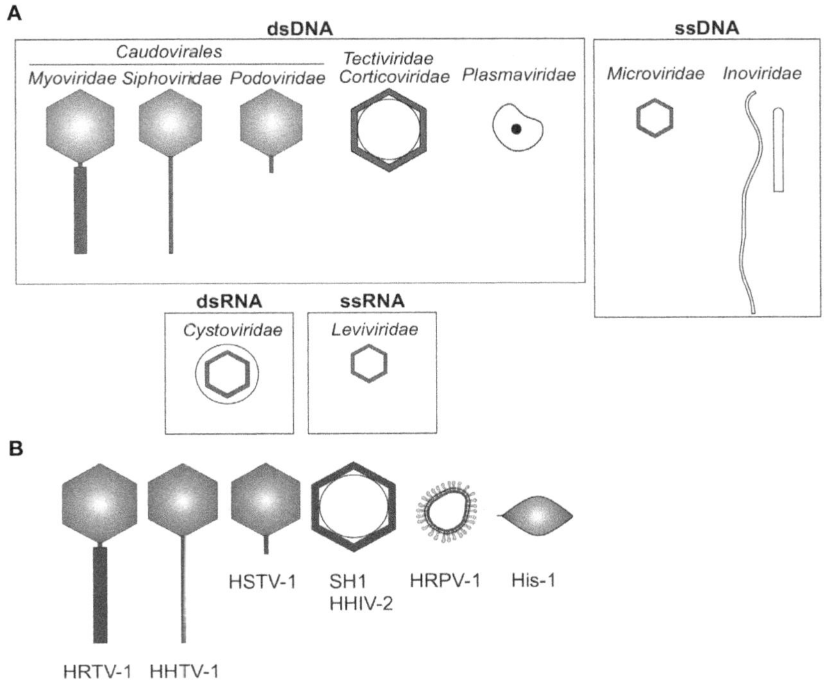

Figure 4.1 Different mcrphotypes of bacteriophages and haloarchaeal viruses. (A) A schematic representation of the d fferent bacteriophage morphotypes. Circles present lipid membranes inside (*Tectiviridae* and *Corticoviridae*) or outside the capsid (*Cystoviridae*). Redrawn and modified from (Ackermann, 2007). (B) A schematic representation of the currently known haloarchaeal virion morphotypes. For each morphotype, at least one example of a haloarchaeal virus species is given.

and seven of those viruses infect haloarchaea. As shown in Table 4.1, the actual number of complete genome sequences of haloarchaeal viruses is 26. All of these numbers are in vast minority when compared to the total number of complete viral genome sequences of almost 3000, and this quantity does not reflect the fact that the vast majority of viruses in the universe infect prokaryotic organisms (Brüssow and Hendrix, 2002; Hendrix, 2002).

Infection cycle

In order to understand the genomics of prokaryotic viruses, it is very useful to keep in mind their life cycle and the different modes of infection. That is, viruses come with at least two different kinds of infection strategies: virulent and temperate (Weinbauer, 2004). Virulent viruses infect the host cell and produce the progeny that are usually released from the infected cell by lysis i.e. break down of the host cells. Temperate viruses, on the other hand, can either go through the same type of life cycle as virulent viruses or postpone the production of progeny viruses often by inserting the viral genome into the host genome (Weinbauer, 2004). In this way, the viral genome becomes an integral part of the host and is passed on to the next generations of the host cells that have become lysogens. Temperate viruses are common not only among tailed viruses, but also among filamentous morphotypes (Brüssow *et al.*, 2004).

Table 4.1 List of haloarchaeal viruses

Virus	Nomenclature/host/morphotype	Reference
Myoviruses		
φCh1	*Natrialba magadii*	Witte *et al.* (1997)
φH	*Halobacterium salinarum*	Schnabel *et al.* (1982b)
HF1	*Haloferax lucentense*	Nuttall and Dyall-Smith (1993)
HF2	*Halorubrum coriense/Halorubrum saccharovorum*	Nuttall and Dyall-Smith (1993)
Hs1	*Halobacterium salinarum*	Torsvik and Dundas (1974)
Ja1	*Halobacterium salinarum*	Wais *et al.* (1975)
S41	*Halobacterium salinarum*	Daniels and Wais (1998)
S50.2	*Halobacterium salinarum*	Daniels and Wais (1998)
S4100	*Halobacterium salinarum*	Daniels and Wais (1990)
S5100	*Halobacterium salinarum*	Daniels and Wais (1990)
HATV-1	*Haloarcula* head-tail virus 1	Atanasova *et al.* (2012)
HATV-2	*Haloarcula* head-tail virus 2	Atanasova *et al.* (2012)
HGTV-1	*Halogranum* head-tail virus 1	Atanasova *et al.* (2012)
HJTV-1	*Haloarcula japonica* head-tail virus 1	Atanasova *et al.* (2012)
HJTV-2	*Haloarcula japonica* head-tail virus 1	Atanasova *et al.* (2012)
HRTV-1	*Halorubrum* head-tail virus 1	Kukkaro and Bamford (2009)
HRTV-2	*Halorubrum* head-tail virus 2	Atanasova *et al.* (2012)
HRTV-3	*Halorubrum* head-tail virus 3	Atanasova *et al.* (2012)
HRTV-5	*Halorubrum* head-tail virus 5	Atanasova *et al.* (2012)
HRTV-6	*Halorubrum* head-tail virus 6	Atanasova *et al.* (2012)
HRTV-7	*Halorubrum* head-tail virus 7	Atanasova *et al.* (2012)
HRTV-8	*Halorubrum* head-tail virus 8	Atanasova *et al.* (2012)
HRTV-9	*Halorubrum* head-tail virus 9	Atanasova *et al.* (2012)
HRTV-10	*Halorubrum* head-tail virus 10	Atanasova *et al.* (2012)
HRTV-11	*Halorubrum* head-tail virus 11	Atanasova *et al.* (2012)
HRTV-12	*Halorubrum* head-tail virus 12	Atanasova *et al.* (2012)
HSTV-2	*Halorubrum sodomense* head-tail virus 2	Atanasova *et al.* (2012)
HSTV-3	*Halorubrum sodomense* head-tail virus 3	Atanasova *et al.* (2012)
HTV-1	Halophilic head-tail virus 1	Atanasova *et al.* (2012)
Siphoviruses		
BJ1	*Halorubrum saccharovorum*	Pagaling *et al.* (2007)
B10	*Halobacterium* sp.	Torsvik (1982)
φN	*Halobacterium halobium*	Vogelsang-Wenke and Oesterhelt (1988)
HCTV-1	'Haloarcula californiae' head-tail virus 1	Kukkaro and Bamford (2009)
HCTV-2	'Haloarcula californiae' head-tail virus 2	Atanasova *et al.* (2012)
HCTV-5	'Haloarcula californiae' head-tail virus 5	Atanasova *et al.* (2012)
Hh-1	*Halobacterium salinarum*	Pauling (1982)

Table 4.1 continued

Virus	Nomenclature/host/morphotype	Reference
Hh-3	*Halobacterium salinarum*	Pauling (1982)
HHTV-1	*Haloarcula hispanica* head-tail virus 1	Kukkaro and Bamford, (2009)
HHTV-2	*Haloarcula hispanica* head-tail virus	Atanasova *et al*. (2012)
HRTV-4	*Halorubrum* head-tail virus 4	Atanasova *et al*. (2012)
HVTV-1	*Haloarcula vallismortis* head-tail virus 1	Atanasova *et al*. (2012)
HVTV-2	*Haloarcula vallismortis* head-tail virus 2	
SNJ1	*Natrinema* sp.	Mei *et al*. (2007)[a]
S45	*Halobacterium salinarum*	Daniels and Wais (1984)
Podoviruses		
HSTV-1	'Haloarcula sinaiiensis' head-tail virus 1	Atanasova *et al*. (2012)
Spindle shaped		
His1	*Haloarcula hispanica*/spindle-shaped virus	Bath and Dyall-Smith (1998)
His2	*Haloarcula hispanica*/spindle-shaped virus	Bath *et al*. (2006)
Unclassified icosahedral		
SH1		Porter *et al*. (2005)
HHIV-2	*Haloarcula hispanica* icosahedral virus 2	Atanasova *et al*. (2012)
SNJ1	*Natrinema* sp.	Zhang *et al*. (2012)[a]
Unclassified pleomorphic		
HHPV-1	*Haloarcula hispanica* pleomorphic virus 1	Roine *et al*. (2010)
HGPV-1	*Halogeometricum* pleomorphic virus 1	Atanasova *et al*. (2012)
HRPV-1	*Halorubrum* pleomorphic virus 1	Pietilä *et al*. (2009)
HRPV-2	*Halorubrum* pleomorphic virus 2	Atanasova *et al*. (2012)
HRPV-3	*Halorubrum* pleomorphic virus 3	Atanasova *et al*. (2012)
HRPV-6	*Halorubrum* pleomorphic virus 6	Pietilä *et al* (2012b)

[a]SNJ1, originally characterized as a tailed halovirus (Mei *et al*., 2007) was recently reported to be icosahedral (Zhang *et al*., 2012).

These two basic modes of infection are not to be considered the same as the two main exit strategies: lysis of the cells or extrusion/budding out without breaking the host cell membrane integrity. All tailed viruses known to date, both virulent and temperate, lyse their host cells in the end of the productive infection cycle. Enveloped viruses take advantage of their hosts by obtaining the membrane from the host cellular membrane in the process of budding out of the cell (Garoff *et al.*, 1998). On the other hand, filamentous bacteriophages (family of *Inoviridae*) such as M13 or *Pseudomonas* Ff-phages produce new progeny viruses by protrusion through the viral encoded outer membrane pore, so called secretin that is similar to the ones found in type II and type III secretion pathways (Hobbs and Mattick, 1993). The filamentous viral particles, however, do not obtain a lipid envelope, but the virion coat proteins are assembled in a controlled fashion into the capsid concomitantly with the assembly and release from the infected cell (Rakonjac *et al.*, 2011; Russel, 1991). Yet another exception is the bacteriophage Φ6, for which the lipid envelope is assembled in the cytoplasm of the host cell prior to lysis of the cells (Bamford *et al.*, 1976; Mindich and Lehman, 1979). All of the known icosahedral viruses that contain a membrane inside the capsid structure (e.g. PRD1, Bam35, PM2, P23-77, SH1, HHIV-2, SSIV-1) lyse the cells in exit (Aalto *et al.*, 2012; Cota-Robles *et al.*, 1968; Jaakkola *et al.*, 2012; Jaatinen *et al.*, 2008; Kivelä *et al.*, 1999; Porter *et al.*, 2005; Roine and Oksanen, 2011).

Salt concentration influences the adsorption and infection rates of halophilic viruses in different ways (Aalto *et al.*, 2012; Daniels and Wais, 1990; Kukkaro and Bamford, 2009; Torsvik and Dundas, 1980). The adsorption of viruses from both low salt and hypersaline environments was systematically studied by Kukkaro and Bamford (2009), who showed that changes in salt concentration affect the adsorption rate. However, no generalizations could be made. Although the lowest adsorption rate constants were found for haloarchaeal viruses (2.9×10^{-13} ml/min for HHTV-1), some haloarchaeal viruses exhibited adsorption rate constants similar to bacteriophages from low salt environments (2.0×10^{-10} ml/min for HHPV-1 and 4.4×10^{-10} ml/min for PRD1). A recent study on tailed haloarchaeal viruses HVTV-1 and HSTV-2 showed that their infectivity was recovered when the particles were first incubated in buffer devoid of salt and then brought back to the concentration of salt required for infection (Pietilä *et al.*, 2013).

Salt concentration also influences the infection strategy (Aalto *et al.*, 2012; Bettarel *et al.*, 2011). The haloarchaeal viruses Hs1 and S5100 were shown to change from lytic infection to lysogenic strategy in increasing concentrations of NaCl (Daniels and Wais, 1990; Torsvik and Dundas, 1980). Similar observations were made in a study using an environmental approach (Bettarel *et al.*, 2011). Opposite results were obtained for the halophage *Salisaeta* icosahedral phage 1 (SSIP-1; Aalto *et al.*, 2012) for which at least 19% artificial salt water (SW; Halohandbook, 2009) concentration was required for plaque production. Temperate viruses in the studied cases have adjusted the switch between different life cycle modes to conditions when the host cell experiences stress (UV-induction, mitomycin C treatment). In halophilic organisms changes of extracellular salt concentrations may be such a stress factor consequently leading to a switch between lytic or lysogenic strategy of the infection.

Related or not related, that's the problem

Prokaryotic classification is in great turmoil due to the whole genome sequencing that reveals the great genomic divergence even between organisms that were considered closely related (Doolittle, 1999; Doolittle and Papke, 2006; Fraser *et al.*, 2009; Rhodes *et al.*, 2011; Thompson *et al.*, 2005). On one hand it explains the confusing results obtained by gene sequencing indicating that organisms having exactly the same 16S rRNA gene sequence can still have less than 70% of their genomic DNA hybridized (Stackebrandt and Goebel, 1994). On the other hand, it also muddles our perception regarding which genes should be used for classification (de la Haba *et al.*, 2012; Fraser *et al.*, 2009; Graham *et al.*, 2000; Papke *et al.*, 2007). With viruses the perplexity is even greater, because of the lack of a universal gene such as the one encoding 16S rRNA in cellular organisms. Before the era of whole genome sequencing the higher level classification of viruses has traditionally been based on the viral genome and replication types (Baltimore, 1971), the host type (prokaryotic/eukaryotic) and the morphology of the virion (King *et al.*, 2012). In prokaryotic viruses the sequence divergence can happen in relatively short evolutionary time due to their fast reproduction that has been estimated to occur at 10^{23} infections every second (Hendrix, 2003; Suttle,

2007). Especially viruses isolated from new environmental niches often show no significant similarity even at the level of translated amino acid sequences of predicted open reading frames (ORFs) let alone at the nucleotide sequence level (for example Hendrix *et al.*, 1999; Krupovič *et al.*, 2011; Senčilo *et al.*, 2012, 2013). Apart from the genes encoding the basic structural proteins (see below) the functions of the vast majority of predicted ORFs remain unknown (for example Baudoux *et al.*, 2012; Senčilo *et al.*, 2013). In addition to sequence divergence occurring through point mutations, short deletions and insertions, recombination is an important mechanism generating the genomic mosaicism of viral genomes.

Genomic mosaicism

Along the whole genome sequencing of viruses it has become clear that the genomes of prokaryotic viruses are mosaic in structure (Hatfull and Hendrix, 2011; Hendrix *et al.*, 1999; Juhala *et al.*, 2000). The genes encoding products involved in the same function, e.g. capsid formation or tail structure, are clustered into modules (Hendrix *et al.*, 1999), and the order of the different genes in a module is often conserved (Hendrix, 2003; Krupovič *et al.*, 2011; Lawrence *et al.*, 2002). The genes found clustered on a chromosome (module) may show similar evolutionary histories but different modules, for example the capsid encoding genes and genes encoding the components of the tail structure, can have different evolutionary histories (Juhala *et al.*, 2000; Lawrence *et al.*, 2002). Viral components involved in host recognition (e.g. tail fibres of tailed viruses and spike proteins of icosahedral viruses) are also under different selection pressure than the components required for the capsid formation (Brüssow and Desiere, 2001; Lubbers *et al.*, 1995; Saren *et al.*, 2005). Tailed viruses have been classified according to the tail type (contractile/non-contractile etc.), which can be seen in the relatedness of the tail-encoding genes also at the sequence level (Lawrence *et al.*, 2002). Modular structure of the bacteriophage genomes exists in all virus morphotypes where comparative genomics has been conducted to analyse them (Krupovič *et al.*, 2011; Lawrence *et al.*, 2002).

What are the mechanisms behind the observed mosaicism? It has been suggested that both homologous (HR) and non-homologous recombination (NHR) are involved in the exchange of the different modules between viruses that are infecting the same cell (Krupovič *et al.*, 2011 and references therein). Naturally, HR is involved in recombination of related genomes (Arbiol *et al.*, 2010) and NHR of non-related genomes (Juhala *et al.*, 2000; Pedulla *et al.*, 2003). Although it is clear that many of the new genomic combinations created through NHR are not viable, it has been estimated, that even NHR is involved in the production of 10^9 to 10^{10} functional viral genomes every second (Hendrix, 2009).

It is easy to imagine that recombination can also occur between an infecting virus and an integrated temperate virus. The temperate viruses are thus in this respect more prone to the exchange of modules than the virulent viruses (Lawrence *et al.*, 2002). Indeed, analyses of virulent T4 family bacteriophages showed that there is an extensive set of core genes that are inherited vertically (Comeau *et al.*, 2007; Filée *et al.*, 2006). In theory, all different kinds of recombination events can occur, but the successful combinations are those that retain the ability to produce infective viral particles (Hatfull and Hendrix, 2011; Hendrix *et al.*, 2000).

Structure based virus classification

Since virion is the hallmark of a virus it has been proposed that the viral architecture and the 3D structure of the major capsid protein should serve as the higher order classification criteria (Abrescia *et al.*, 2012; Bamford, 2003; Bamford *et al.*, 2002; Benson *et al.*, 2004). This suggestion was based on the initial observation that the major capsid proteins of the icosahedral bacteriophage PRD1 and the icosahedral adenovirus both adopt the so called double jelly roll fold (Benson *et al.*, 1999). In further studies several putative lineages have been proposed that consist of viruses infecting hosts from different domains of life, but sharing the same major capsid protein structure (Abrescia *et al.*, 2010, 2012; Bamford *et al.*, 2002; Benson *et al.*, 2004). To date, the PRD1-adenovirus -lineage contains the most members. The hosts for these viruses represent different domains of life: bacteriophages PRD1 and PM2 infect bacteria, adenovirus a eukaryote and crenarchaeal viruses *Sulfolobus* turreted icosahedral virus (STIV (Khayat *et al.*, 2005) and *Sulfolobus* turreted icosahedral virus 2 (STIV2) hyperthermophilic Archaea (Happonen *et al.*, 2010). It is a well-documented phenomenon and an appreciated fact that within an evolutionary time period two homologues of a protein can diverge in amino acid sequence to the point where no identity can be detected at the primary sequence level while the protein 3D structure is still conserved (Orengo and Thornton, 2005, and references therein). It should not then come as a surprise if this has also happened during the viral evolution where related viruses have adopted different hosts after the divergence of the proposed last universal common ancestor (LUCA) of cellular life forms to different domains (Archaea, Bacteria, Eukarya; Woese *et al.*, 1990). Although there is a possibility that similar folds have appeared through convergent evolution, in case of viruses also other features such as similar clustering of packaging ATPases support the hypothesis of a common ancestor (Abrescia *et al.*, 2012).

In conclusion, the 3D structure of the major capsid protein of a virus can reveal deep evolutionary relationships extending beyond the divergence of the cellular life forms (Abrescia *et al.*, 2012; Bamford *et al.*, 2005a; Benson *et al.*, 2004). Viruses related at the highest order of classification do not necessarily show relatedness at the genomic level. Both prokaryotic cellular and viral genomes are mosaics or dynamic networks containing genes from different origins (Hatfull and Hendrix, 2011; Koonin *et al.*, 2009; Papke and Gogarten, 2012). Some genes, however, are more often inherited vertically than other genes. No matter how small this proportion is in comparison with the whole genome content, it still is the part that has the highest potential to reveal the true ancestral cellular or viral lineage.

The influence of viral genes on their hosts

Owing to the high infection and recombination rates combined with the intimate relationship of the prokaryotic viruses with their hosts, it is not surprising to see the many ways prokaryotic viruses influence their hosts: viral encoded genes have turned many benign bacteria into serious pathogens (Brüssow *et al.*, 2004; Casjens, 2003). These genes are often the so called 'morons' (from words 'more DNA': Hendrix *et al.*, 2000; Juhala *et al.*, 2000), genes or gene cassettes inserted between the different gene modules that do not seem to be important for basic viral functions. Examples include the phage-encoded virulence factors such as Shiga- and cholera toxin genes (Brüssow *et al.*, 2004). These toxin-encoding genes have become members of the host genome after integration of a temperate virus genome. In these cases the genes introduced by viruses are beneficial for hosts.

There are also examples where host genes carried by the viruses have been accommodated to serve the purposes of the virus infection. Such examples include the cyanobacterial genes involved in photosynthesis (Lindell *et al.*, 2005; Mann *et al.*, 2003, 2005; Millard *et al.*, 2004; Sharon *et al.*, 2009; Sullivan *et al.*, 2005). These genes and their products have been suggested to enhance the viral infection by maintaining photosynthesis rate of the host during the infection and increasing the cyanophage fitness (Lindell *et al.*, 2005; Mann *et al.*, 2003). The cyanophage encoded photosynthesis genes are also evolving rapidly and provide a reservoir for photosystem evolution for their hosts (Sullivan *et al.*, 2006).

New functions of viral structures

Not only single viral genes can have considerable effects on the host biology, but also sets of viral genes encoding structural components (Veesler and Cambillau, 2011). Examples of these spin-offs are the type VI secretion pathway of Gram-negative bacteria (Leiman *et al.*, 2009; Pell *et al.*, 2009), the macromolecular bacteriocins such as carotovoricins (Nguyen *et al.*, 1999; Yamada *et al.*, 2006) and pyocins R and F (Nakayama *et al.*, 2000) as well as gene transfer agents (GTAs; Lang and Beatty, 2007). In Gram-negative bacteria the virulence related type VI secretion system (T6SS) components VgrG and Hcp proteins can form oligomeric structures that look like bacteriophage spike protein and tail tube, respectively (Basler *et al.*, 2012; Leiman *et al.*, 2009; Pukatzki *et al.*, 2007; Veesler and Cambillau, 2011). The 3D protein structures are also similar to the corresponding structures of gp27 of T4 and gpV_N of bacteriophage λ (Leiman *et al.*, 2009; Pell *et al.*, 2009).

All three known macromolecular bacteriocins, i.e. carotovoricins and pyocins R and F, have similar morphology to either myoviral or siphoviral tail structures and the genes encoding them are related to bacteriophage tail-encoding genes (Nakayama *et al.*, 2000; Yamada *et al.*, 2006). Also haloarchaea produce toxins comparable with the bacteriocins, the so-called halocins (O'Connor and Shand, 2002). Although the genes encoding known halocins do not show homology to viral tail genes, the possibility of finding new macromolecular halocins that are related to bacteriophage tail structures cannot be ruled out.

Gene transfer agents (GTAs) are particles that look like viral particles, but most of the time they contain host DNA (Lang and Beatty, 2007). GTAs have been described frequently among the members of the Alphaproteobacteria and gene regions involved in their production have been identified in the host genomes (Lang and Beatty, 2007; McDaniel *et al.*, 2010). The genes include homologues for terminase, portal, capsid and tail encoding genes of tailed viruses, but their regulation is different suggesting that these genes may have originated from tailed viruses, but have diverged and become part of the host biology and are not just proviral genes any more (Lang and Beatty, 2007; McDaniel *et al.*, 2010).

Haloarchaeal viruses

Tailed haloarchaeal viruses

As mentioned above, the first haloarchaeal virus was the tailed Hs1 (Porter *et al.*, 2008; Torsvik and Dundas, 1974). Subsequently, many other tailed viruses were isolated and studied in terms of their infection biology (Porter *et al.*, 2008). The first genomic sequences of tailed haloarchaeal viruses were obtained from ϕH, which still to date is among the best characterized haloarchaeal viruses, although the genome is not completely sequenced (Gropp *et al.*, 1992; Porter *et al.*, 2008). The genome of ϕH was shown to be very instable due to varying number of insertion elements ISH1.8 and ISH50 as well as deletions (Reiter *et al.*, 1988; Schnabel *et al.*, 1982a). ϕCh1 is a temperate tailed virus that infects haloalkaliphilic

Natrialba magadii (Witte *et al.*, 1997). In addition to the determined structural proteins and the dsDNA genome, the virion also contains RNA species of cellular origin (Witte *et al.*, 1997). The complete genome sequence of ϕCh1 shows high similarity with the sequenced parts of ϕH (Klein *et al.*, 2002) even though the host of ϕCh1 is a haloalkaliphile. The close relatedness between ϕH and ϕCh1 is peculiar also because of the instability of ϕH genome. This instability may directly reflect the genomic instability of the host *Halobacterium salinarum* (Reiter *et al.*, 1988) whereas instability of *Natrialba magadii* genome has not been reported. The 58.5 kb genome of ϕCh1, however, contains a region that can undergo inversion and generate variation in the C-terminal parts of the predicted ORFs34 and 36 polypeptides (Klein *et al.*, 2002; Rössler *et al.*, 2004). The ORF34 has now been shown to encode a tail fibre protein that is subject to phase variation (Klein *et al.*, 2012).

The isolation of two myoviruses, HF1 and HF2, was reported in 1993 (Nuttall and Dyall-Smith, 1993). The genomic sequences of HF1 (Tang *et al.*, 2004) and HF2 (Tang *et al.*, 2002) showed high similarity throughout most parts of the genome (Fig. 4.2). The only exception is the region possibly containing genes encoding tail fibre proteins (approximately between nucleotides 20,000 and 30,000; Fig. 4.2). Since tail fibre proteins are responsible for the host recognition, it makes sense that their sequences are usually the most diverged part of the genome (Brüssow and Desiere, 2001; Lubbers *et al.*, 1995; Saren *et al.*, 2005). Indeed, HF1 was found to have a host range different from HF2 (Nuttall and Dyall-Smith, 1993). BJ1, a temperate tailed haloarchaeal virus was isolated from a hypersaline lake in Inner Mongolia (Pagaling *et al.*, 2007) and shown to infect a *Halorubrum* host (16S rRNA sequence 98% identical to *Halorubrum saccharovorum*). The 42,271 bp genome is terminally redundant and possibly circularly permuted containing 70 predicted ORFs (Pagaling *et al.*, 2007).

In our recent study we have analysed complete genomes (Senčilo *et al.*, 2013) of four myoviruses and six siphoviruses of the isolated haloarchaeal tailed viruses described in Atanasova *et al.* (2012). The sizes of the genomes range between 35 and 144 kb yielding in total of 760 kb of new viral sequence. Analyses show that myoviruses HRTV-5 and HRTV-8 are closely related to HF1 and HF2 even at the nucleotide level whereas the relatedness between HCTV-1 and HCTV-5 can mostly be seen at the amino acid level of the predicted ORFs (Senčilo *et al.*, 2013). Most of the predicted ORFs do not show significant similarity to any sequences in the databases (higher than 32.5% identity at the aa level or BlastP E-value of 10^{-50} or less; Cresawn *et al.*, 2011; Senčilo *et al.*, 2013). The HVTV-1 genome reported by Pietilä and others (Pietilä *et al.*, 2013) is closely related to HCTV-5, and the HSTV-2 is remotely related to the HRTV-5, HRTV-8, HF1 and HF2 cluster of related viruses (Pietilä *et al.*, 2013; Roine, unpublished).

In recent study by Garcia-Heredia and others (Garcia-Heredia *et al.*, 2012), 42 new genomic sequences of haloviruses were reported. They were retrieved by cloning viral DNA

Figure 4.2 (opposite page) Genomic organization of the haloarchaeal tailed viruses HCTV-1 and HCTV-5. Nucleotide sequence similarity is depicted by shading between the genomes according to colour spectrum determined in Phamerator (Cresawn *et al.*, 2011) with purple representing the highest identity. Protein families (pfams) as determined in Phamerator are numbered above each ORF. RnR, ribonucleotide reductase; Thy, thymidylate; AdoMet-Mtase, S-adenosylmethionine-dependent methyltransferase; RFC, replication factor; TMP, thymidine monophosphate; TerL, terminase large subunit. Figure reproduced from Senčilo *et al.* (2013). A colour version of this figure is located in the plate section at the back of the book.

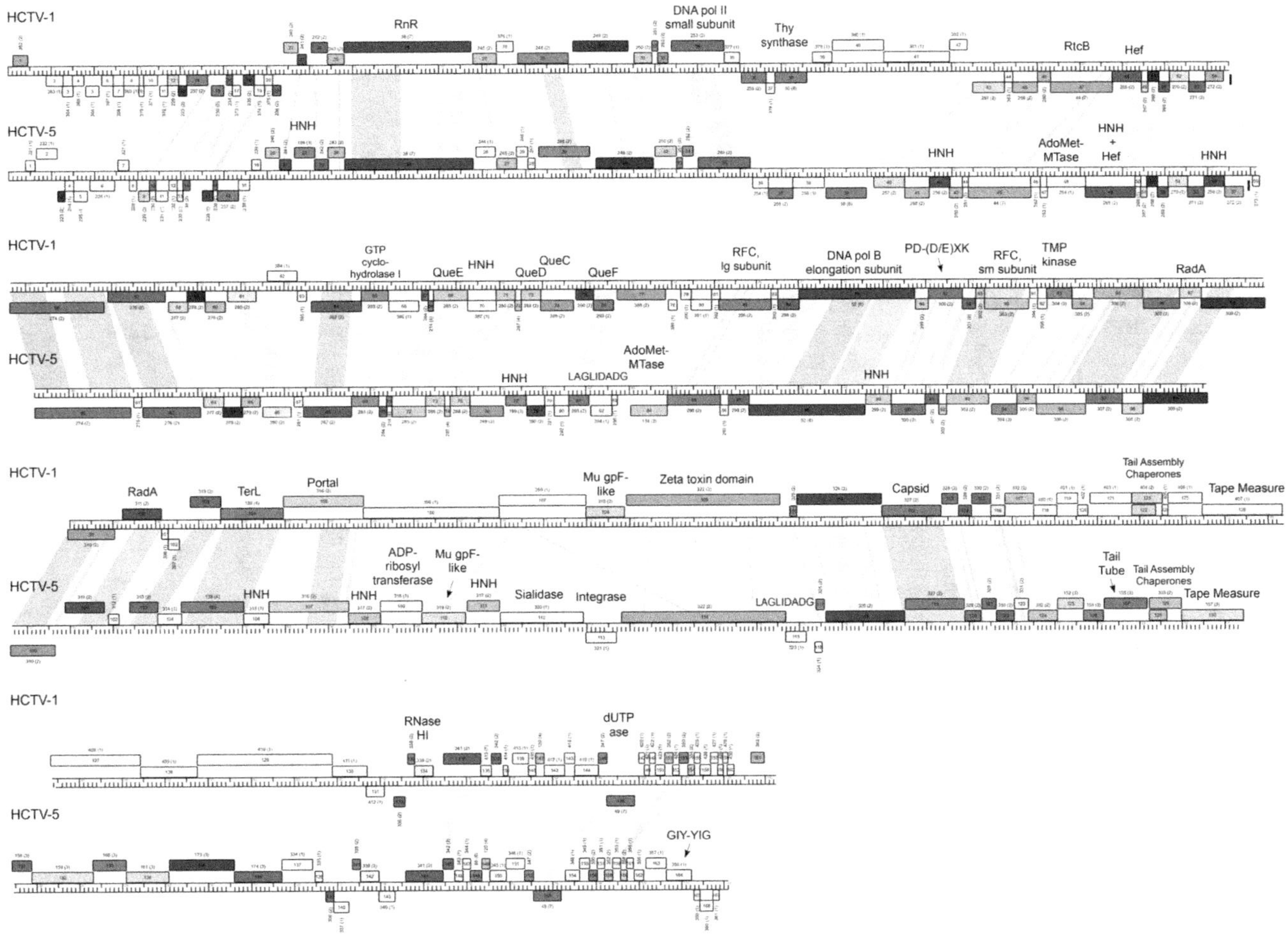

HCTV-1
RnR
DNA pol II
small subunit
Thy
synthase
RtcB
Hef
HCTV-5
HNH
HNH
AdoMet-
MTase
HNH
+
Hef
HNH
HCTV-1
GTP
cyclo-
hydrolase I
QueE
HNH
QueC
QueD
QueF
RFC,
lg subunit
DNA pol B
elongation subunit
PD-(D/E)XK
RFC,
sm subunit
TMP
kinase
RadA
HCTV-5
HNH
LAGLIDADG
AdoMet-
MTase
HNH
HCTV-1
RadA
TerL
Portal
Mu gpF-
like
Zeta toxin domain
Capsid
Tail Assembly
Chaperones
Tape Measure
HCTV-5
HNH
ADP-
ribosyl
transferase
Mu gpF-
like
HNH
HNH
Sialidase
Integrase
LAGLIDADG
Tail
Tube
Tail Assembly
Chaperones
Tape Measure
HCTV-1
RNase
HI
dUTP
ase
HCTV-5
GIY-YIG

isolated and purified from a concentrated water sample. They all contained the hallmark genes for tailed viruses i.e. terminase and/or portal-like genes.

Genomic structure of tailed haloarchaeal viruses

The genome maps of HCTV-1 and HCTV-5 are shown in Fig. 4.2. The genomic sequence analyses of haloarchaeal tailed viruses show that they are in many aspects similar to the tailed bacteriophages (Klein *et al.*, 2002; Senčilo *et al.*, 2013; Tang *et al.*, 2004). This can be seen in the complete genome sequences of isolated tailed haloarchaeal viruses as well as in the partial genomes of the so-called environmental viruses recently reported (Garcia-Heredia *et al.*, 2012; Klein *et al.*, 2002; Pagaling *et al.*, 2007; Senčilo *et al.*, 2013; Tang *et al.*, 2002). Even though the overall nucleotide sequence similarity is practically non-existing, searches with some predicted gene products, can give meaningful hits to proteins in the sequence database. As in bacteriophage genomes, the gene encoding large subunit of terminase can usually be recognized in the genomes of tailed haloarchaeal viruses (Casjens, 2003; Pagaling *et al.*, 2007; Tang *et al.*, 2002, 2004). This protein forms heteromultimeric complexes with the small subunit of terminase (Casjens, 2011), and is involved in packaging of the genome. The genes encoding portal proteins can also often be recognized through matches in the sequence database. The gene encoding tail tape measure protein (Katsura and Hendrix, 1984) can often be identified as a relatively long ORF located in the cluster of genes encoding the tail structure. Enzymes involved in genome replication can also be recognized (Pagaling *et al.*, 2007; Tang *et al.*, 2002) and the different modules encoding identified structural components are organized in the same way: terminase and portal (assembly)–major coat protein (head structure)–tail tube protein (tail structure) (Casjens, 2003; Krupovič *et al.*, 2011; Lawrence *et al.*, 2002). The other parts of the genome, often also transcribed in the opposite direction, contain genes required for genome replication but for most no clear function can be assigned (Hatfull, 2010).

tRNAs in the genomes of tailed viruses

As in many tailed bacteriophages, the genomes of tailed haloarcheal viruses also carry genes encoding tRNAs. The amount of tRNAs varies from none (for example HCTV-2) to 36 in the approximately 144-kb-long genome of the HGTV-1 myovirus (Pagaling *et al.*, 2007; Senčilo *et al.*, 2013; Tang *et al.*, 2002, 2004). It has been hypothesized that tRNA genes in bacteriophage genomes would play a role in enhancing the translational efficiency of viral genes in hosts with codon usage different from the most expressed viral genes (Daniel *et al.*, 1970; Enav *et al.*, 2012; Kunisawa, 2000; Limor-Waisberg *et al.*, 2011). Also, a positive correlation was reported between virulent life style and the abundance of tRNAs in viral genomes (Bailly-Bechet *et al.*, 2007). For cyanophages Limor-Waisberg *et al.* (2011) showed that analysed myoviruses with larger genomes and wider host range seemed to adapt by maintaining tRNA genes whereas podoviruses that had narrow host ranges and smaller genomes had been adapted to host genomic conditions by adjusting their GC-content. In haloarchaeal tailed viruses it is difficult to assess the role of tRNAs before the other aspects of viral infection such as virus–host interactions (virulent/temperate) and the host range has been properly studied. As there also seems to be correlation between the morphotype of the tailed virus (myo-, podo- or siphovirus) and the tRNA content (Enav *et al.*, 2012; Limor-Waisberg *et al.*, 2011) the proper characterization of the viruses and their hosts cannot be emphasized too much.

Other predicted genes

Ten of the newly described genomes encode predicted freestanding homing endonucleases (HE; Pietilä *et al.*, 2013; Senčilo *et al.*, 2013). The most remarkable example is the HCTV-5 genome that contains 14 putative HEs (Senčilo *et al.*, 2013). Some of these HEs flank both the large subunit of terminase (TerL) as well as the portal encoding genes (Senčilo *et al.*, 2013). It is typical for virus-encoded HEs that they are inserted within or near functionally critical virus genes (Edgell *et al.*, 2010).

One family of HEs is the so called PD-(D/E)XK endonucleases that can be found in seven of the newly studied viruses (Pietilä *et al.*, 2013; Senčilo *et al.*, 2013) as well as in HF1 and HF2 (Senčilo *et al.*, 2013). These predicted proteins show homology to RecB exonucleases and give hits to the Cas4 superfamily proteins in the conserved domain database (Senčilo *et al.*, 2013). Cas4 proteins are part of the newly described CRISPR/Cas system (clustered regularly interspaced short palindromic repeats) that is involved in adaptive immunity against foreign viral and plasmid nucleic acid (Barrangou *et al.*, 2007; Bolotin *et al.*, 2005). It is roughly comparable to the RNAi system in eukaryotic cells but it is not homologous and many differences also exist (Bhaya *et al.*, 2011; Goren *et al.*, 2012). Cas4 proteins can be found in the type I and type II systems (Makarova *et al.*, 2011) and have been suggested to be RecB family exonucleases involved in the spacer acquisition pathway (Bhaya *et al.*, 2011). A simple alignment of a subset of the Cas4-like proteins with the Cas4 protein of the closest host homologue (Fig. 4.3) shows that the three C-terminal cysteins conserved in Cas4 proteins (Jansen *et al.*, 2002) can be aligned although the viral encoded Cas4 proteins are longer and vary in amino acid sequence from the host encoded proteins especially in the middle of the polypeptide (Senčilo *et al.*, 2013). The function of these *cas4*-like genes in the viral genomes is currently unknown. We have hypothesized that if used for the benefit of the viruses these Cas4 homologues could interfere with the normal function of the host encoded Cas4 complexes (Senčilo *et al.*, 2013). Whether the viral encoded proteins serve in the normal functions of replication or recombination, or whether there is a benefit for the haloarchaeal virus in counter-defence remains an open question and a subject for future studies.

The newly described viral genomes HCTV-1, HCTV-5 and HVTV-1 contain predicted genes that encode putative zeta toxin domains (Pietilä *et al.*, 2013; Senčilo *et al.*, 2013). The zeta toxins studied before are shown to be fosforyl-transferases and part of the proteinaceous toxin–antitoxin (TA) systems in bacteria (Hayes and Van Melderen, 2011; Mutschler *et al.*, 2011). The toxin–antitoxin systems are involved in many important cellular processes such as programmed cell death, stress response or phage defence through abortive infection (Hayes and Van Melderen, 2011). It is peculiar that in the viral genomes no counterpart of antitoxin can be recognized and that the predicted gene products carrying the zeta toxin domain are more than 100 kDa in size suggesting that in addition to zeta toxin, some other functions are encoded as well. In addition, the predicted genes reside in the module encoding structural proteins of the viruses.

Icosahedral viruses

Currently there are two published complete genomes of icosahedral viruses that both infect *Haloarcula hispanica*: SH1 (Porter *et al.*, 2005) and HHIV-2 (Jaakkola *et al.*, 2012). Recently, Zhang *et al.* (2012) reported that the previously known plasmid pHH205 is actually proviral genome that can be induced in its host, *Natrinema* sp., to produce icosahedral

```
HCTV-1_ORF87        MTEET-------------------------------RSVPQDVRE---HPLSASRVKKFAQ
HCTV-5_ORF90        MTKE--------------------------------RYVPDELFD---DYLSASRVKKFAQ
Hqr.walsbyi_Cas4    MKPVTS------------------------------------IADHDELVHVSALNEYLY
HF1_HalHV1gp046     MSKALTDDGEE----------------------IDLTVPEEHVENGLEHVSKSRVKTYLQ
HF2p052             MSKALTDDGEE----------------------IDLTVPEEHVENGLEHVSKSRVKTYLQ
HRTV-5_ORF62        MSKALTDDGEE----------------------IDLTVPDEHVENGLEHVSKSRVKTYLQ
HRTV-7_ORF51        MSRYNSHDATEGGEATTRAKPSKAEALPGIEPGTTLEIMPEVAEGGMPYISKSRIKTFVQ
HRTV-8_ORF65        MSKALTDDGRE----------------------IDLTVPDDHAENGLEHISKSRIKTYLQ
                    *.                                               :  * :: :

HCTV-1_ORF87        CPLSWWYNYVKEETRTKPGEGYLGLGNAVHDSIEAELK------EQGGTPNSSTLAHRLK
HCTV-5_ORF90        CPLSWWFDYARDEDRTKPEKGYREMGTAVHEAIEEVLLDDESIRDSGIL------SHRFK
Hqr.walsbyi_Cas4    CPRRFYYQRYHDE-MGTP--YELLDGRSKHENA---------------------------
HF1_HalHV1gp046     CPRKFYYSYWCGN-R-TPGSYHTEKGSQIHRAYEDFHLNL--------------------
HF2p052             CPRKFYYSYWCGN-R-TPGSYHTEKGSQIHRAYEDFHLNL--------------------
HRTV-5_ORF62        CPRKFYYSYWCGN-R-TPSSYHTEKGSEIHRAYEDFHLNL--------------------
HRTV-7_ORF51        CPAKFYWKYWCGE-R-GPGSYYTEKGSRLHETFEKFHLNL--------------------
HRTV-8_ORF65        CPRKFLYSYWMDN-R-TPGSYHTEKGSQIHRAYEDFHLNL--------------------
                    **  : :.    :    *       *   *

HCTV-1_ORF87        RRYREEDPDIPEWMYD------------------------KGLK-CCDNAAKFFAEYD--
HCTV-5_ORF90        ERYREKNPDVPEWMYE------------------------NGLD-CCDNAAKYIEKFG--
Hqr.walsbyi_Cas4    -------AHRGGWISE-----------------------RYL----CDTSL---------
HF1_HalHV1gp046     IEYVEEHGERPETYAELMGPWEDWAQWLEPHIRNFWQFEDKRWELALD-AAQRLDAS-KP
HF2p052             IEYVEEHGERPETYAELMGPWEDWAQWLEPHIRNFWQFEDKRWELALD-AAQRLDAS-KP
HRTV-5_ORF62        IEYVQEHGERPEWYADVMGPWEDYAQWLHPHIENFWKFEDKRWELACDYAAAKFRALDDP
HRTV-7_ORF51        FDYLEVNDSRPDRFTDLLPHWRNYSQWLD-QVGAFFLFEERRWQQSVH-EAAKTDAM-MP
HRTV-8_ORF65        IEYVEEHGERPEWYADVMGPWEDYAQWLHPHIENFWKFEDKRWELACDYAAAKFRALDDP
                                   :                               .

HCTV-1_ORF87        --------------DLNI-REIEAEHRYA---------------VKGS-----------V
HCTV-5_ORF90        -------------ADMTF-RGFEVEHQYH---------------VGGE-----------V
Hqr.walsbyi_Cas4    ------------------------------------------------------------
HF1_HalHV1gp046     -EDSKPVDIEAEALNLWLPVGVEVEGRLE----------------GDEIPIGNIPWMGYA
HF2p052             -EDSKPVDIEAEALNLWLPVGVEVEGRLE----------------GDEIPIGNIPWMGYA
HRTV-5_ORF62        -RDGKTV--MEYALDAWLPIGVEVEGRLE----------------GDDIPIGNIPWMGYA
HRTV-7_ORF51        YSASRSL--DDYIHDLWVPVEVEAEAWLGEPPESWVEANGEPDYVSGEPPVGDAPWMGRA
HRTV-8_ORF65        -RDGKTV--MEYALDAWLPLGVEVEGRLE----------------GDDIPIGQLPWMGYA


HCTV-1_ORF87        NAMFNAK-M-DVTTDRF--IIDWKTGNAHDSDGNIRDYR---------IRDELIQGMVYA
HCTV-5_ORF90        NKGFNAK-M-DIVTEEG--VLDWKTGKRKDSDGNVRDYR---------KRDELIQGMVYA
Hqr.walsbyi_Cas4    --GLHG-KIDVIESEDGVLT--PIERKRAESG----GY--------YTNDEIQLAG--YC
HF1_HalHV1gp046     DALLHAATVPGIEADEGVVILDYKTGKVQDP-----KYR---------HKGIYLEGEFYG
HF2p052             DALLHAATVPGIEADEGVVILDYKTGKVQDP-----KYR---------HKGIYLEGEFYG
HRTV-5_ORF62        DALLHAATVPGIEADEGVVILDYKTGKVQDP-----KYR---------HKGIYLEGEFYG
HRTV-7_ORF51        DLIVKTQSLPGVDG-NGVTIIDYKTGSAPTV-----RYKDHGMLEQILNEGIYLEGEYYG
HRTV-8_ORF65        DALLHAATVPGIEADEGVVIVDYKTGKVQDE-----KYR---------HKGIYLEGEFYG
                       .:   :  :              .          *               : *  *

HCTV-1_ORF87        GAYLDRYGEYPDKVIFVYLGDG-TVRRSDPSQDRWEEMKQYARSLLQAMDA------ENF
HCTV-5_ORF90        GAYLNKYGEYPEYVTFVYLGDG-EIANRIPDEEQWSQMKQYARSLLQAMGA------GEF
Hqr.walsbyi_Cas4    MLLSHVINE-PVNFGYIYLYSTDQRHSIRITEDHRQAVNKIVEQIQSMS-A------KNI
HF1_HalHV1gp046     WLFENDLDYEIAGVAGYYPQEDELVVSPYPDEDRRHIIRKAVLGMQMMPEV------ENY
HF2p052             WLFENDLDYEIAGVAGYYPQEDELVVSPYPDEDRRHIIRKAVLGMQMMPEV------ENY
HRTV-5_ORF62        WLFENDLDYEIAGVAGYYPQEDELVVSPYPDEDRRHIIRKAVLGMQMKPEV------ENY
HRTV-7_ORF51        WLFENF--YDVDAVAGYYPGDDELVVSPYPNKDRRRLIKRAVIGMQREPDVEGNGPPENY
HRTV-8_ORF65        WLFENELDYEIAGVAGYYPQEDELVVSPYPDEDRRHIIRKAVLGMHMKPTE------ENY
                        .        .   *  .          :::   :.: .  :             :

HCTV-1_ORF87        PA--KTGGHCG----FCDYQFVCPAQDHS-------------------------MADVSY
HCTV-5_ORF90        PA--KPGGHCG----FCDYEFVCPAQETS-------------------------MANLSY
Hqr.walsbyi_Cas4    PPLTDNPSKCEA----CSAREYCMPE----------------ETAMLE---PEKARGTGW
HF1_HalHV1gp046     DI--DTGPLCHYGHGKCFFYDQCPSSWGKKGGEGYHGFAEPDGSTKPKDEITDHKRAKNW
HF2p052             DI--DTGPLCHYGHGKCFFYDQCPSSWGKKGGEGYHGFAEPDGSTKPKDEITDHKRAKNW
HRTV-5_ORF62        DI--DTGPLCHYGHGKCFFYDQCPSSWGKKGGEGYHGFAEPDGSTKPKDEITQHKRDKNW
HRTV-7_ORF51        PH--EEQPLCNYSSGNCHFYNICDS---------------------------------TW
HRTV-8_ORF65        EL--NTGPLCHYGHGKCFFYDECASTWGKKGGEGYHGFAEPDGSTKPKDEITNHKRTKNW
                        .    *      *     * .                                  :

HCTV-1_ORF87        W-KY
HCTV-5_ORF90        W-KY
Hqr.walsbyi_Cas4    EGEI
HF1_HalHV1gp046     Y-PY
HF2p052             Y-PY
HRTV-5_ORF62        Y-PY
HRTV-7_ORF51        T-PK
HRTV-8_ORF65        Y-PY
```

Figure 4.3 Comparison of the predicted viral Cas4 proteins with Cas4 of *Haloquadratum walsbyi* C23 Cas4A homologue. Alignment of the HCTV-1, HCTV-5, HRTV-5, HRTV-7, HRTV-8, HF1 (HalHV1gp046, NP_861634) and HF2 (HF2p052, accession NP_542546) Cas4-like protein sequences with Cas4A homologue of *Hqw* C23 (YP_005839106.1). The C-terminal conserved cysteine residues are marked in grey. Figure reproduced from Senčilo *et al.* (2013).

virions. This is in contrast to the previous report that the virus in question, SNJ1, was a tailed halovirus (Mei *et al.*, 2007).

Preliminary virion structure analysis of HHIV-2 (Jaakkola *et al.*, 2012) and a detailed study on SH1 (Jäälinoja *et al.*, 2008) show that both virions contain an internal membrane underneath the capsid. The dsDNA genomes are almost of the same size (30,889 bp of SH1 and 30,578 of HHIV-2) and the overall sequence similarity is 59.1% (Jaakkola *et al.*, 2012). Both genomes are linear dsDNA molecules with inverted terminal repeats (ITRs) of 309 bp (SH1) and 305 bp (HHIV-2) in length (Fig. 4.4; Bamford *et al.*, 2005b; Jaakkola *et al.*, 2012). There is strong evidence suggesting that the 5′-ends of the SH1 genome are covalently attached to terminal proteins (Porter and Dyall-Smith, 2008) that are responsible for the priming of genome replication (Salas, 1991) by host polymerases. The genomes of SH1 and HHIV-2 contain 56 and 43 predicted ORFs, respectively (Fig. 4.4). There are 15 (SH1) and 16 (HHIV-2) structural proteins in the virions, of which 11 for SH1 and 10 for HHIV-2 have been identified using N-terminal sequencing and/or mass spectrometry (Bamford *et al.*, 2005b; Jaakkola *et al.*, 2012; Kivelä *et al.*, 2006). As in the bacteriophage PRD1, the genes encoding structural proteins reside mainly in the middle of the genome in both viruses (Fig. 4.4B). Exceptions are the VP1 encoding gene in both viruses and the VP18 encoding gene in HHIV-2 (Bamford *et al.*, 2005b; Jaakkola *et al.*, 2012). Also, the region encoding the spike proteins VP3 and VP6 of SH1 (Jäälinoja *et al.*, 2008) shows no similarity with the HHIV-2 genome encoding predicted ORFs 22 to 26 (Jaakkola *et al.*, 2012). As discussed above for the tailed viruses, this is a clear example of closely related viruses in which the region encoding host recognition functions are less related. It also suggests that even though both viruses infect the same host, the receptors for these two viruses may be different.

In both viruses there are also two genes encoding major capsid proteins yielding products of approximately 25 kDa (VP4) and 20 kDa (VP7) in size (Fig. 4.4A). In SH1 these proteins have been suggested to fold into single β-barrel structures (Jäälinoja *et al.*, 2008). Corresponding proteins of HHIV-2 are 84% (VP4) and 72% (VP7) identical in amino acid sequence (Jaakkola *et al.*, 2012) to the SH1 proteins suggesting strongly similar folding of the proteins. Both SH1 and HHIV-2 contain a predicted ORF encoding a putative homologue for a packaging ATPase that has been shown to share features similar to the other packaging ATPases of icosahedral viruses (Jaakkola *et al.*, 2012; Strömsten *et al.*, 2005). It has been proposed that SH1 and HHIV-2 belong to the PRD1-adenovirus – lineage of icosahedral viruses (Bamford *et al.*, 2005b; Jaakkola *et al.*, 2012; Jäälinoja *et al.*, 2008; Kivelä *et al.*, 2006; Porter *et al.*, 2005). There are several features in common especially with bacteriophage PRD1 such as the icosahedral morphotype of the virion, an internal membrane underneath the capsid, similar packaging ATPase, linear dsDNA genome with terminal proteins and the genomic organization of the genes, which support the hypothesis (Bamford *et al.*, 2005b; Jäälinoja *et al.*, 2008; Kivelä *et al.*, 2006; Krupovič and Bamford, 2008; Porter and Dyall-Smith, 2008; Porter *et al.*, 2005; Strömsten *et al.*, 2005). Although the 3D structure of the major capsid proteins of SH1 and HHIV-2 have not been solved to high resolution, it is clear that instead of the double jelly roll fold structure found in the major structural proteins of the other members of the PRD1-adenovirus –lineage, there are two separate proteins both folding possibly to single β-barrel structures (Jäälinoja *et al.*, 2008). Therefore, it has been suggested that the SH1 and HHIV-2 viruses together with a thermophilic bacteriophage P23–77 (Jaatinen *et al.*, 2008) would form a related group of viruses, the so-called single β-barrel lineage (Jaakkola *et al.*, 2012; Krupovič and Bamford, 2008).

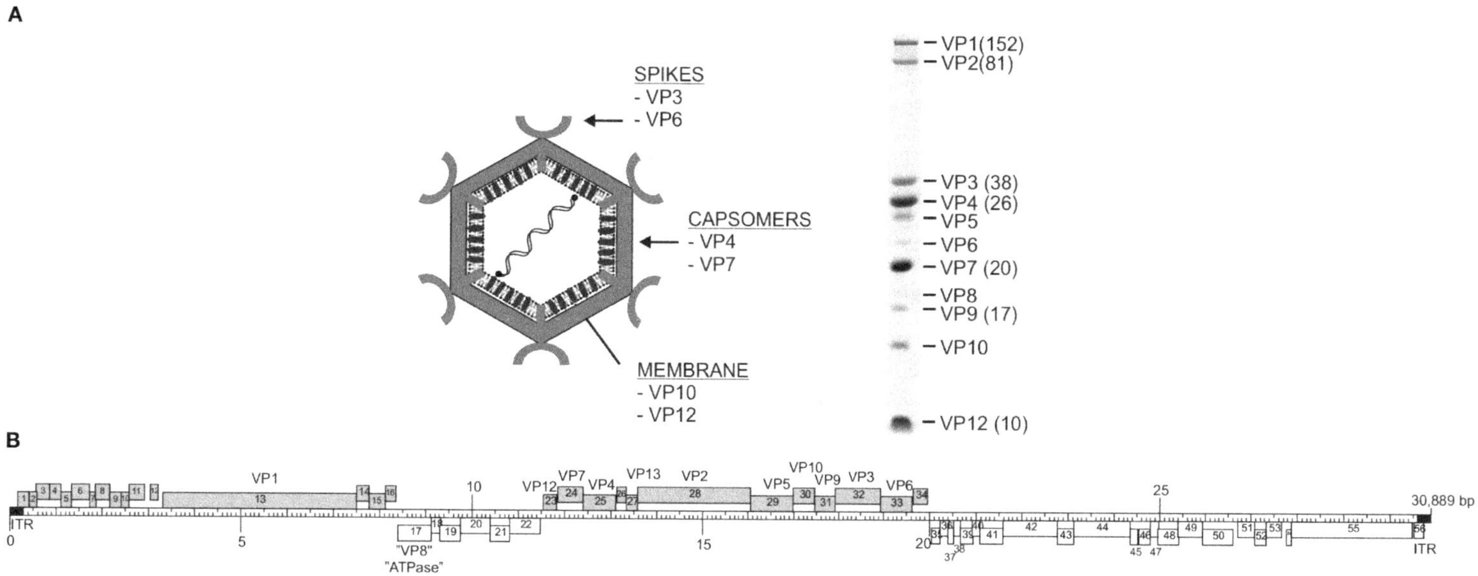

Figure 4.4 Icosahedral haloarchaeal virus SH1. (A) A schematic of the icosahedral SH1 virion (on the left). The specific locations of some structural proteins are indicated (Jäälinoja *et al*., 2008; Kivelä *et al*., 2006). On the right: protein analysis (14% modified tricine-SDS-PAGE, see (Kivelä *et al*., 2006) of the SH1 structural proteins. All indicated proteins except for VP8 have been identified by protein chemistry. Calculated molecular masses (in kDa) for some of the proteins are shown in parentheses. (B) Genomic features of the icosahedral SH1. The predicted ORFs are numbered and the gene products identified by protein chemistry are marked using the protein names (VP=viral proteins) except for the VP8, which has not been identified experimentally. ORFs and genes encoded by the sense strand are coloured in pink and the ORFs encoded by the reverse strand are shown in grey/white (Bamford *et al*., 2005b). The inverted terminal repeats (ITRs) are shown as black boxes.

Relatively few proviral elements have been found in the haloarchaeal host genomes that show relatedness to the icosahedral viral genomes (Jaakkola *et al.*, 2012; Jalasvuori *et al.*, 2009). There are short regions in *Haloarcula marismortui* chromosome that contain homologues for the two major capsid proteins as well as the packaging ATPase (Jalasvuori *et al.*, 2009) and one longer region in *Haladaptatus paucihalophilus* (strain DX253; RefSeq NZ_AEMG01000027.1) which shows similarity along longer genomic region including the genes encoding most of the structural genes (Jaakkola *et al.*, 2012). As discussed above for tailed viruses, this most probably reflects the virulent life styles of the two icosahedral viruses (Jaakkola *et al.*, 2012; Roine and Oksanen, 2011).

Spindle-shaped viruses

Currently there are two haloarchaeal viruses, His1 and His2 that have been classified into the genus *Salterprovirus* (King *et al.*, 2012). The genomes of both viruses are linear dsDNA molecules of 14,464 bp (His1) and 16,067 bp (His2) in length. As in icosahedral viruses the genomes contain inverted terminal repeats and 5′ terminal proteins (Bath *et al.*, 2006; Porter and Dyall-Smith, 2008). The genomes share very little similarity at the nucleotide level and at the amino acid sequence level the only detectable similarity can be found between putative DNA polymerases (Bath *et al.*, 2006). Both His1 and His2 genomes have undergone independent recombination events (Bath *et al.*, 2006). Recently, it has been suggested that instead of *Salterprovirus*, His2 should actually be classified as a member of the haloarchaeal pleomorphic virus group (see below; Pietilä *et al.*, 2012b; Senčilo *et al.*, 2012). In addition, a new study on the structure of His1 virion shows that unlike His2 it does not contain lipids, but the 7.9 kDa major capsid protein VP21 is most probably lipid modified (Pietilä *et al.*, 2012a).

Pleomorphic viruses

Haloarchaeal pleomorphic viruses represent a new group of recently isolated viruses with simple virion architecture (Fig. 4.5A and B; Atanasova *et al.*, 2012; Pietilä *et al.*, 2009, 2012b; Roine *et al.*, 2010; Senčilo *et al.*, 2012). HRPV-1 was the first pleomorphic virus isolated (Pietilä *et al.*, 2009) and currently there are seven members that are suggested to belong to this group of viruses (Table 4.2; Pietilä *et al.*, 2012b; Senčilo *et al.*, 2012). His2 is also suggested to belong to the group of pleomorphic viruses on the basis of the virion morphology, the major structural protein pattern and collinearity of the approximately 7 kb of the His2 genome (Fig. 4.5; Pietilä *et al.*, 2012b; Senčilo *et al.*, 2012). HRPV-1 was also the first archaeal virus shown to contain an ssDNA genome (Pietilä *et al.*, 2009). Pleomorphic viruses are geographically widely distributed: they have been isolated from two different locations in Italy (Trapani and Margherita di Savoia), Israel, Thailand, Spain and Australia (Atanasova *et al.*, 2012; Bath *et al.*, 2006; Pietilä *et al.*, 2012b).

The virion architecture is pleomorphic and consists of a genome enclosed in lipid membrane vesicle (Fig. 4.5B; Pietilä *et al.*, 2010, 2012b). The virions contain two major structural proteins that are associated with the membrane. The larger protein, VP4 and VP4-like proteins, form spike structures, protrusions that are usually responsible for the viral recognition of the host receptor. The spike proteins are C-terminally anchored to the membrane whereas the smaller structural protein, VP3 and VP3-like proteins, are mostly embedded in the viral envelope (Pietilä *et al.*, 2009, 2010, 2012b; Roine *et al.*, 2010). HGPV-1 and His2 viruses have one or two structural proteins in addition to the two major ones (Pietilä *et al.*,

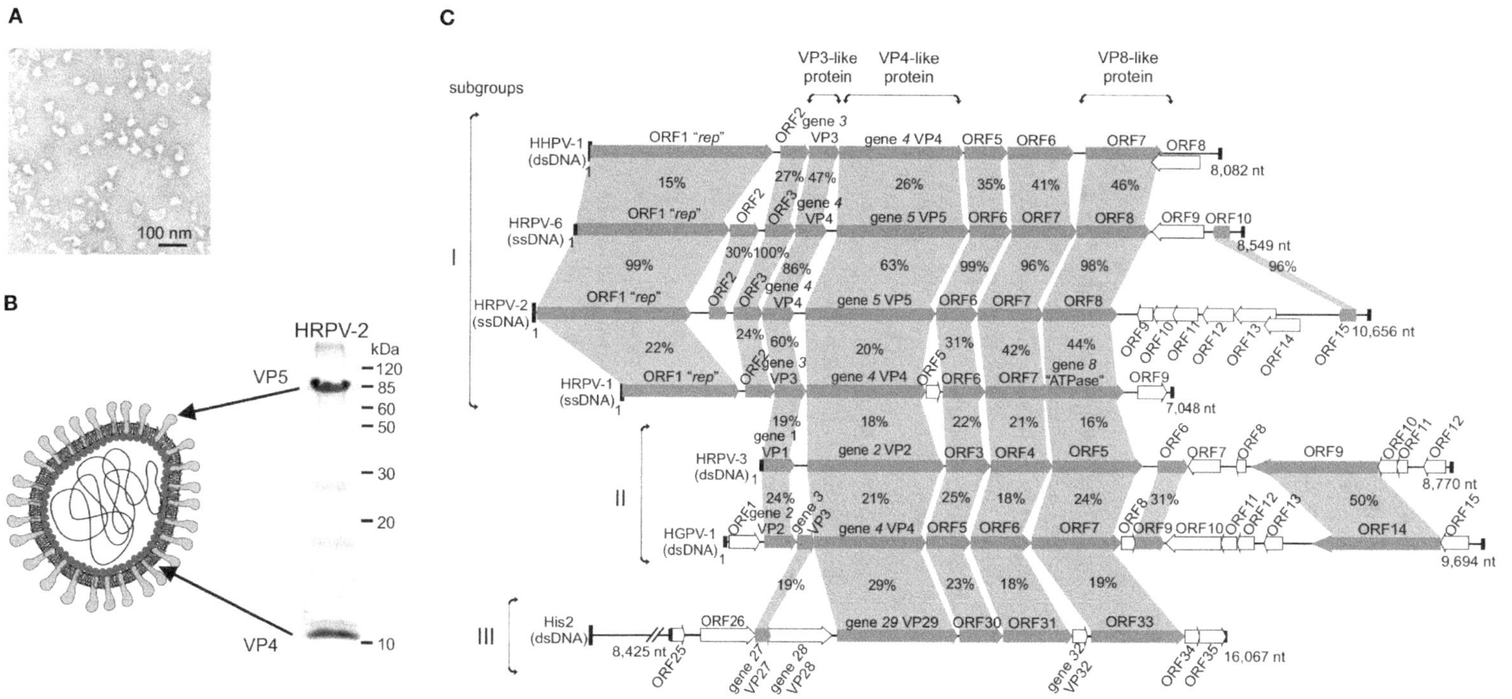

Figure 4.5 New group of haloarchaeal pleomorphic viruses. (A) Transmission electron micrograph of the negative stained (3% ammonium molybdate, pH 7.2) pleomorphic particles of HRPV-2. The picture was taken with Jeol 1200EX electron microscope operating at 80 kV. (B) Schematic representation of the pleomorphic viral particle (on the left) and the analysis of structural proteins (16% SDS-PAGE) of HRPV-2 (on the right). (C) Genome alignment of the pleomorphic viruses (Senčilo *et al.*, 2012). The genomes are divided into subgroups according to the genome content and organization as discussed in the text. Percentages of identical amino acids in homologous putative polypeptides and identified proteins are indicated.

Table 4.2 The molecular weight (in kDa) of the two major structural proteins and their protein names in different pleomorphic viruses

Virus	VP3-like protein	VP4-like protein[a]
HHPV-1	14.4 (VP3)	51.6 (VP4)
HRPV-1	13.9 (VP3)	53.3 (VP4)
HRPV-2	14.3 (VP4)	56.5 (VP5)
HRPV-3	14.6 (VP1)	57.3 (VP2)
HRPV-6	14.5 (VP4)	57.3 (VP5)
HGPV-1	9.5 (VP2)	48.5 (VP4)
His2		53.6 (VP29)[b]

[a]The molecular weight of the protein calculated in its processed form is shown. [b]In addition to VP4-like protein, His2 virion contains three other major structural proteins: VP27, VP28 and VP32 (Pietilä *et al.*, 2012b). Neither of the membrane proteins VP27 or VP32 are related to the VP3-like proteins.

2012b). HRPV-1 VP4 is N-glycosylated comprising a pentasaccharide containing glucose, glucuronic acid, mannose, sulfated glucuronic acid and a terminal 5-N-formyl-legionaminic acid residue (Kandiba *et al.*, 2012). The virion structure with a host acquired lipid envelope suggests an exit strategy similar to the one of many animal viruses i.e. by budding without lysis of the host cells, and there is evidence that the cells producing viral progeny are alive, but grow slower than the non-infected cells (Pietilä *et al.*, 2009, 2012b; Roine *et al.*, 2010). HRPV-1 has been reported to establish an infection persisting through several successive passages of culturing (Pietilä *et al.*, 2009).

Genomic structure of pleomorphic viruses

The genome sizes of the pleomorphic viruses vary from the 7 kb of HRPV-1 to the 16.1-kb of His2 (Senčilo *et al.*, 2012). These viruses are characterized by a conserved set of five genes (Fig. 4.5B) starting from the genes encoding the two major structural proteins and ending after the predicted gene encoding putative P-loop NTPase. At the nucleotide level significant sequence similarity can only be found between the HRPV-2 and HRPV-6 genomes (Senčilo *et al.*, 2012), the two viruses isolated from the same solar saltern in Thailand (Atanasova *et al.*, 2012; Pietilä *et al.*, 2012b). Although related, the amino acid sequences of the conserved cluster genes are not similar enough in order to conduct proper phylogenetic analyses of the different gene products. Preliminary analyses suggest, however, that at least VP3-like, VP8-like and ORF7-like products may have a common evolutionary history. Thus, the genes and predicted ORFs in the conserved cluster may be a uniform module. These genomes, however, also contain variable regions that do not show significant similarity (Fig. 4.5; see below).

Based on the genome organization and gene content in the variable regions, these viruses can be further divided into three subgroups (Fig. 4.5C). The first predicted open reading frame in the first subgroup members is a putative replication initiation protein (Rep) of rolling circle replication (RCR). There is also a short homologous ORF (ORF2 in HRPV-1) just upstream of the gene encoding the smaller major structural protein. The second subgroup members differ from the first subgroup members by showing relatively low amino acid sequence identity of the VP3-like proteins both within this subgroup and between the subgroups 1 and 2. In addition, instead of the putative RCR Rep, they contain a putative conserved ORF in the variable part of the genome that is transcribed to the direction opposite from the conserved gene cluster and is predicted to contain a winged helix–turn–helix motif (wHTH) in the C-terminal part of the protein (Senčilo *et al.*, 2012). His2 with linear dsDNA genome is, at the moment, the only representative of the third subgroup. The approximately 7 kb region at the 3′ end of His2 genome harbours the cluster of genes conserved in the pleomorphic viruses (Dyall-Smith *et al.*, 2011; Pietilä *et al.*, 2009; Senčilo *et al.*, 2012).

The most remarkable feature of the pleomorphic viruses is that although conserved in the virion architecture and co-linearity of the genome, the HRPV-1, HRPV-2 and HRPV-6 genomes are circular single-stranded DNA (ssDNA) whereas the HHPV-1 is circular double-stranded DNA (dsDNA; Fig 4.5C). The genomes of HRPV-3 and HGPV-1 are also circular dsDNA molecules, but discontinuous (see below). The His2 genome, on the other hand, is linear dsDNA molecule containing 5′ covalently attached terminal proteins (Bath *et al.*, 2006; Porter and Dyall-Smith, 2008) comparable to those in SH1 and HHIV-2. Thus, these related viruses, containing different genome types, violate the rules of current viral classification schemes (King *et al.*, 2012). It is also perhaps an extreme example of genomic plasticity among viruses.

Discontinuous regions of HRPV-3 genome

The pleomorphic virus group further violates conventional observations and wisdom because HRPV-3 and HGPV-1, members of the subgroup 2 viruses, contain a rare type of genome that is mostly dsDNA, but with short stretches of ssDNA. We were able to show that in the genome of HRPV-3 the ssDNA stretch coincided with specific pentanucleotide motif GCCCA (Senčilo *et al.*, 2012). There are 27 GCCCA pentanucleotide motifs found in the genome of HRPV-3, and in 13 cases there was an associated ssDNA region adjacent to it. The location of these motifs also correlated with the transcriptional direction of the genes (Senčilo *et al.*, 2012). As these ssDNA stretches could not be repaired using ligase alone, but required the combined activities of polymerase and a ligase, we concluded that the discontinuity is not only a nick, but a genuine short stretch of ssDNA (Senčilo *et al.*, 2012). In case of the HGPV-1 genome, it has been more difficult to distinguish a similar type of motif and the studies are ongoing. Discontinuous dsDNA genomes can be considered rare at least among the prokaryotic viruses, although some tailed bacteriophage genomes are reported to contain nicks (Kulakov *et al.*, 2009; Wang *et al.*, 2005). Among eukaryotic viruses the Human hepatitis virus B (HBV) and its relative Woodchuck hepatitis virus (WHV) have been shown to contain partially ssDNA genomes (Summers *et al.*, 1975, 1978). In all of these cases the ultimate reason for discontinuous genomic regions remains open. Upon the infection of HBV, the genome is converted into covalently closed circular form (cccDNA; Tuttleman *et al.*, 1986) by host enzymes (Köck and Schlicht, 1993). Although similar initial steps may also occur for the haloarchaeal genomes, the rest of the genome replication cycle does not need to be reminiscent of the elaborated genome replication scheme of HBV (Ganem and Varmus, 1987). It is tempting to speculate, however, that in all of these cases the discontinuous viral genomes attract the host machinery involved in DNA repair and replication, and eventually take advantage of it.

Proviral regions related to pleomorphic viruses

In addition to the pleomorphic viral genomes characterized, there are related proviral elements found frequently in many of the haloarchaeal host genomes (Roine and Oksanen, 2011; Senčilo *et al.*, 2012). These proviral regions are usually longer than the characterized viral genomes and many of them also contain ORFs predicted to encode putative integrase and ϕH repressor-like proteins. In most of them a tRNA gene could also be found in the vicinity (Roine *et al.*, 2010; Roine and Oksanen, 2011; Senčilo *et al.*, 2012). Related genetic elements have also been reported as plasmids (Holmes *et al.*, 1995; Roine *et al.*, 2010). Thus, they seem to represent typical complete proviral elements, although there is no evidence that they can be induced to produce functional virions. Comparison of the proviral elements with the isolated pleomorphic viruses shows that the viral genomes are indeed smaller than the proviral regions. To date we have not been able to isolate a single pleomorphic virus the genome of which would contain an ORF predicted to encode a putative integrase. Future studies will hopefully show whether this tendency continues and has some meaningful biology behind it.

Viruses and plasmids, related genetic elements

The genomes of pleomorphic viruses are in many ways similar to plasmids. Furthermore, pleomorphic viruses do not lyse the cells, but retard the host growth and exit most probably through budding (Pietilä *et al.*, 2009, 2012b; Roine *et al.*, 2010). High copy-number plasmids also retard the growth of a cell. Since plasmids and viral genomes often are very close to each other as genetic elements we have to admit that there may be a 'twilight zone' between viruses and plasmids. As mentioned before, the manifestation of a virus is the ability to produce particles packaged with the genetic information for its production. Thus, we consider the pleomorphic viruses to be viruses. An alternative concept, a plasmid that has an autonomous vehicle for its delivery from one cell to another, would need even more imagination and maybe additional examples in the real life to support its existence. Among haloarchaea this may not be so far-fetched, however, because there are no reports of existence of traditional conjugative plasmids. In comparison, thermophilic crenarchaeal plasmids have been characterized that carry genes encoding TraG and TrbE homologues, proteins involved in conjugational transfer of plasmids in bacteria (Lipps, 2006; Schröder and Lanka, 2005).

Conclusions and future prospects

The field of haloarchaeal virus genomics is in its infancy. Currently, there are approximately 100 haloarchaeal viruses and less than 30 complete viral genomic sequences published. As in other prokaryotic viruses, the genomes of haloarchaeal viruses very often show little if any similarity to the database deposited nucleotide sequences. The second level of information, the primary amino acid sequence may reveal more relationships to existing data. Yet, for more than half of the predicted genes in viral genomes a putative function on the basis of homologues found in the database cannot be assigned. On the basis of the 'higher level' observations we can say that haloarchaeal viruses resemble bacteriophages: tailed viruses are common and we can also find icosahedral viruses that seem to be related with the icosahedral bacteriophages. The genomes of both of these viral classes also resemble the corresponding bacteriophage genomes. Pleomorphic viruses have been suggested to be related to *Plasmaviridae* family of bacteriophages such as L172 (Dybvig *et al.*, 1985; Pietilä *et al.*, 2009; Roine and Oksanen, 2011), but the closer characterization of L172 virion architecture and genome sequence is missing. The genomes of analysed haloarchaeal viruses show similar type of mosaicism already seen in bacteriophage genomes. The mosaicism of viral genomes is most apparent in the genomes of viruses with complex virion architectures such as tailed viruses, but also in the more simple ones, such as the pleomorphic viruses, there is clear modularity. In conclusion, the new haloarchaeal viruses and information of their genomes further support the suggested classification scheme in which the structural features of the virion as well as the atomic structure of the major structural proteins are used in the higher order classification. For further division into lower levels of classification, the features of the rest of the genome can be utilized.

In addition to the direct influence of viruses as parasites of their hosts, there are examples of the indirect effects viruses can have (see 'Influence of viral genes on their hosts', above) on their host biology. For example, part of the type VI secretion pathway (Leiman *et al.*, 2009; Pell *et al.*, 2009) and macromolecular toxins (Nakayama *et al.*, 2000; Yamada *et al.*, 2006) are encoded by homologues of viral genes encoding the tail structure, and toxin encoding genes carried by viral genomes have been incorporated into the host genomes for the benefit of the host (Brüssow *et al.*, 2004; Casjens, 2003). These examples nicely light up the close relationships and dynamics between the viral and host genomes. Does this also happen in haloarchaea and what are the outcomes of the interplay between haloarchaea and their viruses remains to be answered in future studies.

Characterization of the genomes of new haloarchaeal viruses, especially the pleomorphic viruses has further reinforced the dichotomy of viral functions. As already discussed above, the genes encoding for the virion structural components experience the selective pressure for virion formation whereas the genes involved in genome replication, transcription and translation need to be compatible with the same processes of the host. The nature of the genome itself may also be a reflection of the conditions of the host. One of the early examples was the genomic instability of ϕH (Reiter *et al.*, 1988; Schnabel *et al.*, 1982a) which was in accordance with the instability of the host genome (Pfeifer *et al.*, 1981; Reiter *et al.*, 1988; Sapienza *et al.*, 1982; Simsek *et al.*, 1982). The pleomorphic viruses, although clearly related, contain different types of genomes (see above). This may also reflect the nature of the replication process in their host cells. For example, it may be that in different host cells the rates of different replication steps vary leading to packaging of either single stranded or double stranded DNA genomes. Genomic instability has also been reported for the crenarchaeal virus SIRV1 (Prangishvili *et al.*, 1999). Although the mechanisms for this instability were also not studied further, the interplay between unidentified factors in the viral genome together with a certain host was suggested (Prangishvili *et al.*, 1999).

Taken together, there is relatively small amount of complete prokaryotic viral genome sequences published and even smaller amount of haloarchaeal viral genomes. Considering the fact that prokaryotic viruses, including the haloarchaeal viruses, are the most abundant life forms on earth we still have many new things to discover. In addition, the already shown influence the prokaryotic viruses have on their hosts warrants proper efforts to be put in the characterization of them. Not only new genome sequences, but also the functional testing of the gene content is required. The most interesting findings in the studies of viral genomes come from careful characterization of both the viruses and their hosts. Unfortunately, it is not an easy, cheap or fast process even during these modern times of molecular biology. Yet, only by detailed studies of the viral–host interactions will we be able to find out the basic rules, if any, behind the interplay between the viral and host genomes resulting in the dynamic genomes of prokaryotic viruses.

Acknowledgements

This review is dedicated to the memory of Kielo Haahtela, Professor in Microbiology, who gave me the opportunity to try my best. This work was supported by the Helsinki University Three Year Grant (2010–2012).

References

Aalto, A.P., Bitto, D., Ravantti, J.J., Bamford, D.H., Huiskonen, J.T., and Oksanen, H.M. (2012). Snapshot of virus evolution in hypersaline environments from the characterization of a membrane-containing *Salisaeta* icosahedral phage 1. Proc. Natl. Acad. Sci. U.S.A. *109*, 7079–7084.

Abrescia, N.G.A., Grimes, J.M., Ravantti, J.J., Bamford, D.H., and Stuart, D.I. (2010). What does it take to make a virus: the concept of the viral "self". In Emerging Topics in Physical Virology, P.G. Stockley, and R. Twarock, eds. (Imperial College Press, London, UK), pp. 35–58.

Abrescia, N.G., Bamford, D.H., Grimes, J.M., and Stuart, D.I. (2012). Structure unifies the viral universe. Annu. Rev. Biochem. *81*, 795–822.

Ackermann, H.-W. (2007). 5500 Phages examined in the electron microscope. Arch. Virol. *152*, 227–243.

Arbiol, C., Comeau, A.M., Kutateladze, M., Adamia, R., and Krisch, H.M. (2010). Mobile regulatory cassettes mediate modular shuffling in T4-type phage genomes. Genome Biol. Evol. *2*, 140–152.

Atanasova, N.S., Roine, E., Oren, A., Bamford, D.H., and Oksanen, H.M. (2012). Global network of specific virus–host interactions in hypersaline environments. Environ. Microbiol. *14*, 426–440.

Bailly-Bechet, M., Vergassola, M., and Rocha, E. (2007). Causes for the intriguing presence of tRNAs in phages. Genome Res. *17*, 1486–1495.

Baltimore, D. (1971). Expression of animal virus genomes. Bacteriol. Rev. *35*, 235–241.

Bamford, D.H. (2003). Do viruses form lineages across different domains of life? Res. Microbiol. *154*, 231–236.

Bamford, D.H., Palva, E.T., and Lounatmaa, K. (1976). Ultrastructure and life cycle of the lipid-containing bacteriophage ϕ6. J. Gen. Virol. *32*, 249–259.

Bamford, D.H., Burnett, R.M., and Stuart, D.I. (2002). Evolution of viral structure. Theoret. Popul. Biol. *61*, 461–470.

Bamford, D.H., Grimes, J.M., and Stuart, D.I. (2005a). What does structure tell us about virus evolution? Curr. Opin. Struct. Biol. *15*, 655–663.

Bamford, D.H., Ravantti, J.J., Rönnholm, G., Laurinavičius, S., Kukkaro, P., Dyall-Smith, M., Somerharju, P., Kalkkinen, N., and Bamford, J.K. (2005b). Constituents of SH1, a novel lipid-containing virus infecting the halophilic euryarchaeon *Haloarcula hispanica*. J. Virol. *79*, 9097–9107.

Barrangou, R., Fremaux, C., Deveau, H., Richards, M., Boyaval, P., Moineau, S., Romero, D.A., and Horvath, P. (2007). CRISPR provides acquired resistance against viruses in prokaryotes. Science *315*, 1709–1712.

Basler, M., Pilhofer, M., Henderson, G.P., Jensen, G.J., and Mekalanos, J.J. (2012). Type VI secretion requires a dynamic contractile phage tail-like structure. Nature *483*, 182–186.

Bath, C., and Dyall-Smith, M.L. (1998). His1, an archaeal virus of the *Fuselloviridae* family that infects *Haloarcula hispanica*. J. Virol. *72*, 9392–9395.

Bath, C., Cukalac, T., Porter, K., and Dyall-Smith, M.L. (2006). His1 and His2 are distantly related, spindle-shaped haloviruses belonging to the novel virus group, *Salterprovirus*. Virology *350*, 228–239.

Baudoux, A.-C., Hendrix, R.W., Lander, G.C., Bailly, X., Podell, S., Paillard, C., Johnson, J.E., Potter, C.S., Carragher, B., and Azam, F. (2012). Genomic and functional analysis of *Vibrio* phage SIO-2 reveals novel insights into ecology and evolution of marine siphoviruses. Env. Microbiol. *14*, 2071–2086.

Benson, S.D., Bamford, J.K., Bamford, D.H., and Burnett, R.M. (1999). Viral evolution revealed by bacteriophage PRD1 and human adenovirus coat protein structures. Cell *98*, 825–833.

Benson, S.D., Bamford, J.K., Bamford, D.H., and Burnett, R.M. (2004). Does common architecture reveal a viral lineage spanning all three domains of life? Mol. Cell *16*, 673–685.

Bettarel, Y., Bouvier, T., Bouvier, C., Carre, C., Desnues, A., Domaizon, I., Jacquet, S., Robin, A., and Sime-Ngando, T. (2011). Ecological traits of planktonic viruses and prokaryotes along a full-salinity gradient. FEMS Microbiol. Ecol. *76*, 360–372.

Bhaya, D., Davison, M., and Barrangou, R. (2011). CRISPR-Cas systems in Bacteria and Archaea: versatile small RNAs for adaptive defense and regulation. Annu. Rev. Genet. *45*, 273–297.

Bolotin, A., Quinquis, B., Sorokin, A., and Ehrlich, S.D. (2005). Clustered regularly interspaced short palindrome repeats (CRISPRs) have spacers of extrachromosomal origin. Microbiology *151*, 2551–2561.

Brüssow, H., and Desiere, F. (2001). Comparative phage genomics and the evolution of *Siphoviridae*: insights from dairy phages. Mol. Microbiol. *39*, 213–222.

Brüssow, H., and Hendrix, R.W. (2002). Phage genomics: small is beautiful. Cell *108*, 13–16.

Brüssow, H., Canchaya, C., and Hardt, W.D. (2004). Phages and the evolution of bacterial pathogens: from genomic rearrangements to lysogenic conversion. Microbiol. Mol. Biol. Rev. *68*, 560–602.

Carver, T.J., Rutherford, K.M., Berriman, M., Rajandream, M.A., Barrell, B.G., and Parkhill, J. (2005). ACT: the Artemis Comparison Tool. Bioinformatics *21*, 3422–3423.

Casjens, S. (2003). Prophages and bacterial genomics: what have we learned so far? Mol. Microbiol. *49*, 277–300.

Casjens, S.R. (2011). The DNA-packaging nanomotor of tailed bacteriophages. Nat. Rev. Microbiol. *9*, 647–657.

Comeau, A.M., Bertrand, C., Letarov, A., Tetart, F., and Krisch, H.M. (2007). Modular architecture of the T4 phage superfamily: a conserved core genome and a plastic periphery. Virology *362*, 384–396.

Cota-Robles, E., Espejo, R.T., and Haywood, P.W. (1968). Ultrastructure of bacterial cells infected with bacteriophage PM2, a lipid-containing bacterial virus. J. Virol. *2*, 56–68.

Cresawn, S.G., Bogel, M., Day, N., Jacobs-Sera, D., Hendrix, R.W., and Hatfull, G.F. (2011). Phamerator: a bioinformatic tool for comparative bacteriophage genomics. BMC Bioinformatics *12*, 395.

Daniel, V., Sarid, S., and Littauer, U.Z. (1970). Bacteriophage induced transfer RNA in *Escherichia coli*. New transfer RNA molecules are synthesized on the bacteriophage genome. Science *167*, 1682–1688.

Daniels, L.L., and Wais, A.C. (1984). Restriction and modification of halophage S45 in *Halobacterium*. Curr. Microbiol. *10*, 133–136.

Daniels, L.L., and Wais, A.C. (1990). Ecophysiology of bacteriophage S5100 infecting *Halobacterium cutirubrum*. Appl. Environ. Microbiol. *56*, 3605–3608.

Daniels, L.L., and Wais, A.C. (1998). Virulence in phage populations infecting *Halobacterium curtirubrum*. FEMS Microbiol. Ecol. *25*, 129–134.

Doolittle, W.F. (1999). Phylogenetic classification and the universal tree. Science *284*, 2124–2129.

Doolittle, W.F., and Papke, R.T. (2006). Genomics and the bacterial species problem. Genome Biol. *7*, 116.

Dyall-Smith, M., Tang, S.L., and Bath, C. (2003). Haloarchaeal viruses: how diverse are they? Res. Microbiol.*154*, 309–313.

Dyall-Smith, M.L., Pfeiffer, F., Klee, K., Palm, P., Gross, K., Schuster, S.C., Rampp, M., and Oesterhelt, D. (2011). *Haloquadratum walsbyi*: limited diversity in a global pond. PLoS One *6*, e20968.

Dybvig, K., Nowak, J.A., Sladek, T.L., and Maniloff, J. (1985). Identification of an enveloped phage, mycoplasma virus L172, that contains a 14-kilobase single-stranded DNA genome. J. Virol. *53*, 384–390.

Edgell, D.R., Gibb, E.A., and Belfort, M. (2010). Mobile DNA elements in T4. Virol. J. *7*, 290.

Enav, H., Béjà, O., and Mandel-Gutfreund, Y. (2012). Cyanophage tRNAs may have a role in cross-infectivity of oceanic *Prochlorococcus* and *Synechococcus* hosts. ISME J. *6*, 619–628.

Filée, J., Bapteste, E., Susko, E., and Krisch, H.M. (2006). A selective barrier to horizontal gene transfer in the T4-type bacteriophages that has preserved a core genome with the viral replication and structural genes. Mol. Biol. Evol. *23*, 1688–1696.

Fraser, C., Alm, E.J., Polz, M.F., Spratt, B.G., and Hanage, W.P. (2009). The bacterial species challenge: making sense of genetic and ecological diversity. Science *323*, 741–746.

Ganem, D., and Varmus, H.E. (1987). The molecular biology of the hepatitis B viruses. Annu. Rev. Biochem. *56*, 651–693.

Garcia-Heredia, I., Martin-Cuadrado, A.B., Mojica, F.J., Santos, F., Mira, A., Antón, J., and Rodriguez-Valera, F. (2012). Reconstructing viral genomes from the environment using fosmid clones: the case of haloviruses. PLoS One *7*, e33802.

Garoff, H., Hewson, R., and Opstelten, D.J. (1998). Virus maturation by budding. Microbiol. Mol. Biol. Rev. *62*, 1171–1190.

Goren, M., Yosef, I., Edgar, R., and Qimron, U. (2012). The bacterial CRISPR/Cas system as analog of the mammalian adaptive immune system. RNA Biol. *9*, 549–554.

Graham, D.E., Overbeek, R., Olsen, G.J., and Woese, C.R. (2000). An archaeal genomic signature. Proc. Natl. Acad. Sci. U.S.A. *97*, 3304–3308.

Gropp, F., Grampp, B., Stolt, P., Palm, P., and Zillig, W. (1992). The immunity-conferring plasmid pϕHL from the *Halobacterium salinarium* phage ϕH: nucleotide sequence and transcription. Virology *190*, 45–54.

Guixa-Boixareu, N., Calderón-Paz, J.I., Heldal, M., Bratbak, G., and Pedrós-Alió, C. (1996). Viral lysis and bacterivory as prokaryotic loss factors along a salinity gradient. Aquat. Microb. Ecol. *11*, 215–227.

de la Haba, R.R., Márquez, M C., Papke, R.T., and Ventosa, A. (2012). Multilocus sequence analysis of the family *Halomonadaceae*. Int. J. Syst. Evol. Microbiol. *62*, 520–538.

Halohandbook, 2009. Protocols for Halobacterial genetics version 7.2. M. Dyall-Smith, ed. 144 pp. Available at: http://www.haloarchaea.com

Happonen, L.J., Redder, P., Peng, X., Reigstad, L.J., Prangishvili, D., and Butcher, S.J. (2010). Familial relationships in hyperthermo- and acidophilic archaeal viruses. J. Virol. *84*, 4747–4754.

Hatfull, G.F. (2010). Mycobacteriophages: genes and genomes. Annu. Rev. Microbiol. *64*, 331–356.

Hatfull, G.F., and Hendrix, R.W. (2011). Bacteriophages and their genomes. Curr. Opin. Virol. *1*, 298–303.

Hayes, F., and Van Melderen, L. (2011). Toxins–antitoxins: diversity, evolution and function. Crit. Rev. Biochem. Mol. Biol. *46*, 386–408.

Hendrix, R.W. (2002). Bacteriophages: evolution of the majority. Theor. Popul. Biol. *61*, 471–480.

Hendrix, R.W. (2003). Bacteriophage genomics. Curr. Opin. Microbiol. *6*, 506–511.

Hendrix, R.W. (2009). Jumbo bacteriophages. In Current Topics in Microbiolology and Immunology: The Jumbo Phages, J.L. Van Etten, ed. (Springer, Berlin), pp. 229–240.

Hendrix, R.W., Smith, M.C., Burns, R.N., Ford, M.E., and Hatfull, G.F. (1999). Evolutionary relationships among diverse bacteriophages and prophages: all the world's a phage. Proc. Natl. Acad. Sci. U.S.A. *96*, 2192–2197.

Hendrix, R.W., Lawrence, J.G., Hatfull, G.F., and Casjens, S. (2000). The origins and ongoing evolution of viruses. Trends Microbiol. *8*, 504–508.

Hobbs, M., and Mattick, J.S. (1993). Common components in the assembly of type 4 fimbriae, DNA transfer systems, filamentous phage and protein-secretion apparatus: a general system for the formation of surface-associated protein complexes. Mol. Microbiol. *10*, 233–243.

Holmes, M.L., Pfeifer, F., and Dyall-Smith, M.L. (1995). Analysis of the Halobacterial plasmid pHK2 minimal replicon. Gene *153*, 117–121.

Jaakkola, S.T., Penttinen, R.K., Vilén, S.T., Jalasvuori, M., Rönnholm, G., Bamford, J.K., Bamford, D.H., and Oksanen, H.M. (2012). Closely related archaeal *Haloarcula hispanica* icosahedral viruses HHIV-2 and SH1 have nonhomologous genes encoding host recognition functions. J. Virol. *86*, 4734–4742.

Jäälinoja, H.T., Roine, E., Laurinmäki, P., Kivelä, H.M., Bamford, D.H., and Butcher, S.J. (2008). Structure and host–cell interaction of SH1, a membrane-containing, halophilic euryarchaeal virus. Proc. Natl. Acad. Sci. U.S.A. *105*, 8008–8013.

Jaatinen, S.T., Happonen, L.J., Laurinmäki, P., Butcher, S.J., and Bamford, D.H. (2008). Biochemical and structural characterisation of membrane-containing icosahedral dsDNA bacteriophages infecting thermophilic *Thermus thermophilus*. Virology 379, 10–19.

Jalasvuori, M., Jaatinen, S.T., Laurinavičius, S., Ahola-Iivarinen, E., Kalkkinen, N., Bamford, D.H., and Bamford, J.K. (2009). The closest relatives of icosahedral viruses of thermophilic bacteria are among viruses and plasmids of the halophilic archaea. J. Virol. *83*, 9388–9397.

Jansen, R., Embden, J.D., Gaastra, W., and Schouls, L.M. (2002). Identification of genes that are associated with DNA repeats in prokaryotes. Mol. Microbiol. *43*, 1565–1575.

Juhala, R.J., Ford, M.E., Duda, R.L., Youlton, A., Hatfull, G.F., and Hendrix, R.W. (2000). Genomic sequences of bacteriophages HK97 and HK022: pervasive genetic mosaicism in the lambdoid bacteriophages. J. Mol. Biol. *299*, 27–51.

Kandiba, L., Aitio, O., Helin, J., Guan, Z., Permi, P., Bamford, D.H., Eichler, J., and Roine, E. (2012). Diversity in prokaryotic glycosylation: an archaeal-derived N-linked glycan contains legionaminic acid. Mol. Microbiol. *84*, 578–593.

Katsura, I., and Hendrix, R.W. (1984). Length determination in bacteriophage lambda tails. Cell 39, 691–698.

Khayat, R., Tang, L., Larson, E.T., Lawrence, C.M., Young, M., and Johnson, J.E. (2005). Structure of an archaeal virus capsid protein reveals a common ancestry to eukaryotic and bacterial viruses. Proc. Natl. Acad. Sci. U.S.A. *102*, 18944–18949.

King, A.M.Q., Adams, M.J., Carstens, E.B., and Lefkowitz, E.J., eds. (2012). Virus Taxonomy: Ninth Report of the International Committee on Taxonomy of Viruses (Elsevier, Oxford, UK).

Kivelä, H.M., Männistö, R.H., Kalkkinen, N., and Bamford, D.H. (1999). Purification and protein composition of PM2, the first lipid-containing bacterial virus to be isolated. Virology *262*, 364–374.

Kivelä, H.M., Roine, E., Kukkaro, P., Laurinavičius, S., Somerharju, P., and Bamford, D.H. (2006). Quantitative dissociation of archaeal virus SH1 reveals distinct capsid proteins and a lipid core. Virology *356*, 4–11.

Klein, R., Baranyi, U., Rössler, N., Greineder, B., Scholz, H., and Witte, A. (2002). *Natrialba magadii* virus ϕCh1: first complete nucleotide sequence and functional organization of a virus infecting a haloalkaliphilic archaeon. Mol. Microbiol. *45*, 851–863.

Klein, R., Rössler, N., Iro, M., Scholz, H., and Witte, A. (2012). Haloarchaeal myovirus ϕCh1 harbours a phase variation system for the production of protein variants with distinct cell surface adhesion specificities. Mol. Microbiol. *83*, 137–150.

Köck, J., and Schlicht, H.J. (1993). Analysis of the earliest steps of hepadnavirus replication: genome repair after infectious entry into hepatocytes does not depend on viral polymerase activity. J. Virol. *67*, 4867–4874.

Koonin, E.V., Wolf, Y.I., Nagasaki, K., and Dolja, V.V. (2009). The complexity of the virus world. Nat. Rev. Microbiol. *7*, 250.

Krupovič, M., and Bamford, D.H. (2008). Virus evolution: how far does the double β-barrel viral lineage extend? Nat. Rev. Microbiol. *6*, 941–948.

Krupovič, M., Prangishvili, D., Hendrix, R.W., and Bamford, D.H. (2011). Genomics of bacterial and archaeal viruses: dynamics within the prokaryotic virosphere. Microbiol. Mol. Biol. Rev. *75*, 610–635.

Kukkaro, P., and Bamford, D.H. (2009). Virus–host interactions in environments with a wide range of ionic strengths. Environ. Microbiol. Rep. *1*, 71–77.

Kulakov, L.A., Ksenzenko, V.N., Shlyapnikov, M.G., Kochetkov, V.V., Del Casale, A., Allen, C.C., Larkin, M.J., Ceyssens, P.J., and Lavigne, R. (2009). Genomes of "phiKMV-like viruses" of *Pseudomonas aeruginosa* contain localized single-strand interruptions. Virology *391*, 1–4.

Kunisawa, T. (2000). Functional role of mycobacteriophage transfer RNAs. J. Theor. Biol. *205*, 167–170.

Lang, A.S., and Beatty, J.T. (2007). Importance of widespread gene transfer agent genes in α-proteobacteria. Trends Microbiol. *15*, 54–62.

Lawrence, J.G., Hatfull, G.F., and Hendrix, R.W. (2002). Imbroglios of viral taxonomy: genetic exchange and failings of phenetic approaches. J. Bacteriol. *184*, 4891–4905.

Leiman, P.G., Basler, M., Ramagopal, U.A., Bonanno, J.B., Sauder, J.M., Pukatzki, S., Burley, S.K., Almo, S.C., and Mekalanos, J.J. (2009). Type VI secretion apparatus and phage tail-associated protein complexes share a common evolutionary origin. Proc. Natl. Acad. Sci. U.S.A. *106*, 4154–4159.

Limor-Waisberg, K., Carmi, A., Scherz, A., Pilpel, Y., and Furman, I. (2011). Specialization versus adaptation: two strategies employed by cyanophages to enhance their translation efficiencies. Nucleic Acids Res. *39*, 6016–6028.

Lindell, D., Jaffe, J.D., Johnson, Z.I., Church, G.M., and Chisholm, S.W. (2005). Photosynthesis genes in marine viruses yield proteins during host infection. Nature *438*, 86–89.

Lipps, G. (2006). Plasmids and viruses of the thermoacidophilic crenarchaeote *Sulfolobus*. Extremophiles *10*, 17–28.

Lubbers, M.W., Waterfield, N.R., Beresford, T.P., Le Page, R.W., and Jarvis, A.W. (1995). Sequencing and analysis of the prolate-headed lactococcal bacteriophage c2 genome and identification of the structural genes. Appl. Environ. Microbiol. *61*, 4348–4356.

McDaniel, L.D., Young, E., Delaney, J., Ruhnau, F., Ritchie, K.B., and Paul, J.H. (2010). High frequency of horizontal gene transfer in the oceans. Science *330*, 50.

Makarova, K.S., Haft, D.H., Barrangou, R., Brouns, S.J., Charpentier, E., Horvath, P., Moineau, S., Mojica, F.J., Wolf, Y.I., Yakunin, A.F., van der Oost, J., and Koonin, E.V. (2011). Evolution and classification of the CRISPR-Cas systems. Nat. Rev. Microbiol. *9*, 467–477.

Mann, N.H., Cook, A., Millard, A., Bailey, S., and Clokie, M. (2003). Marine ecosystems: bacterial photosynthesis genes in a virus. Nature *424*, 741.

Mann, N.H., Clokie, M.R., Millard, A., Cook, A., Wilson, W.H., Wheatley, P.J., Letarov, A., and Krisch, H.M. (2005). The genome of S-PM2, a "photosynthetic" T4-type bacteriophage that infects marine *Synechococcus* strains. J. Bacteriol. *187*, 3188–3200.

Mei, Y., Chen, J., Sun, D., Chen, D., Yang, Y., Shen, P., and Chen, X. (2007). Induction and preliminary characterization of a novel halophage SNJ1 from lysogenic *Natrinema* sp. F5. Can. J. Microbiol. *53*, 1106–1110.

Millard, A., Clokie, M.R., Shub, D.A., and Mann, N.H. (2004). Genetic organization of the *psbAD* region in phages infecting marine *Synechococcus* strains. Proc. Natl. Acad. Sci. U.S.A. *101*, 11007–11012.

Mindich, L., and Lehman, J. (1979). Cell wall lysin as a component of the bacteriophage ϕ6 virion. J. Virol. *30*, 489–496.

Mutschler, H., Gebhardt, M., Shoeman, R.L., and Meinhart, A. (2011). A novel mechanism of programmed cell death in bacteria by toxin–antitoxin systems corrupts peptidoglycan synthesis. PLoS Biol. *9*, e1001033.

Nakayama, K., Takashima, K., Ishihara, H., Shinomiya, T., Kageyama, M., Kanaya, S., Ohnishi, M., Murata, T., Mori, H., and Hayashi, T. (2000). The R-type pyocin of *Pseudomonas aeruginosa* is related to P2 phage, and the F-type is related to lambda phage. Mol. Microbiol. *38*, 213–231.

Nguyen, A.H., Tomita, T., Hirota, M., Sato, T., and Kamio, Y. (1999). A simple purification method and morphology and component analyses for carotovoricin Er, a phage-tail-like bacteriocin from the plant pathogen *Erwinia carotovora* Er. Biosci. Biotech. Biochem. *63*, 1360–1369.

Nuttall, S.D., and Dyall-Smith, M.L. (1993). HF1 and HF2: novel bacteriophages of halophilic archaea. Virology *197*, 678–684

O'Connor, E.M., and Shand, R.F. (2002). Halocins and sulfolobicins: the emerging story of archaeal protein and peptide antibiotics. J. Indust. Microbiol. Biotechnol. *28*, 23–31.

Oren, A. (2008). Microbial life at high salt concentrations: phylogenetic and metabolic diversity. Saline Syst. *4*, 2.

Orengo, C.A., and Thornton, J.M. (2005). Protein families and their evolution-a structural perspective. Annu. Rev. Biochem. *74*, 867–900.

Pagaling, E., Haigh, R.D., Grant, W.D., Cowan, D.A., Jones, B.E., Ma, Y., Ventosa, A., and Heaphy, S. (2007). Sequence analysis of an archaeal virus isolated from a hypersaline lake in Inner Mongolia, China. BMC Genomics *8*, 410.

Papke, R.T., and Gogarten, J.P. (2012). Ecology. How bacterial lineages emerge. Science *336*, 45–46.

Papke, R.T., Zhaxybayeva, O., Feil, E.J., Sommerfeld, K., Muise, D., and Doolittle, W.F. (2007). Searching for species in haloarchaea. Proc. Natl. Acad. Sci. U.S.A. *104*, 14092–14097.

Park, J.S., Kim, H., Choi, D.H., and Cho, B.C. (2003). Active flagellates grazing on prokaryotes in high salinity waters of a solar saltern. Aquat. Microbial Ecol. 33, 173–179.

Pauling, C. (1982). Bacteriophages of *Halobacterium halobium*: isolated from fermented fish sauce and primary characterization. Can. J. Microbiol. *28*, 916–921.

Pedrós-Alió, C., Calderón-Paz, J.I., MacLean, M.H., Medina, G., Marrasé, C., Gasol, J.M., and Guixa-Boixereu, N. (2000). The microbial food web along salinity gradients. FEMS Microbiol. Ecol. 32, 143–155.

Pedulla, M.L., Ford, M.E., Houtz, J.M., Karthikeyan, T., Wadsworth, C., Lewis, J.A., Jacobs-Sera, D., Falbo, J., Gross, J., and Pannunzio, N.R. (2003). Origins of highly mosaic mycobacteriophage genomes. Cell *113*, 171–182.

Pell, L.G., Kanelis, V., Donaldson, L.W., Howell, P.L., and Davidson, A.R. (2009). The phage lambda major tail protein structure reveals a common evolution for long-tailed phages and the type VI bacterial secretion system. Proc. Natl. Acad. Sci. U.S.A. *106*, 4160–4165.

Pfeifer, F., Weidinger, G., and Goebel, W. (1981). Genetic variability in *Halobacterium halobium*. J. Bacteriol. *145*, 375–381.

Pietilä, M.K., Roine, E., Paulin, L., Kalkkinen, N., and Bamford, D.H. (2009). An ssDNA virus infecting archaea: a new lineage of viruses with a membrane envelope. Mol. Microbiol. 72, 307–319.

Pietilä, M.K., Laurinavičius, S., Sund, J., Roine, E., and Bamford, D.H. (2010). The single-stranded DNA genome of novel archaeal virus *Halorubrum* pleomorphic virus 1 is enclosed in the envelope decorated with glycoprotein spikes. J. Virol. *84*, 788–798.

Pietilä, M.K., Atanasova, N.S., Oksanen, H.M., and Bamford, D.H. (2012a). Modified coat protein forms the flexible spindle-shaped virion of haloarchaeal virus His1. Env. Microbiol. *15*, 1674–1686.

Pietilä, M.K., Atanasova, N.S., Manole, V., Liljeroos, L., Butcher, S.J., Oksanen, H.M., and Bamford, D.H. (2012b). Virion architecture unifies globally distributed pleolipoviruses infecting halophilic archaea. J. Virol. *86*, 5067–5079.

Pietilä, M.K., Laurinmäki, P., Russell, D.A., Ko, C.C., Jacobs-Sera, D., Butcher, S.J., Bamford, D.H., and Hendrix, R.W. (2013). Insights into head-tailed viruses infecting extremely halophilic Archaea. J. Virol. *87*, 3248–3260.

Pina, M., Bize, A., Forterre, P., and Prangishvili, D. (2011). The archeoviruses. FEMS Microbiol. Rev. 35, 1035–1054.

Plemenitaš, A., and Gunde-Cimerman, N. (2011). Molecular mechanisms of adaptations to high salt concentration in the extremely halotolerant black yeast *Hortea werneckii*. In Halophiles and Hypersaline Environments, A. Ventosa, A. Oren, and Y. Ma, eds. (Springer, Berlin Heidelberg), pp. 137–151.

Porter, K., and Dyall-Smith, M. (2006). The isolation and study of viruses of halophilic microorganisms. In Methods in Microbiology, Extremophiles, F.A. Rainey, and A. Oren, eds. (Elsevier/Academic Press, London), pp. 681–702.

Porter, K., and Dyall-Smith, M.L. (2008). Transfection of haloarchaea by the DNAs of spindle and round haloviruses and the use of transposon mutagenesis to identify non-essential regions. Mol. Microbiol. *70*, 1236–1245.

Porter, K., Kukkaro, P., Bamford, J.K., Bath, C., Kivelä, H.M., Dyall-Smith, M.L., and Bamford, D.H. (2005). SH1: a novel, spherical halovirus isolated from an Australian hypersaline lake. Virology *335*, 22–33.

Porter, K., Russ, B.E., and Dyall-Smith, M.L. (2007). Virus–host interactions in salt lakes. Curr. Opin. Microbiol. *10*, 418–424.

Porter, K., Russ, B.E., Thorburn, A.N., and Dyall-Smith, M.L. (2008). Viruses infecting Euryarchaea. In Encyclopedia of Virology, B.W.J. Mahy, and M.H.V. van Regenmortel, eds. (Elsevier, Oxford, UK), pp. 411–423.

Prangishvili, D., Arnold, H.P., Gotz, D., Ziese, U., Holz, I., Kristjansson, J.K., and Zillig, W. (1999). A novel virus family, the Rudiviridae: structure, virus–host interactions and genome variability of the *Sulfolobus* viruses SIRV1 and SIRV2. Genetics *152*, 1387–1396.

Pukatzki, S., Ma, A.T., Revel, A.T., Sturtevant, D., and Mekalanos, J.J. (2007). Type VI secretion system translocates a phage tail spike-like protein into target cells where it cross-links actin. Proc. Natl. Acad. Sci. U.S.A. *104*, 15508–15513.

Rakonjac, J., Bennett, N.J., Spagnuolo, J., Gagic, D., and Russel, M. (2011). Filamentous bacteriophage: biology, phage display and nanotechnology applications. Curr. Issues Mol. Biol. *13*, 51–76.

Reiter, W.D., Zillig, W., and Palm, P. (1988). Archaebacterial viruses. Adv. Virus Res. *34*, 143–188.

Rhodes, M.E., Spear, J.R., Oren, A., and House, C.H. (2011). Differences in lateral gene transfer in hypersaline versus thermal environments. BMC Evol. Biol. *11*, 199.

Roine, E., and Oksanen, H.M. (2011). Viruses from the hypersaline environment. In Halophiles and Hypersaline Environments, A. Ventosa, A. Oren, and Y. Ma, eds. (Springer, Berlin Heidelberg), pp. 153–172.

Roine, E., Kukkaro, P., Paulin, L., Laurinavičius, S., Domanska, A., Somerharju, P., and Bamford, D.H. (2010). New, closely related haloarchaeal viral elements with different nucleic acid types. J. Virol. *84*, 3682–3689.

Rössler, N., Klein, R., Scholz, H., and Witte, A. (2004). Inversion within the haloalkaliphilic virus ϕCh1 DNA results in differential expression of structural proteins. Mol. Microbiol. *52*, 413–426.

Russel, M. (1991). Filamentous phage assembly. Mol. Microbiol. *5*, 1607–1613.

Salas, M. (1991). Protein-priming of DNA replication. Annu. Rev. Biochem. *60*, 39–71.

Santos, F., Yarza, P., Parro, V., Meseguer, I., Rosselló-Móra, R., and Antón, J. (2012). Culture-independent approaches for studying viruses from hypersaline environments. Appl. Environ. Microbiol. *78*, 1635–1643.

Sapienza, C., Rose, M.R., and Doolittle, W.F. (1982). High-frequency genomic rearrangements involving archaebacterial repeat sequence elements. Nature *299*, 182–185.

Saren, A.M., Ravantti, J.J., Benson, S.D., Burnett, R.M., Paulin, L., Bamford, D.H., and Bamford, J.K. (2005). A snapshot of viral evolution from genome analysis of the *Tectiviridae* family. J. Mol. Biol. *350*, 427–440.

Schnabel, H., Schramm, E., Schnabel, R., and Zillig, W. (1982a). Structural variability in the genome of phage ϕH of *Halobacterium halobium*. Mol. Gen. Genet. *188*, 370–377.

Schnabel, H., Zillig, W., Pfaffle, M., Schnabel, R., Michel, H., and Delius, H. (1982b). *Halobacterium halobium* phage ϕH. EMBO J. *1*, 87–92.

Schröder, G., and Lanka, E. (2005). The mating pair formation system of conjugative plasmids – a versatile secretion machinery for transfer of proteins and DNA. Plasmid *54*, 1–25.

Senčilo, A., Paulin, L., Kellner, S., Helm, M., and Roine, E. (2012). Related haloarchaeal pleomorphic viruses contain different genome types. Nucleic Acids Res. *40*, 5523–5534.

Senčilo, A., Jacobs-Sera, D., Russell, D.A., Ko, C.-C., Bowman, C.A., Atanasova, N.S., Österlund, E., Oksanen, H.M., Bamford, D.H., Hatfull, G.F., *et al.* (2013). Snapshot of haloarchaeal tailed virus genomes. RNA Biol. *10*, 803–816.

Sharon, I., Alperovitch, A., Rohwer, F., Haynes, M., Glaser, F., Atamna-Ismaeel, N., Pinter, R.Y., Partensky, F., Koonin, E.V., Wolf, Y.I., *et al.* (2009). Photosystem I gene cassettes are present in marine virus genomes. Nature *461*, 258–262.

Simsek, M., DasSarma, S., RajBhandary, U.L., and Khorana, H.G. (1982). A transposable element from *Halobacterium halobium* which inactivates the bacteriorhodopsin gene. Proc. Natl. Acad. Sci. U.S.A. *79*, 7268–7272.

Stackebrandt, E., and Goebel, B.M. (1994). Taxonomic note: A place for DNA–DNA reassociation and 16S rRNA sequence analysis in the present species definition in bacteriology. Int. J. Syst. Bacteriol. *44*, 846–849.

Strömsten, N.J., Bamford, D.H., and Bamford, J.K. (2005). In vitro DNA packaging of PRD1: a common mechanism for internal-membrane viruses. J. Mol. Biol. *348*, 617–629.

Sullivan, M.B., Coleman, M.L., Weigele, P., Rohwer, F., and Chisholm, S.W. (2005). Three *Prochlorococcus* cyanophage genomes: signature features and ecological interpretations. PLoS Biol. 3, e144.

Sullivan, M.B., Lindell, D., Lee, J.A., Thompson, L.R., Bielawski, J.P., and Chisholm, S.W. (2006). Prevalence and evolution of core photosystem II genes in marine cyanobacterial viruses and their hosts. PLoS Biol. *4*, e234.

Summers, J., O'Connell, A., and Millman, I. (1975). Genome of hepatitis B virus: restriction enzyme cleavage and structure of DNA extracted from Dane particles. Proc. Natl. Acad. Sci. U.S.A. *72*, 4597–4601.

Summers, J., Smolec, J.M., and Snyder, R. (1978). A virus similar to human hepatitis B virus associated with hepatitis and hepatoma in woodchucks. Proc. Natl. Acad. Sci. U.S.A. *75*, 4533–4537.

Suttle, C.A. (2005). Viruses in the sea. Nature *437*, 356–361.

Suttle, C.A. (2007). Marine viruses – major players in the global ecosystem. Nat. Rev. Microbiol. *5*, 801–812.

Tang, S.L., Nuttall, S., Ngui, K., Fisher, C., Lopez, P., and Dyall-Smith, M. (2002). HF2: a double-stranded DNA tailed haloarchaeal virus with a mosaic genome. Mol. Microbiol. *44*, 283–296.

Tang, S.L., Nuttall, S., and Dyall-Smith, M. (2004). Haloviruses HF1 and HF2: evidence for a recent and large recombination event. J. Bacteriol. *186*, 2810–2817.

Thompson, J.R., Pacocha, S., Pharino, C., Klepac-Ceraj, V., Hunt, D.E., Benoit, J., Sarma-Rupavtarm, R., Distel, D.L., and Polz, M.F. (2005). Genotypic diversity within a natural coastal bacterioplankton population. Science *307*, 1311–1313.

Torsvik, T. (1982). Characterization of four bacteriophages for *Halobacterium*, with a special emphasis on phage Hs1. In *Archaebacteria*, O. Kandler, ed. (Gustav Fischer, Stuttgart, Germany), p. 351.

Torsvik, T., and Dundas, I.D. (1974). Bacteriophage of *Halobacterium salinarium*. Nature *248*, 680–681.

Torsvik, T., and Dundas, I.D. (1980). Persisting phage infection in *Halobacterium salinarium* str. 1. J. Gen. Virol. *47*, 29–36.

Veesler, D., and Cambillau, C. (2011). A common evolutionary origin for tailed-bacteriophage functional modules and bacterial machineries. Microbiol. Mol. Biol. Rev. *75*, 423–433.

Vogelsang-Wenke, H., and Oesterhelt, D. (1988). Isolation of a Halobacterial phage with a fully cytosine-methylated genome. Mol. Gen. Genet. *211*, 407–414.

Wais, A.C., Kon, M., MacDonald, R.E., and Stollar, B.D. (1975). Salt-dependent bacteriophage infecting *Halobacterium cutirubrum* and *H. halobium*. Nature *256*, 314–315.

Wang, J., Jiang, Y., Vincent, M., Sun, Y., Yu, H., Bao, Q., Kong, H., and Hu, S. (2005). Complete genome sequence of bacteriophage T5. Virology *332*, 45–65.

Weinbauer, M.G. (2004). Ecology of prokaryotic viruses. FEMS Microbiol. Rev. *28*, 127–181.

Witte, A., Baranyi, U., Klein, R., Sulzner, M., Luo, C., Wanner, G., Kruger, D.H., and Lubitz, W. (1997). Characterization of *Natronobacterium magadii* phage ϕCh1, a unique archaeal phage containing DNA and RNA. Mol. Microbiol. *23*, 603–616.

Woese, C.R., Kandler, O., and Wheelis, M.L. (1990). Towards a natural system of organisms: proposal for the domains Archaea, Bacteria, and Eucarya. Proc. Natl. Acad. Sci. U.S.A. *87*, 4576–4579.

Yamada, K., Hirota, M., Niimi, Y., Nguyen, H.A., Takahara, Y., Kamio, Y., and Kaneko, J. (2006). Nucleotide sequences and organization of the genes for carotovoricin (Ctv) from *Erwinia carotovora* indicate that Ctv evolved from the same ancestor as *Salmonella typhi* prophage. Biosci. Biotech. Biochem. *70*, 2236–2247.

Zhang, Z., Liu, Y., Wang, S., Yang, D., Cheng, Y., Hu, J., Chen, J., Mei, Y., Shen, P., Bamford, D.H., *et al.* (2012). Temperate membrane-containing halophilic archaeal virus SNJ1 has a circular dsDNA genome identical to that of plasmid pHH205. Virology *434*, 233–241.

Microbial Adaptation to Saline Environments: Lessons from the Genomes of *Natranaerobius thermophilus* and *Halobacillus halophilus*

5

Noha M. Mesbah, Inga Hänelt, Baisuo Zhao and Volker Müller

Abstract

The ability of extremophiles to survive and multiply under extreme conditions is of great importance for microbial physiology, evolution and industry. Whole genome sequencing has provided significant insight into mechanisms used by extremophiles for adaptation to extreme environments. This chapter reviews the current knowledge on the adaptation of two halophiles, the anaerobic alkalithermophilic *Natranaerobius thermophilus*, and the aerobic *Halobacillus halophilus*, to their extreme environments, with emphasis on traits delineated from their genome sequences. *N. thermophilus* and *H. halophilus* have developed different mechanisms for adaptation to their extreme environments. *N. thermophilus* faces multiple extremes and consequently employs different and diverse adaptive mechanisms, including accumulation of compatible solutes to counteract high salinity, multiple cation/proton antiporters for intracellular pH and ion regulation and changing intracellular amino acid content in response to high temperature. *H. halophilus* faces predominantly salt stress and has developed a hybrid strategy for adaptation involving accumulation of a mixture of compatible solutes in addition to accumulation of molar concentrations of chloride and probably potassium inside its cells. Intracellular solute composition in *H. halophilus* is strictly regulated to adjust to changing extracellular osmolarity. Genomic diversity of *N. thermophilus* indicates the presence of complex regulatory mechanisms necessary for survival under multiple extreme conditions.

Introduction

Several ecosystems on earth are extreme from a human perspective. Such environments may have extremely high or low pH, high or low temperatures, high salinity, high pressure and various combinations thereof. These environments are not 'sterile' but inhabited by specialized microorganisms that are adapted to them. Since these microorganisms live under extreme conditions, they are termed 'extremophiles'. Extremophiles are present in all three domains of life, the Archaea, Bacteria and Eukarya. Often, extremophiles are challenged by more than one extreme condition, these are termed 'polyextremophiles' (Mesbah and

Wiegel, 2008). Examples of multiple extreme conditions are hot, alkaline springs, hypersaline, alkaline lakes, and hot, acidic springs.

Microorganisms adapted to life at high salinities were the first extremophiles discovered. These extremophiles, termed halophiles, grow optimally in the presence of extremely high salinities, up to 5 M NaCl. They accumulate intracellular KCl in concentrations higher than the external NaCl concentration to maintain a turgor pressure. This strategy, called the 'salt-in' strategy, is found, in the *Halobacteriales* (Archaea) and the bacterium *Salinibacter ruber.* The 'salt-in' strategy requires that the cellular machineries are adapted to high KCl concentrations. In general, halophiles using the 'salt-in' strategy are restricted to growth at high salt. A more flexible strategy is found in moderately halophilic bacteria that grow over a wide range of salinities (typically 0.5–3 M NaCl). This strategy, the 'low-salt-in' strategy, relies on accumulation of high concentrations of organic compatible solutes and retains a cytoplasm iso-osmotic or slightly hyperosmotic with its surrounding.

This chapter summarizes current knowledge on the molecular basis of adaptation of two moderate halophiles, the anaerobic polyextremophile *Natranaerobius thermophilus,* and the aerobic moderate halophile *Halobacillus halophilus,* to their environments. Special emphasis is on the discussion of the metabolic traits delineated from the genome sequence.

The anaerobic polyextremophile *Natranaerobius thermophilus*

Strains of *Natranaerobius thermophilus* were isolated from the highly saline soda lakes of the Wadi An Natrun, Egypt (Mesbah *et al.*, 2007b). The alkaline brines of the Wadi An Natrun have pH values ranging between 9 and 11 (Imhoff *et al.*, 1979; Mesbah *et al.*, 2007a; Taher, 1999). These alkaline brines are rich in carbonate and chloride, and lack significant quantities of the divalent cations magnesium and calcium. *N. thermophilus* grows optimally at 3.3–3.9 M Na^+ (derived from both NaCl and $Na_2CO_3/NaHCO_3$), pH 9.5 (measured at 55°C) and 53°C. *N. thermophilus* is obligately anaerobic and chemoorganotrophic, producing acetate and formate when grown in the presence of sucrose, yeast extract and tryptone (Mesbah *et al.*, 2007b).

N. thermophilus is phylogenetically classified within the order *Natranaerobiales,* Class *Clostridia,* Phylum *Firmicutes* (Mesbah *et al.*, 2007b). The order *Natranaerobiales* consists exclusively of extremely halophilic, alkaliphilic and moderately thermophilic anaerobes. Among the *Natranaerobiales, N. thermophilus* has been the most extensively studied physiologically and biochemically (Krulwich *et al.*, 2009; Mesbah *et al.*, 2007b, 2009; Mesbah and Wiegel, 2011, 2012).

General features of the *Natranaerobius thermophilus* genome

The genome of *N. thermophilus* strain JW/NM-WN-LF^T (= DSM 18059^T = ATCC BAA-1301^T) was sequenced at the Joint Genome Institute (Walnut Creek, CA, USA), using a combination of 3-, 8-, and 40 kbp insert libraries and random shotgun sequencing (Zhao *et al.*, 2011). The *N. thermophilus* genome consists of a single, circular chromosome (3.2 Mbp) and two circular plasmids, pNTHE01 and pNTHE02 (17 and 8.6 kbp, respectively) (Fig. 5.1). The average G+C content of the three elements was 36.4% for the chromosome, 34.1% for pNTHE01, and 35.7% for pNTHE02 (Zhao *et al.*, 2011). On the basis of the G+C ratio

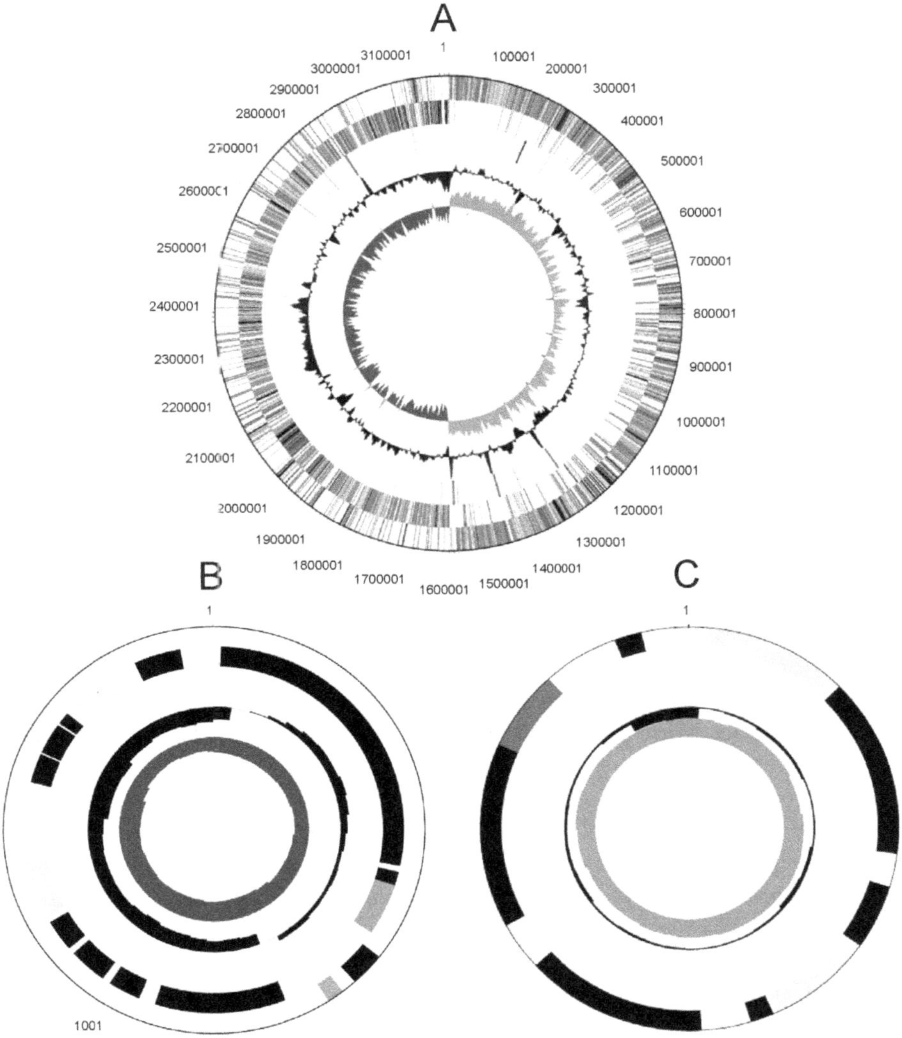

Figure 5.1 Circular representation of the *N. thermophilus* chromosome (A), plasmid pNTHE01 (B) and plasmid pNTHE02 (C). The distribution of coding sequences is depicted by the coloured boxes according to functional categories. The first two circles from the outside represent the predicted coding sequences on the plus and minus strands, respectively. The two innermost circles represent the per cent G+C and G+C skew, respectively. Colours represent the following: dark grey, hypothetical proteins; light grey, conserved hypothetical and unknown function; brown, general function prediction; red, replication and repair; green, energy metabolism; blue, carbon and carbohydrate metabolism; cyan, lipid metabolism; magenta, transcription; yellow, translation; orange, amino acid metabolism; pink, metabolism of cofactors and vitamins; light red, purine and pyrimidine metabolism; lavender, signal transduction; sky blue, cellular processes. The third circle represents red for rRNA and green for tRNA genes. A colour version of this figure is located in the plate section at the back of the book.

and G-C skew, the origin of replication was estimated to be upstream of the gene coding for the chromosomal replication initiator protein (DnaA), and the site of termination of replication is nearly 1.6 Mbp from the origin. However, a gene encoding the replication termination protein was not found.

Coding sequences covered 88.3% of the chromosome, 49.9% of pNTHE01, and 25.7% of pNTHE02. The chromosome encodes for 2882 predicted proteins, 71.4% of which were assigned putative functions. The remainder was identified as conserved proteins of unknown function. The plasmid pNTHE01 encoded for 14 proteins, five of which were assigned putative functions and plasmid pNTHE02 encoded for ten predicted proteins, five with putative functions. Fifty annotated pseudogenes were also identified on the chromosome, and one pseudogene was identified on pNTHE01. Proteins of *N. thermophilus* predicted from the 2882 coding regions were estimated to range in length from 31 to 1474 amino acids.

The chromosome of *N. thermophilus* contains four rRNA operons, each consisting of 16S, 23S and 5S rRNAs. Fifty-one transfer RNAs (tRNAs), representing 36 different anticodons, are encoded in the genome.

Long-term survival: endospore formation

Spore formation in Gram-positive bacteria is a complex developmental process that takes about 6–8 hours for completion (Errington, 1993; Stragier and Losick, 1996). Sporulation requires intricate networks of temporal and compartmental regulation and, in *Bacillus subtilis*, at least 75 genes must act sequentially (Paidhungat *et al.*, 2001). Therefore, if a component involved in a sequential process is missing, spore formation will not be observed. As a result, asporogenic phenotypes occur (Onyenwoke *et al.*, 2004). This appears to be the case in *N. thermophilus*. Many genes involved in the sporulation process were identified in the genome, including Spo0A, a DNA-binding protein shown to be the master regulator for entry into sporulation in *B. subtilis* (Molle *et al.*, 2003). In addition, many genes involved in post-septation and post-engulfment are present in the *N. thermophilus* genome. However, spores have not been observed in cells of *N. thermophilus*, neither in ultrathin sections using electron microscopy nor by light microscopy after heat treatment (10 minutes at 80°C) (Mesbah *et al.*, 2007b). A similar phenomenon has been reported in the halophilic and thermophilic bacterium *Halothermothrix orenii*. The genome sequence of *H. orenii* contains homologues for the majority of the genes involved in the sporulation process but the strain is apparently asporogenic (Mavromatis *et al.*, 2009).

Metabolism

N. thermophilus is an anaerobic chemoorganotroph, growing best in a complex medium supplied with 0.2% yeast extract and 0.2% tryptone. No growth was observed in minimal medium, indicating that it is unable to synthesize necessary components, such as essential amino acids, fatty acids, and purine and pyrimidine and nucleotides. *N. thermophilus* can utilize xylose, ribose, glucose, fructose, sucrose, cellobiose, trehalose, pyruvate, casamino acids, acetate, and peptone as carbon and energy sources (Mesbah *et al.*, 2007b). The chromosome of *N. thermophilus* contains putative genes for a complete Embden–Meyerhof pathway, a non-oxidative pentose phosphate pathway and a reductive branch of the tricarboxylic acid cycle (Fig. 5.2). The genes for a putative pyruvate formate lyase (catalysing the non-oxidative cleavage of pyruvate to formate and acetyl-CoA) and the phosphate acetyltransferase (converting acetyl CoA to acetylphosphate, the acetate kinase forming acetate and ATP)

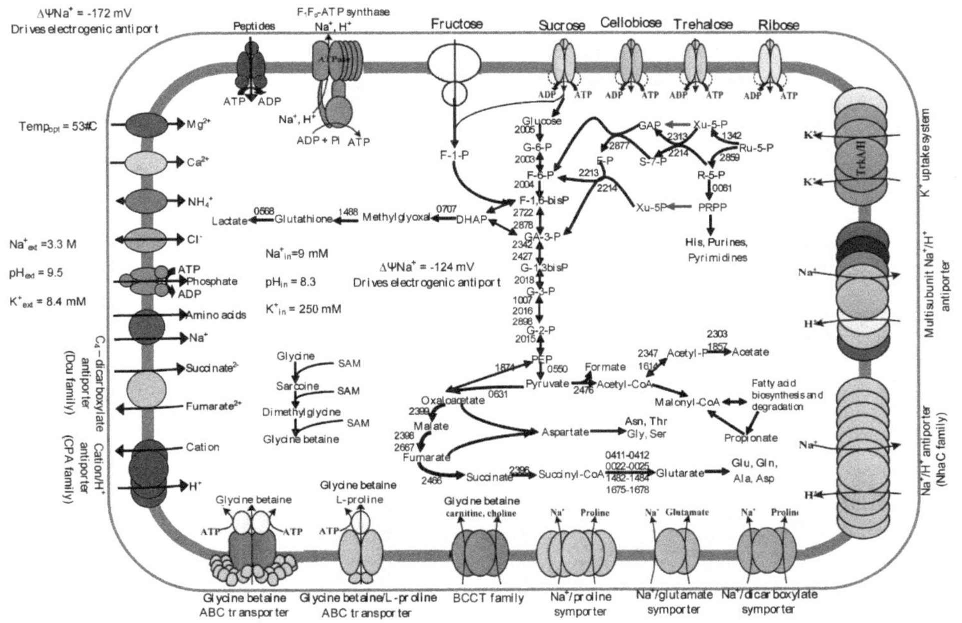

Figure 5.2 Overview of putative transport systems and central metabolism predicted in the chromosome of *N. thermophilus*. Abbreviations: G-6-P, glucose-6-phosphate; F-6-P, fructose-6-phosphate; F-1,6-bisP, fructose-1,6-bisphosphate; GA-3-P, glyceraldehyde-3-phosphate; G-1,3-bisP, 1,3-biophosphoglycerate; G-3-P, 3-biophosphoglycerate; G-2-P, 2-biophosphoglycerate; PEP, phosphoenolpyruvate; Ru-5-P, ribulose-5-phosphate; Xu-5-P, xylulose-5-phosphate; R-5-P, ribose-5-phosphate; GAP, glyceraldehyde-3-phosphate; S-7P, sedoheptulose-7-phosphate; E-P, erythrose-4-phosphate; DHAP, dihydroxyacetone phosphate; PRPP, phosphoribosyl pyrophosphate.

were also identified. Formate and acetate were the major fermentation products when *N. thermophilus* was grown on sucrose (Mesbah *et al.*, 2007b). *N. thermophilus* possesses genes for a phosphenolpyruvate synthase and fructose-1,6-bisphosphatase, but a gene encoding for glucose-6-phosphatase was not found. Therefore, *N. thermophilus* is not genetically capable of gluconeogenesis. Genes for the oxidative branch of the pentose phosphate pathway and the Entner–Doudoroff pathway were not found.

The chromosome of *N. thermophilus* contains six genes for nitrate reductases, a gene encoding for an anaerobic sulfide reductase and an anaerobic ribonucleoside triphosphate reductase. *N. thermophilus* was able to utilize nitrate and thiosulfate as electron acceptors (Mesbah *et al.*, 2007b). *N. thermophilus* possesses a complete F-type ATPase complex. In addition, two clusters of genes with high similarity to the ion-motive Rnf electron transport complex were found (Biegel *et al.*, 2011).

Long-term adaptation: adaptation to hypersaline conditions

N. thermophilus was isolated from a mixed sediment water sample from the alkaline, hypersaline Lake Fazda, Wadi An Natrun, Egypt (Mesbah *et al.*, 2007b). The water of this lake typically evaporates during the summer and fills during the winter due to a small amount of rainfall. As a result, the halophilic *N. thermophilus* must have the ability to rapidly adjust to fluctuating salinity.

Analysis of the proteome of *N. thermophilus* showed that, consistent with being an extreme halophile, the isoelectric point of proteins is predominantly acidic, ranging between 4 and 5. Thus the majority of intracellular proteins are negatively charged. This resembles the profiles mainly observed in organisms that use the 'salt-in' strategy (Elevi Bardavid and Oren, 2012). Analysis of the amino acid composition of *N. thermophilus* showed predominance for the amino acids glutamate as well as the non-polar leucine and isoleucine. It thus would be expected that *N. thermophilus* uses the 'salt-in' strategy for osmotic adaptation.

However, analysis of the intracellular ion content in energized cells of *N. thermophilus* showed that, at the optimal growth conditions of 53°C, $pH^{55°C}$ 9.5 and 3.5 M Na^+, intracellular K^+ concentration was only 250 mM. Intracellular K^+ concentration did not vary with changes in the K^+ concentration of the culture medium. The concentration remained constant below extracellular $pH^{55°C}$ 9.5, but increased sharply at more alkaline extracellular pH values reaching 540 mM at $pH^{55°C}$ 10.6, the maximum growth pH for *N. thermophilus*. These values were still too low to combat external salinity but consistent with those measured in exponentially growing cells under the same growth conditions. The intracellular Na^+ concentration in both exponentially growing and energized cell suspensions was only 8 mM at extracellular $pH^{55°C}$ 9.5, and increased to 33 mM at $pH^{55°C}$ 10.5 (Mesbah *et al.*, 2009). Concentrations of other ions (Mg^{2+}, Mn^{2+}, Li^+) were in the nanomolar range. These data indicate that *N. thermophilus* does not rely solely on the 'salt-in' strategy for osmotic adaptation.

Compatible solutes are usually amino acids and their derivatives, methylamines, polyols and sugars (Roberts, 2005). In many microorganisms, the compatible solute glycine betaine can be synthesized *de novo* or from exogenously provided choline. Synthesis *de novo* of glycine betaine is by way of glycine; it involves 3-fold methylation of glycine at the N position with S-adenosylmethionine as a methyl donor. These methylation reactions are catalysed by two enzymes, glycine sarcosine methyltransferase (GSMT) and sarcosine dimethylglycine methyltransferase (SDMT) (Nyyssola *et al.*, 2001). The chromosome of *N. thermophilus* contains genes encoding putative GSMT and SDMT. Thus, *N. thermophilus* has the genetic ability to synthesize glycine betaine. Homologues of genes involved in the synthesis of glycine betaine from choline were not found. The chromosome of *N. thermophilus* also contains 15 genes for potential glycine betaine ABC transporters, four genes for glycine betaine/L-proline ABC transporters, and three for betaine/carnitine/choline transporters (Zhao *et al.*, 2011) (Fig. 5.2). The delineated ability to accumulate glycine betaine by *de novo* synthesis and/or uptake is consistent with physiological data. *N. thermophilus* grown at 52°C, $pH^{55°C}$ 9.5 and 3.3 M NaCl had an internal glycine betaine concentration of 410 mM that increased to 1.1 M at 4.5 M Na^+, indicating that this compatible solute plays a role in osmotic stress response.

The chromosome of *N. thermophilus* also contains three genes for glutamine synthetase, five genes for Na^+/proline symporters and two for Na^+/glutamate symporters (Fig. 5.2). Collectively, the presence of a large number of compatible solute symporters together with the low intracellular ion content of *N. thermophilus* indicates that this extreme anaerobic halophile uses mainly the 'salt-out' strategy for osmotic adaptation. Noteworthy is the 'acidic' proteome that would argue for a 'salt-in' strategy or a hybrid strategy as discussed below for *H. halophilus*.

Long-term adaptation: sodium bioenergetics and adaptation to alkaline conditions

The bioenergetic problems faced by alkaliphiles are enormous, an inverted ΔpH, suboptimal proton motive force and the need to constantly acidify the cytoplasm while growing in near absence of protons (Krulwich, 1995). Alkaliphiles must have mechanisms for cytoplasmic acidification and/or homeostasis. Many strategies exist, including increased expression and activity of cation/proton antiporters (CPA), changes in cell surface properties and increased production of metabolic acids *via* amino acid deaminases and sugar fermentation. Among these, cation/proton antiporters play an essential and dominant role in cytoplasm pH regulation (Padan *et al.*, 2005; Slonczewski *et al.*, 2009).

The chromosome of *N. thermophilus* contains genes for 11 monovalent cation/proton antiporters of the NhaC type, a gene cluster encoding a multisubunit cation/proton antiporter of the CPA-3 family, four monovalent cation/proton antiporters of the CPA1 and CPA2 family, and one gene encoding a cation/proton antiporter of the NdhF-a family (Krulwich *et al.*, 2009; Mesbah *et al.*, 2009; Zhao *et al.*, 2011) (Fig. 5.2). Cloning and heterologous expression of 12 of these genes revealed that seven had $Na^+(K^+)/H^+$ antiport activity, one was a specific K^+/H^+ antiporter, and four showed K^+ transport, where they complemented a K^+ uptake deficient strain of *Escherichia coli* (Mesbah *et al.*, 2009). The kinetic and biochemical properties of the antiporter proteins correlated well with the intracellular conditions of *N. thermophilus* (Mesbah *et al.*, 2009).

The unusually large number of predicted antiporter loci in the genome of *N. thermophilus* suggests that the size of the antiporter complement of a bacterium is influenced by the number of environmental stressors it faces. There are 17 predicted antiporters in the chromosome of *N. thermophilus*; extremely alkaliphilic, halotolerant and mesophilic *Bacillus halodurans* and *Oceanobacillus iheyensis* have only five and four predicted antiporters, respectively (Takami *et al.*, 2000, 2002). Halophilic and thermophilic *Halothermothrix orenii* has only one predicted antiporter of the NhaC type in its genome (Mavromatis *et al.*, 2009), whereas the haloalkaliphilic *Thioalkalivibrio sulfidiphilus* has only two predicted antiporters (Muyzer et al., 2011). It is possible that extremophiles facing only one extreme require a small number of antiporters specifically adapted to cope with the bioenergetic problems faced by that extreme. On the other hand, poly-extremophiles such as *N. thermophilus* utilize a larger number of antiporters, with different roles to function as defence mechanisms against the large number of bioenergetic difficulties posed by multiple extreme conditions.

N. thermophilus harbours an ATPase with nine predicted subunits. The deduced subunits are similar to those of F_1F_O ATP synthases. In contrast to most other F_1F_O ATP synthases, but similar to the anaerobic bacterium *Acetobacterium woodii* (Fritz and Müller, 2007; Rahlfs *et al.*, 1999; Schmidt *et al.*, 2009), the *atp* operon encodes two different *c* subunits. The *c* subunits of F_1F_O ATP synthases make the rotor of the membrane-embedded motor and are assembled into a ring-like structure. In most cases, the *c* ring is made of multiple copies of just one *c* subunit that has two membrane-integral α-helices that are connected by a small cytoplasmic loop. Each *c* subunit harbours one ion binding site. In V_1V_O ATPases found in organelles of eukarya, the *c* subunit encoding gene underwent a duplication event followed by the loss of one ion binding site. This gave rise to a *c* subunit with four transmembrane helices and one ion binding site and thus c rings that have only half the number of ion binding sites (Manolson *et al.*, 1992). This is seen as the reason for the apparent inability of V_1V_O ATPases to synthesize ATP (Müller and Grüber, 2003). In contrast, this design favours the generation of steep ion gradients driven by ATP hydrolysis. The *atp* operon of *N. thermophilus* harbours a second gene encoding a *c* subunit of the V_1V_O ATPase type. This indicates the presence of a hetero-oligomeric *c* ring, as has been reported in *Acetobacterium woodii* (Fritz *et al.*, 2008). Furthermore, the sequence signature for coordination of Na^+ (Q^{33}, E^{66}, T^{67}, T^{68}, Y^{71}) is present in the *c* subunit of the *N. thermophilus* ATPase. Purification and functional characterization of the enzyme showed that the ATPase is Na^+-coupled, and reported maximal hydrolysis at the alkaline pH of 9.4 and 47°C (Mesbah and Wiegel, 2011). Synthesis of ATP by the enzyme proceeded very slowly, and it was concluded that the enzyme is geared primarily in the hydrolysis direction, where it serves as a pump to expel excess Na^+ from the cytoplasm (Mesbah and Wiegel, 2011). Genes homologous to V-type ATPases were not found in the chromosome of *N. thermophilus*.

Long-term adaptation: adaptation to elevated temperature

An analysis of a subset of genome sequences showed that the major effect of thermophily at the proteome level was a significant reduction in the frequency of thermolabile amino acids: histidine, glutamine and threonine (Hickey and Singer, 2004). Thermophiles also show an increase in both positively charged amino acid residues (arginine and lysine) and negatively charged residues (glutamate and aspartate), suggesting that ionic bonds between oppositely charged residues help to stabilize proteins at high temperature (Tekaia *et al.*, 2002). The proteomes of thermophiles also contain a larger fraction of proteins with isoelectric points in the basic range (Kawashima *et al.*, 2000).

Analysis of the amino acid composition of *N. thermophilus* showed a low abundance of the amino acids cysteine, histidine, methionine and tryptophan. There was a large abundance in the amino acids leucine, isoleucine and glutamate. The proteome of *N. thermophilus* contains proteins with isoelectric points in the acid range (pH 4–5), consistent with being a halophile (Mesbah *et al.*, 2009). Based on genomic analyses, it appears that adaptation mechanisms to high salinity and high temperature conflict. This could explain why *N. thermophilus* is a moderate thermophile, showing no growth at temperatures greater than 56°C (Mesbah *et al.*, 2007b). It is interesting to note that the intracellular concentration of the amino acid glutamate in *N. thermophilus* increases from 61 to 178 mM in response to an increase in growth temperature from 37 to 54°C. Compatible solutes of thermophiles are generally negatively charged (Martins *et al.*, 1997). Therefore the increase in the intracellular concentration of the negatively charged amino acid glutamate indicates that it plays a role in adaptation of *N. thermophilus* to heat.

The genome of *N. thermophilus* contains a number of genes potentially involved in adaptation to life at elevated temperature. The *N. thermophilus* genome contains a homologue for a putative spermine synthase, involved in the synthesis of spermine. Spermine is a positively charged polyamine that has been reported to play critical roles in the stabilization of proteins and nucleic acids during exposure to extreme temperatures (Tabor and Tabor, 1985; Terui *et al.*, 2005). Indeed, analysis of the polyamine content in cells of *N. thermophilus* showed the presence of the polyamine spermine, as well as spermidine, putrescine and two other unidentified polyamines. These polyamines could play a role in stabilization of proteins and nucleic acids in the moderately thermophilic *N. thermophilus*.

The *N. thermophilus* genome also contains four genes encoding orthologous rRNA methyltransferases and three genes encoding orthologous tRNA methyltransferase. It has been suggested that rRNA and tRNA methyltransferases play roles in thermophily, where base methylation provides structural stabilization of DNA and RNA at high temperature. The genome of thermophilic *Bacillus kaustophilus* (optimal growth at 60°C, maximum growth temperature 74°C), contains three tRNA and rRNA methyltransferase genes that are not present in other mesophilic bacilli, *B. subtilis*, *B. halodurans*, *B. anthracis*, *B. cereus* and *Oceanobacillus iheyensis* (Takami *et al.*, 2004). In addition, 12 genes for heat shock proteins are present in the chromosome of *N. thermophilus*.

Combating oxidative stress

Genes encoding for a copper/zinc superoxide dismutase, manganese superoxide dismutase, manganese catalase, as well as a bacterioferritin and peroxiredoxin peroxidase AhpC were identified in the chromosome of *N. thermophilus*. Superoxide dismutase catalyses the dismutation of superoxide into molecular oxygen and hydrogen peroxide, preventing damage from oxygen-mediated free radicals. Bacterioferritin and peroxiredoxins are abundant in a wide variety of cells and play major roles in peroxide detoxification in prokaryotic cells (Seaver and Imlayt, 2001; Wang *et al.*, 2005). Superoxide dismutases and catalases function in antioxidative defence in aerobic and facultative anaerobic microorganisms, but have also been detected in obligate anaerobes such as *H. orenii* (Mavromatis *et al.*, 2009). In general, obligate anaerobes are not uniformly sensitive to oxygen, some species are very sensitive to oxygen and others remain viable for long periods in the presence of oxygen. A relationship between superoxide dismutase activity and aerotolerance has been reported, where high activity was reported in an aerotolerant strain and *vice versa* (Brioukhanov *et al.*, 2006). *N. thermophilus* can tolerate exposure to oxygen for short periods as long as it is not at the optimal temperature for growth. It remains to be determined whether the catalase and peroxidase genes are expressed in *N. thermophilus* and whether the enzyme activities increase in response to oxygen.

Combating radiation

Genes encoding the DNA mismatch repair proteins MutS and MutL were identified in the chromosome of *N. thermophilus*. Homologues for the *recFOR* genes were also detected. The large and small subunits of exodeoxyribonuclease VII and exodeoxyribonuclease III were present, as were two homologues for DNA polymerase V (Pol V). In addition, genes involved in nucleotide excision repair (exinucleases), base excision repair and photorepair (UvrC, DNA repair photolyase) were present. The RecFOR pathway is a pathway of homologous recombination that repairs DNA, it can repair single- or double-stranded gaps (Hiom, 2009). Exodeoxyribonuclease VII is involved in methyl-directed mismatch repair (Burdett *et al.*, 2001). In *E. coli*, Pol V is involved in translesion synthesis, which allows DNA replication machinery to bypass base substitution mutations in DNA due to exposure to ultraviolet light (Friedberg *et al.*, 2005; Wagner *et al.*, 2002). *N. thermophilus* showed remarkable resistance when exposed to the three wavelengths of UV A,B and C radiation (385, 312 and 254 nm resulting in irradiance levels of 0.35 mW/cm^2, 0.38 mW/cm^2 and 0.645 mW/cm^2, respectively), where it had between 40% and 80% survival after 28 h of exposure (J. Blamey, personal communication). Under the same test conditions, *E. coli* showed no survival after 2 hours of exposure. The presence of several putative DNA repair systems in the genome of *N. thermophilus* could explain its resistance to UV radiation. Functional genetic analyses are needed to confirm the role of each gene in the UV resistance of *N. thermophilus*.

Mobile genetic elements

It has been proposed that lateral gene transfer plays a crucial role in adaptation of prokaryotes to extreme environments (Averhoff, 2009). This appears to be the case in *N. thermophilus*. Its genome contains 72 mobile genetic elements, consisting of 39 transposons, 29 integrases and four phage-related genes. A putative phage integrase was encoded on plasmid pNTHE01 and a putative integrase family protein was present on pNTHE02. However, clustered regularly interspaced short palindromic repeat (CRISPR) elements common in Gram-positive bacteria and Archaea were not found (Sorek *et al.*, 2008).

The aerobic moderately halophilic *Halobacillus halophilus*

The rod-shaped, endospore-forming, Gram-positive bacterium *H. halophilus* was isolated from a salt marsh at the North Sea coast of Germany and was originally described as *Sporosarcina halophila* (Claus *et al.*, 1983). Based on 16S rRNA homologies, *H. halophilus* is now phylogenetically classified within the order *Bacillales*, class *Bacilli*, phylum *Firmicutes* (Spring *et al.*, 1996). Being moderately halophilic, *H. halophilus* grows optimally between 0.5 and 2.0 M NaCl but can tolerate NaCl concentrations up to 3.0 M NaCl with a growth rate only 38% of the optimum (Roessler and Müller, 1998). *H. halophilus* adapts to changing environmental salinities that occur due to extensive rain fall or dryness and can survive a broad range of external salt concentrations. This is possible through a fine-tuned adaptation of osmoregulation that is evident in the genome sequence of *H. halophilus* (Saum and Müller, 2008b).

General features of the *Halobacillus halophilus* genome

The genome of *H. halophilus* DSM 2266 consists of one chromosome (4,150,632 bp) and two plasmids of 16,047 bp (PL16) and 3329 bp (PL3) (Fig. 5.3) (Saum *et al.*, 2013). The chromosome has a G+C content of 42% and codes for 4132 proteins with a coding density of 83%. The chromosome shows a typical bacterial GC skew with a minimum at the replication origin (pos 1, upstream of the *dnaA* gene) and a maximum at 2.17 Mbp, the probable replication termination point. A gene for DNA polymerase III epsilon subunit (Hbhal_3177) is close to the proposed termination point. *H. halophilus* has seven rRNA operons and 67 tRNA genes.

The 16 kbp plasmid has a low coding density with only seven ORFs encoding proteins longer than 100 aa. Among those are a phage integrase family protein and a helix–turn–helix (HTH) domain protein. No replication proteins could be identified; there were no sequences with similarity at the nucleotide level in the GenBank database. The 3 kbp plasmid codes for a plasmid replication protein and three short ORFs.

H. halophilus has several genomic islands with significantly altered GC content, typical of prokaryotic genomes. Within these islands, many of the genes have no assigned function. Several are from protein families typically found in plasmids and phages (transposase, integrase, maturase, prophage, DNA replication protein).

Catabolic traits and nutritional versatility

H. halophilus is a chemoorganoheterotrophic, strictly aerobic bacterium with great nutritional versatility. It is able to hydrolyse complex substrates such as casein, gelatin, DNA, starch and pullulan (Claus *et al.*, 1983). Genes encoding two extracellular proteases (Hbhal_5155 and

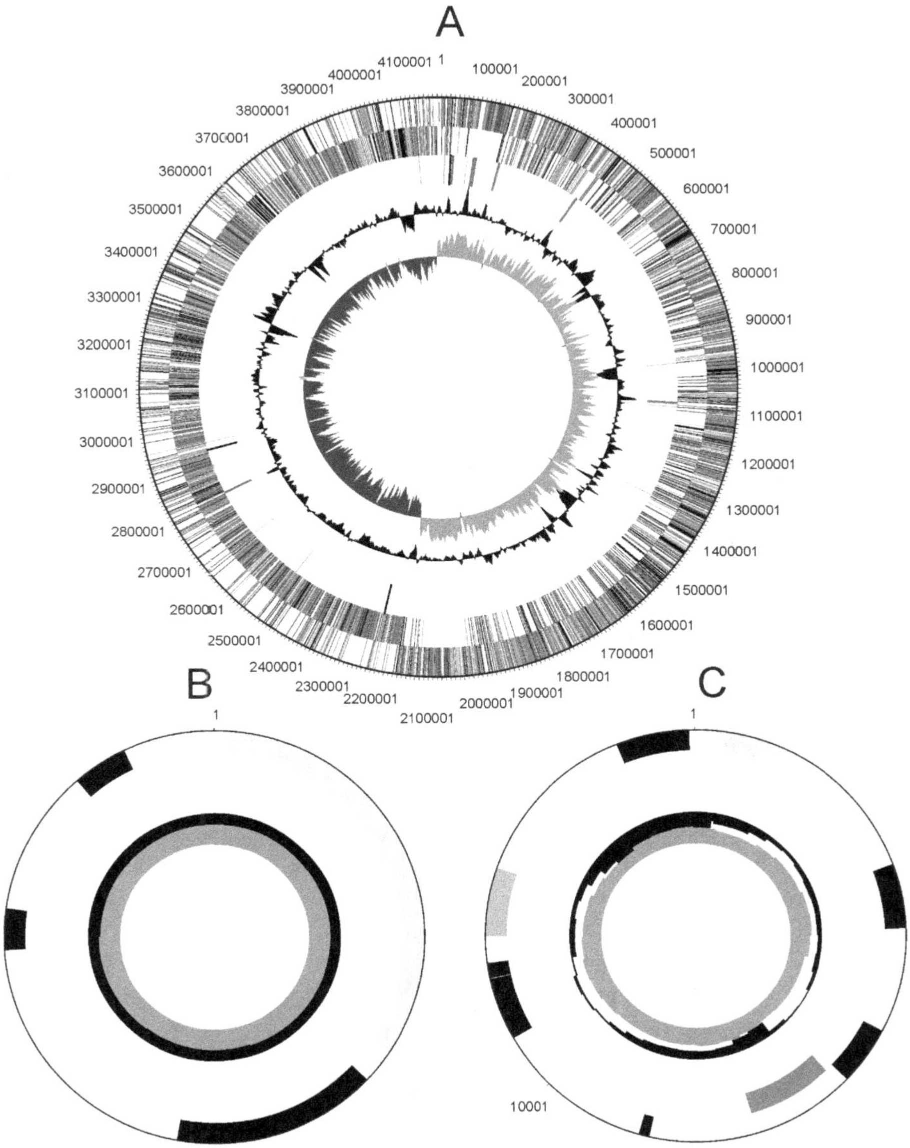

Figure 5.3 Circular representation of the *H. halophilus* chromosome (A), plasmid PL3 (B) and plasmid PL16 (C). The distribution of coding sequences is depicted by the coloured boxes according to functional categories. The first two circles from the outside represent the predicted coding sequences on the plus and minus strands, respectively. The two innermost circles represent the per cent G+C and G+C skew, respectively. Colours represent the following: dark grey, hypothetical proteins; light grey, conserved hypothetical and unknown function; brown, general function prediction; red, replication and repair; green, energy metabolism; blue, carbon and carbohydrate metabolism; cyan, lipid metabolism; magenta, transcription; yellow, translation; orange, amino acid metabolism; pink, metabolism of cofactors and vitamins; light red, purine and pyrimidine metabolism; lavender, signal transduction; sky blue, cellular processes. The third circle represents red for rRNA and green for tRNA genes. A colour version of this figure is located in the plate section at the back of the book.

Hbhal_4449), one amylase (Hbhal_4101) and one pullulanase (Hbhal_2962) were identified in the genome. In addition, *H. halophilus* is able to grow on hexoses such as glucose or fructose (by way of glycolysis) and on amino acids. Noteworthy is the use of carbon sources that are also used as compatible solutes such as glutamate and proline. Proline and glutamate, major compatible solutes of *H. halophilus*, are good growth substrates and may also be used as nitrogen sources, indicating a sophisticated regulatory network balancing different cellular needs. For example, proline is degraded by proline dehydrogenase (ProDH) and Δ^1-pyrroline-5-carboxylate dehydrogenase (P5CDH) to glutamate via Δ^1-pyrroline-5-carboxylate of which *H. halophilus* has two coding isogenes each for *prodh* and *p5cdh*. *prodh2* and *p5cdh2* form an operon (*put* operon) that is involved in the utilization of proline as carbon and energy source whereas ProDH1 and P5CDH1 may be involved in supplying the cell with nitrogen from proline (Saum *et al.*, 2013).

In contrast to *Halomonas elongata*, the genome of *H. halophilus* does not encode for ectoine utilization genes, consistent with the observation that ectoine is only a minor solute in *H. halophilus*.

The chloride modulon

One of the outstanding physiological features of *H. halophilus* is its chloride dependence of growth, gene expression and enzymatic activity. In the first instance, growth of *H. halophilus* was shown to strictly depend on the presence of chloride and to sense the salt concentration of its environment via the external chloride concentration. In absence of chloride the strain does not grow. In consequence of high external salinities, Cl^- is accumulated in the cytoplasm following the 'salt-in' strategy. While the internal Cl^- concentration is negligible at low external Cl^- concentrations, it increases to 50% of the external Cl^- concentration at higher salt concentrations (Roessler and Müller, 1998). This elevated internal chloride in consequence is not only necessary to regulate the germination of endospores (Dohrmann and Müller, 1999) and the motility of the vegetative cell (Roessler and Müller, 2002) but also to control the expression of genes and the enzymatic activity of proteins needed for the halophilic life style of *H. halophilus* (Saum and Müller, 2008b). These proteins are key players of the biosynthesis of compatible solutes which are essential to cope with high and changing salinities. All known processes regulated by the internal chloride concentration are summarized in the chloride modulon (Fig. 5.4).

Since the measured internal chloride concentrations are too high to be in equilibrium with the membrane potential, it has been suggested that chloride is actively accumulated in the cytoplasm of *H. halophilus* (Roessler and Müller, 1998). However, so far no gene could be identified to encode for a chloride transporter or even a simple channel (Saum *et al.*, 2013). Several genes are annotated to encode for potential symporters but their substrate specificity remains to be elucidated. To completely understand the chloride modulon the chloride transporter has to be found and characterized in detail.

Hybrid strategy for long-term adaptation to saline environments

H. halophilus has originally been described as a bacterium that amasses compatible solutes to establish cellular turgor. However, as discussed above, it is now known that it also accumulates molar concentrations of chloride in the cytoplasm. This survival strategy is now seen as a hybrid strategy of the moderate halophilic *H. halophilus* to cope with changing salinities of the environment.

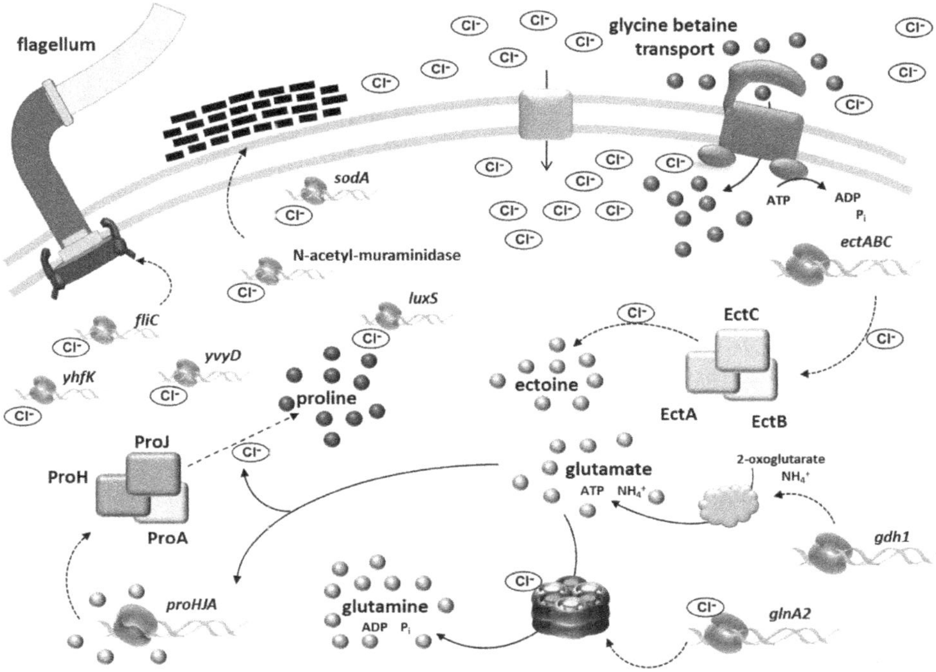

Figure 5.4 The chloride regulon of *Halobacillus halophilus*. For further explanations see text.

A detailed analysis of the proteome deduced from the genome sequence revealed a clear distinction of proteomes from extreme halophiles and non-halophiles with a higher number of acidic proteins in the extreme halophiles. The proteome of *H. halophilus* takes an intermediary position for the number of proteins with a slightly acidic isoelectric point of 6.6. This trend was seen for soluble as well as for membrane proteins. Based on these data and the fact that *H. halophilus* amasses compatible solutes as well as chloride in the cytoplasm, it was postulated that it uses a hybrid strategy of 'solute-in' and 'salt-in' to combat the external salinity (Saum *et al.*, 2013).

The counterion for chloride is not known but is likely to be potassium as shown for many other organisms. It is not yet understood why potassium is the main monovalent intracellular cation and why it is preferred above Na^+ in this function. In bacteria, K^+ might modulate activity and correct folding of proteins more effectively than Na^+. Another explanation for the accumulation of K^+ and the concomitant active extrusion of Na^+ is that this situation enables the cells to develop an inwardly directed electrochemical transmembrane Na^+ gradient. This gradient can then be used for energy-consuming processes like secondary transport or flagella movement (Bakker, 1992). Interestingly, the genome of *H. halophilus* encodes for only one Ktr-type potassium transporter (Hbhal_1246) and two potassium channels (Hbhal_3837, Hbhal_3881) while high-affinity, ATP-dependent transporters like the Kdp system could not be found (Saum *et al.*, 2013). Further studies are needed to elucidate the role of potassium and identify involved uptake systems.

Biosynthesis of compatible solutes

H. halophilus synthesizes a cocktail of different solutes to combat external salinity. The major solutes are glutamate, glutamine and proline but also ectoine, N^{δ}-acetyl ornithine and N^{ε}-acetyl lysine are produced (Saum *et al.*, 2013; Saum and Müller, 2008b). The biosynthesis pathways of glutamate, glutamine, proline and ectoine are predicted based on the genome sequence and studied biochemically (Fig. 5.5) while the pathways of N^{δ}-acetyl ornithine and N^{ε}-acetyl lysine are deduced from the genome sequence.

For the biosynthesis of glutamate and glutamine three main enzymatic reactions are known (Saum *et al.*, 2006). The biosynthesis of glutamate can either be accomplished by the action of a glutamate synthase (GOGAT, a sequence of glutamine synthetase and glutamate synthase) or by the action of a glutamate dehydrogenase (GDH) while glutamine is synthesized by the action of a glutamine synthetase (Gln). The genome contains two putative open reading frames encoding a glutamate synthase (*gdh1* and *gdh2*), only one open reading frame encoding the large subunit of a glutamate synthase (*gltA*) and two open reading frames each potentially encoding the small subunit of a glutamate synthase (*gltB1* and *gltB2*). The glutamine synthetase Gln is encoded by two open reading frames (*glnA1* and *glnA2*). While the former (*glnA1*) clusters with a gene (*glnR*) encoding the regulatory protein GlnR that is known from *B. subtilis* to be essential in nitrogen metabolism, the latter (*glnA2*) lies solitary and is predicted to be regulated by a promoter recognized by σ^B, the general stress σ-factor.

Proline is synthesized from glutamate by a sequence of a glutamate 5-kinase (ProJ), a glutamate 5-semialdehyde dehydrogenase (ProA) and a pyrroline-5-carboxylate reductase

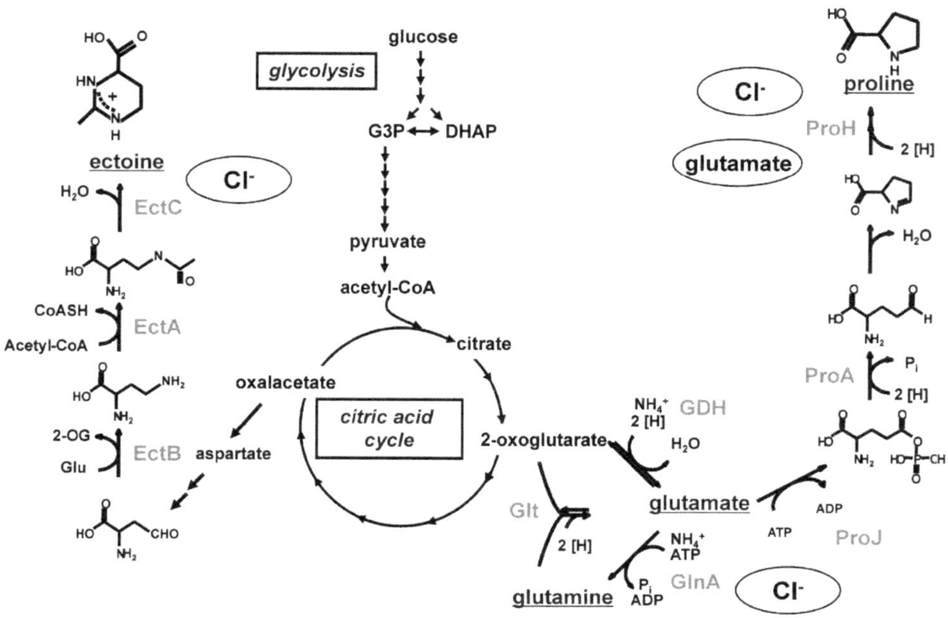

Figure 5.5 Proposed biochemical pathways of the main compatible solutes in *Halobacillus halophilus*.

(ProH) (Saum and Müller, 2007). The enzymes are encoded by a cluster of three genes – *proH, proJ* and *proA,* which are organized in an operon.

Also the three biosynthetic genes (*ectABC*) for the production of ectoine are arranged on one operon (Saum and Müller, 2008a). Ectoine is produced from aspartate semialdehyde by a sequence of a diaminobutyrate-2-oxoglutarate transaminase (EctB), a diaminobutyric acid acetyltransferase (EctA) and an ectoine synthase.

The pathway for the production of N^{δ}-acetyl ornithine still has to be elucidated but presumably ornithine is the direct precursor, for which several pathways are known. All necessary genes encoding the corresponding enzymes were identified on the genome of *H. halophilus* and possible pathways were recently described in detail by Saum and colleagues (Saum *et al.*, 2013).

The last potential compatible solute that is synthesized by *H. halophilus* is N^{ε}-acetyl lysine. However, its role as compatible solute still has to be confirmed by further studies. Based on the now available genome sequence it was assumed that *H. halophilus* is capable of synthesizing lysine using the classical diaminopimelate pathway (Saum *et al.*, 2013). In this pathway aspartate is initially activated by an aspartate kinase. *Halobacillus halophilus* possesses two copies of this enzyme (*dapG1*, Hbhal_3090, *dapG2*, Hbhal_3465) which might be subject to different modes of regulation. The activated aspartyl moiety then becomes reduced to the corresponding semialdehyde [catalysed by an aspartate semialdehyde dehydrogenase (*asd*, Hbhal_3089)] which then undergoes a condensation reaction with one molecule of pyruvate resulting in the formation of 2,3-dihydrodipicolinate. This reaction, which is the first that differs from the ectoine biosynthesis pathway, is catalysed by the dihydrodipicolinate synthase. Three copies of the corresponding gene were identified (*dapA1*, Hbhal_2387, *dapA2*, Hbhal_3091, *dapA3*, Hbhal_5017) which may indicate its critical role in the biosynthesis. It is likely that the different genes are controlled by different demands such as the need for lysine as a compound in protein biosynthesis, the need of N^{ε}-acetyl lysine as osmoprotectant or the need to provide precursors for the biosynthesis of peptidoglycan. The following sequence of reactions leading to lysine is then catalysed by a dihydrodipicolinate reductase (Hbhal_3261), a tetrahydrodipicolinate *N*-acetyltransferase (Hbhal_2790), an acetyltransferase, an *N*-acetyldiaminopimelate deacetylase (Hbhal_2791), a diaminopimelate epimerase (*dapF*, Hbhal_2627) and finally a diaminopimelate decarboxylase (*lysA*, Hbhal_3343). The final acetylation at the ε-amino group requires an acetyltransferase (Hbhal_3877).

Salinity- and growth phase-dependent adaptation of the solute pool

Interestingly, *H. halophilus* switches its osmolyte content depending on salinity (Saum and Müller, 2008b). At intermediate salinities of around 1.5 M NaCl, glutamate and glutamine are the major solutes. Transcription analyses after an osmotic upshock from 0.8 to 2 M NaCl have shown that one of the putative glutamate dehydrogenase genes (*gdh1*) was induced and the mRNA level increased within 1.5 hours to about 4-fold compared to the level before the upshock which enables an increased production of glutamate from 2-oxoglutarate and NH_4^+. In contrast, the transcript levels of the second glutamate hydrogenase gene (*gdh2*) were close to the detection limit likely being involved in nitrogen metabolism rather than osmoregulation. So far, also the glutamate synthase gene (*gltA*) did not seem to be involved in osmoregulation. Glutamine is synthesized by the action of a glutamine synthetase which is encoded by two genes (*glnA1* and *glnA2*) in *H. halophilus.* On a transcriptional level only the expression of *glnA2* was shown to be up-regulated at increasing salt concentrations with a maximal increase of transcripts of about 4-fold (compared to the value at 0.4 M NaCl) at 1.5 M NaCl or higher. The expression of *glnA1* was not affected. Moreover, the expression of *glnA2* and especially the glutamine synthetase activity were shown to be chloride-dependent being increased with increasing chloride concentrations. The maximal enzymatic activity was found at 2.5 M NaCl or higher (Saum *et al.*, 2006). This is in line with the chloride dependence of growth and the accumulation of Cl^- into the cytoplasm as described above. However, it is unknown how chloride modulates the enzymatic activity. Both, a direct interaction of chloride with the glutamine synthetase or the involvement of a regulatory protein that senses the concentration of chloride, are possible.

At high salinities (2.0 M NaCl or higher), glutamine and glutamate pools stay rather constant but proline is produced in addition, and becomes the dominant solute at high salt. It was shown that the transcription of the *pro* operon was increased with increasing salinities with a maximum at 2.5 M. The mRNA level reached a maximum 1.5 hours after an osmotic upshock while the maximal concentration of proline was determined after six hours. Consequently, the increased amount of enzymes led to an increased production of proline (Saum and Müller, 2007). However, not only NaCl but also Na-glutamate had an effect on gene expression. Compared to NaCl Na-glutamate was shown to dramatically increase the *proHJA* mRNA concentration. Since proline is produced from glutamate, and sodium glutamate had a more stimulating effect on gene expression than NaCl, NaCl now is supposed to be the initial signal for the proline production only. Glutamate instead acts as 'second messenger' that further regulates the *pro* operon expression by its internal concentration which increases with increasing salinity (Saum and Müller, 2007).

In addition to the salinity-dependent solute regulation, another layer of regulation is active: proline contents are maximal in exponentially growing cultures but reduced in stationary phase cultures. Under these conditions, ectoine is synthesized. To resolve the time-dependent kinetics of ectoine production *H. halophilus* cells were subjected to an osmotic upshock from 0.8 to 2.0 M NaCl and the biosynthesis of ectoine was measured at the levels of transcription, translation and solute accumulation. Transcripts were readily detectable already at time-point 0 h, but increased dramatically with time and reached a maximum not before three hours after upshock. Most important, expression of *ect* genes was preceded by expression of genes responsible for glutamine, glutamate or proline biosynthesis. The signal leading to *ect* gene transcription is therefore assumed to be an indirect one mediated by one or more yet to be identified factors rather than by the presence of the osmolyte. The production of the ectoine synthase EctC nicely corresponds to the increase of *ectC* transcript. Both were found to increase 2-fold. Surprisingly, four hours after upshock the EctC content again decreased with time and the level reached a value only slightly above the value at the beginning, although the external stress was still present. This decrease, however, was not reflected in the ectoine concentration, which steadily increased and reached a maximum after 18 hours after upshock. Again, this demonstrates a great delay in accumulation compared to proline that reached its maximum already 6 hours after upshock and hints to a role of ectoine not only in the immediate response to osmotic upshock but to a function as a more general protectant in the cell (Saum and Müller, 2008a).

To quickly adapt to changing salinities and growth phases *H. halophilus* regulates the synthesis of solutes by both increasing expression of the enzymes involved and activating the produced enzymes in a chloride-dependent manner. *H. halophilus* also has a gene encoding a potential ectoine hydroxylase, but hydroxyectoine has not been detected (Saum *et al.*, 2013). *H. halophilus* does not have the genetic capacity for *de novo* biosynthesis of glycine betaine, but can take up choline from the environment and oxidize it to glycine betaine (Burkhardt *et al.*, 2009).

Long-term adaptation: sodium bioenergetics and pH regulation?

Growth of *H. halophilus* is strictly sodium ion dependent. Use of Na^+ in primary bioenergetics seems unlikely since the *c* subunit of the F_1F_O ATP synthase does not have the conserved sodium ion binding motif. Likewise, the known redox-driven Na^+ pumps Nqr (Hayashi *et al.*, 2001) and Rnf (Biegel *et al.*, 2011) are absent. Therefore, secondary sodium ion bioenergetics and pH homeostasis is the most likely reason for the observed Na^+ dependence. In contrast to *N. thermophilus*, *H. halophilus* is not an alkaliphile. However, it grows in the range of pH 7 to 9 with an optimum at pH 7.8 indicating a capacity to grow at slightly alkaline pH values and to regulate internal pH. *H. halophilus* has four genes that potentially encode an NhaC-type Na^+/H^+ antiporter, one that encodes a Nhe-type as well as three that encode NhaP-type Na^+/H^+ antiporter (Saum *et al.*, 2013). These may be involved in Na^+ homeostasis (expelling Na^+ to keep the intracellular Na^+ concentration low) and/or pH regulation under alkaline conditions. In addition to the mono-subunit Na^+/H^+ antiporter the genome of *H. halophilus* encodes a multisubunit Mrp-like Na^+/H^+ antiporter that is known to be involved in stress-induced Na^+ export in *B. subtilis* (Ito *et al.*, 1999). The sodium ion gradient established by these Na^+/H^+ antiporters may then be used to drive solute uptake. Indeed, the genome of *H. halophilus* encodes for a multitude of potentially sodium ion-coupled symporters for diverse nutrients as well as a sodium ion-driven flagella motor.

Adjusting membrane lipids

The major lipids produced by *H. halophilus* are sulfoquinovosyldiacylglycerol and phosphatidylglycerol, while cardiolipin is a minor lipid together with phosphatidic acid, alanyl-phosphatidylglycerol and two not yet fully identified lipid components. Upon an increase of the salinity of the culture medium, an increase in the shorter chains and chain unsaturation were observed (Lopalco *et al.*, 2013). These changes in the lipid core structures might compensate for the increase in packing and rigidity of phospholipid and sulfoglycolipid polar heads in high salt medium, therefore contributing to the homeostasis of membrane fluidity and permeability in salt stress conditions.

Long-term survival: endospore formation

In contrast to *N. thermophilus*, *H. halophilus* is known to produce endospores and the genes required for the trait are present in the genome. Interestingly, germination of endospores requires chloride (Dohrmann and Müller, 1999), indicating the existence of a chloride sensor in the germination cascade that has not been identified in the genome.

Combating UV radiation and oxidative stress: the role of carotenoids

Carotenoids are naturally occurring pigments found in a wide variety of plants and microorganisms (Sandmann, 2001). Interestingly, many endospore forming bacteria isolated from saline environments like salt marshes contain carotenoids, whereas their non-halophilic relatives do not (Turner and Jervis, 1963), indicating that carotenoids play a crucial role in salt adaptation in these organisms. *H. halophilus* is pigmented and the major carotenoid was identified as a C30 methyl glucosyl-3,4-dehydro-apo-8′-lycopenoate (Fig. 5.6) (Köcher *et al.*, 2009). Based on the genomic data as well as on spectroscopic evidence, biosynthesis starts from geranyl-pyrophosphate and geranylgeranyl-pyrophosphate via apo-8′-phytoene, methyl-hydroxy-3,4-dehydro-apo-8′-lycopinoate and glycosyl-3,4-dehydro-apo-8′-lycopene to a C30 methyl glucosyl-3,4-dehydro-apo-8′-lycopenoate.

The common feature of carotenoids is their function as lipophilic antioxidants. For several bacteria, the presence of the antioxidative enzyme catalase was shown to be important for osmoprotection (Cho *et al.*, 2000; Lee *et al.*, 2005). In contrast, growth of *H. halophilus* at conditions where the synthesis of coloured carotenoids was inhibited was comparable to the non-inhibited culture. However, when oxidative conditions were applied that allowed about 50% growth of the non-inhibited culture, the culture devoid of coloured carotenoids did not grow. This result indicates that the carotenoids produced by *H. halophilus* allow it to

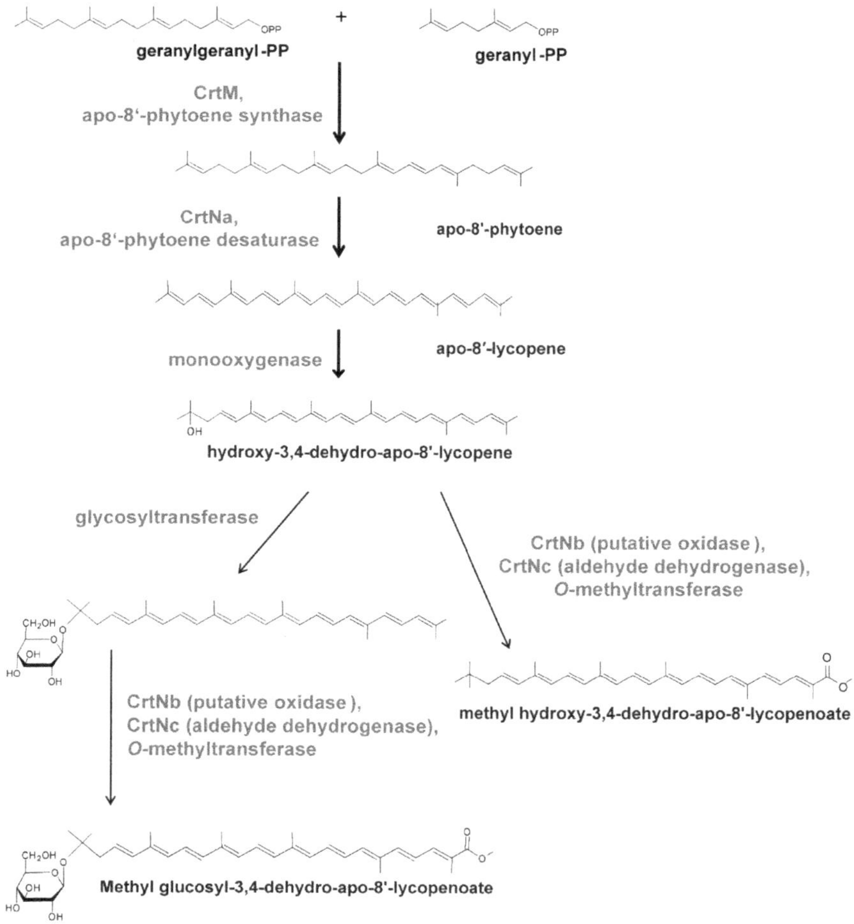

Figure 5.6 Proposed carotenoid biosynthetic pathway in *Halobacillus halophilus* DSM 2266. Gene products catalysing the individual reactions are indicated in grey next to the arrows. Substrates, intermediates and products are shown in black.

cope with oxidative stress. This is in accordance with the fact that the main carotenoid produced by *H. halophilus* (methyl glucosyl-3,4-dehydro-apo-8′-lycopenoate) is a very potent antioxidant (Shindo *et al.*, 2008).

Evolutive traits: gene transfer

As mentioned above the genome of *H. halophilus* harbours several genomic islands that typically are gained by horizontal or lateral gene transfer. Horizontal gene transfer is seen as essential for adaptation to life in extreme environments (Averhoff, 2009). It thus is very likely, that *H. halophilus* has a natural competence as is predicted from the genome sequence. However, so far no growth condition has been identified in which the uptake of naked DNA was observed. Possible reasons are diverse: the right condition to develop natural competence has not yet been found, genes are mutated resulting in inactive proteins or genes essential for competence are missing in *H. halophilus*. Potential candidates for the latter are the missing genes *comX*, *comS*, *rok* and *nucA*, respectively, whose products have been shown to be involved in natural competence in other organisms (Saum *et al.*, 2013).

Synopsis

The two examples *N. thermophilus* and *H. halophilus* highlighted in this chapter give a good impression of the genetic variety that enables microbial adaptation to saline environments. Both have developed independent strategies to cope with changing external salt concentration. While *N. thermophilus* probably follows the 'salt-out' strategy, *H. halophilus* has developed a hybrid strategy of both 'salt-in' and 'salt-out'. In *N. thermophilus* neither the accumulation of Na^+ and K^+ nor of any other cation was observed. *H. halophilus* in contrast accumulates up to molar concentrations of Cl^- and probably K^+ as counter ion. Another variation becomes obvious by comparing the major compatible solutes accumulated for osmoregulation. *N. thermophilus* is hypothesized to synthesize glycine betaine while the major solutes in *H. halophilus* are glutamate and glutamine as well as proline and ectoine. This is likely to have evolved due to the supply from the environment which also explains why *H. halophilus* can use the prevalent solutes as carbon and nitrogen sources. Since *H. halophilus* faces mainly salt stress as the changing extreme its solute regulation is strictly adapted to cope with changing osmolarities in an energetically optimized way. *N. thermophilus* was shown to be much more diverse because it faces different extremes namely alkalinity, salinity and elevated temperatures. It needs to adapt to different changes which requires among others different compatible solutes, changing amino acid contents and varying cation/proton antiporters. The regulatory mechanisms of *N. thermophilus* are not yet elucidated in detail but the genomic diversity suggests a much more complex regulatory concept for the survival of this 'polyextremophile'.

References

Averhoff, B. (2009). Shuffling genes around in hot environments: the unique DNA transporter of *Thermus thermophilus*. FEMS Microbiol. Rev. 33, 611–626.

Bakker, E.P. (1992). Cellular K^+ and K^+ transport systems in prokaryotes (CRC Press, Boca Raton, FL).

Biegel, E., Schmidt, S., Gonzalez, J.M., and Müller, V. (2011). Biochemistry, evolution and physiological functions of the Rnf complex, a novel ion-motive electron transport complex in prokaryotes. Cell. Mol. Life Sci. 68, 613–634.

Brioukhanov, A.L., Netrusov, A.I., and Eggen, R.I.L. (2006). The catalase and superoxide dismutase genes are transcriptionally up-regulated upon oxidative stress in the strictly anaerobic archaeon *Methanosarcina barkeri*. Microbiology *152*, 1671–1677.

Burdett, V., Taitinger, c., Viswanathan, M., Lovett, S.T., and Modrich, P. (2001). *In vivo* requirement for RecJ, ExoVII, ExoI, and ExoX in methyl-directed mismatch repair. Proc. Natl. Acad. Sci. U.S.A. *98*, 6765–6770.

Burkhardt, J., Sewald, X., Bauer, B., Saum, S.H., and Müller, V. (2009). Synthesis of glycine betaine from choline in the moderate halophilic *Halobacillus halophilus*: co-regulation of two divergent, polycistronic operons. Environ. Microbiol. Rep. *1*, 38–43.

Cho, Y.H., Lee, E.J., and Roe, J.H. (2000). A developmentally regulated catalase required for proper differentiation and osmoprotection of *Streptomyces coelicolor*. Mol. Microbiol. *35*, 150–160.

Claus, D., Fahmy, F., Rolf, H.J., and Tosunoglu, N. (1983). *Sporosarcina halophila* sp. nov., an obligate, slightly halophilic bacterium from salt marsh soils. System. Appl. Microbiol. *4*, 496–506.

Dohrmann, A.B., and Müller, V. (1999). Chloride dependence of endospore germination in *Halobacillus halophilus*. Arch. Microbiol. *172*, 264–267.

Elevi Bardavid, R., and Oren, A. (2012). Acid-shifted isoelectric point profiles of the proteins in a hypersaline microbial mat – an adaptation to life at high salt concentrations? Extremophiles *16*, 787–792.

Errington, J. (1993). *Bacillus subtilis* sporulation: regulation of gene expression and control of morphogenesis. Microbiol. Mol. Biol. Rev. *57*, 1–33.

Friedberg, E.C., Lehmann, A.R., and Fuchs, R.P.P. (2005). Trading places: How do DNA polymerases switch during translesion DNA synthesis? Mol. Cell *18*, 499–505.

Fritz, M., and Müller, V. (2007). An intermediate step in the evolution of ATPases – the F_1F_O-ATPase from *Acetobacterium woodii* contains F-type and V-type rotor subunits and is capable of ATP synthesis. FEBS J. *274*, 3421–3428.

Fritz, M., Klyszejko, A.L., Morgner, N., Vonch, J., Brutschy, B., Müller, D.J., Meier, T., and Müller, V. (2008). An intermediate step in the evolution of ATPases – a hybrid F_O–V_O rotor in a bacterial Na^+ F_1F_0 ATP synthase. FEBS J. *275*, 1999–2007.

Hayashi, M., Nakayama, Y., and Unemoto, T. (2001). Recent progress in the Na^+-translocating NADH-quinone reductase from the marine *Vibrio alginolyticus*. Biochim. Biophys. Acta *1505*, 37–44.

Hickey, D.A., and Singer, G.A.C. (2004). Genomic and proteomic adaptations to growth at high temperature. Genome Biol. *5*, 117.

Hiom, K. (2009). DNA repair: common approaches to fixing double-strand breaks. Curr. Biol. *19*, R523–R525.

Imhoff, J.F., Sahl, H.G., Soliman, G.S., and Trüper, H.G. (1979). The Wadi Natrun: Chemical composition and microbial mass developments in alkaline brines of eutrophic desert lakes. Geomicrobiol. J. *1*, 219–234.

Ito, M., Guffanti, A.A., Oudega, B., and Krulwich, T.A. (1999). *mrp*, a multigene, multifunctional locus in *Bacillus subtilis* with roles in resistance to cholate and to Na^+ and in pH homeostasis. J. Bacteriol. *181*, 2394–2402.

Kawashima, T., Amano, N., Koike, H., Makino, S., Higuchi, S., Dawashima-Ohya, Y., Watanabe, K., Yamazaki, M., Kanehori, K., Kawamoto, T. *et al.* (2000). Archaeal adaptation to higher temperature revealed by genomic sequence of *Thermoplasma volcanium*. Proc. Natl. Acad. Sci. U.S.A. *97*, 14257–14262.

Köcher, S., Breitenbach, J., Müller, V., and Sandmann, G. (2009). Structure, function and biosynthesis of carotenoids in the moderately halophilic bacterium *Halobacillus halophilus*. Arch. Microbiol. *191*, 95–104.

Krulwich, T.A. (1995). Alkaliphiles: 'Basic' molecular problems of pH tolerance and bioenergetics. Mol. Microbiol. *15*, 403–410.

Krulwich, T.A., Hicks, D.B., and Ito, M. (2009). Cation/proton antiporter complements of bacteria: why so large and diverse? Mol. Microbiol. *74*, 257–260.

Lee, J.S., Heo, Y.J., Lee, J.K., and Cho, Y.H. (2005). KatA, the major catalase, is critical for osmoprotection and virulence in *Pseudomonas aeruginosa* PA14. Infect. Immun. *73*, 4399–4403.

Lopalco, P., Angelini, R., Lobasso, S., Kocher, S., Thompson, M., Müller, V., and Corcelli, A. (2013). Adjusting membrane lipids under salt stress: the case of the moderate halophilic organism *Halobacillus halophilus*. Environ. Microbiol. *15*, 1078–1087.

Manolson, M.F., Proteau, D., and Jones, E.W. (1992). Evidence for a conserved 95–120kDa subunit associated with and essential for activity of V-ATPases. J. Exp. Biol. *172*, 105–112.

Martins, L.O., Huber, R., Huber, H., Stetter, K.O., Da Costa, M., and Santos, H. (1997). Organic solutes in hyperthermophilic Archaea. Appl. Environ. Microbiol. *63*, 896–902.

Mavromatis, K., Ivanova, N., Anderson, I., Lykidis, A., Hooper, S.D., Sun, H., Kunin, V., Lapidus, A., Hugenholtz, P., Patel, B., *et al.* (2009). Genome analysis of the anaerobic, thermohalophilic bacterium *Halothermothrix orenii*. PLoS One *4*, e4192.

Mesbah, N.M., and Wiegel, J. (2008). Life at extreme limits: the anaerobic halophilic alkalithermophiles. Ann. N. Y. Acad. Sci. *1125*, 44–57.

Mesbah, N.M., and Wiegel, J. (2011). The Na^+-translocating F_1F_O-ATPase from the halophilic, alkalithermophile *Natranaerobius thermophilus*. Biochim. Biophys. Acta. *1807*, 1133–1142.

Mesbah, N.M., and Wiegel, J. (2012). Life under multiple extreme conditions: diversity and physiology of the halophilic alkalithermophiles. Appl. Environ. Microbiol. *78*, 4074–4082.

Mesbah, N.M., Abou-El-Ela, S.H., and Wiegel, J. (2007a). Novel and unexpected prokaryotic diversity in water and sediments of the alkaline, hypersaline lakes of the Wadi An Natrun, Egypt. Microb. Ecol. *54*, 598–617.

Mesbah, N.M., Hedrick, D.B., Peacock, A.D., Rohde, M., and Wiegel, J. (2007b). *Natranaerobius thermophilus* gen. nov., sp. nov., a halophilic, alkalithermophilic bacterium from soda lakes of the Wadi An Natrun, Egypt, and proposal of *Natranaerobiaceae* fam. nov. and *Natranaerobiales* ord. nov. Int. J. Syst. Evol. Microbiol. *57*, 2507–2512.

Mesbah, N.M., Cook, G.M., and Wiegel, J. (2009). The halophilic alkalithermophile *Natranaerobius thermophilus* adapts to multiple environmental extremes using a large repertoire of $Na^+(K^+)/H^+$ antiporters. Mol. Microbiol. *74*, 270–281.

Molle, V., Fujita, M., Jensen, S.T., Eichenberger, P., Gonzalez-Pastor, J.E., Liu, J.S., and Losick, R. (2003). The Spo0A regulon of *Bacillus subtilis*. Mol. Microbiol. *50*, 1683–1704.

Müller, V., and Grüber, G. (2003). ATP synthases: structure, function and evolution of unique energy converters. Cell. Mol. Life Sci. *60*, 474–494.

Muyzer, G., Sorokin, D.Y., Mavromatis, K., Lapidus, A., Clum, A., Ivanova, N., Pati, A., d'Haeseleer, P., Woyke, T., and Kyrpides, N.C. (2011). Complete genome sequence of '*Thioalkalivibrio sulfidophilus*' HL-EbGr7. Stand. Genomic Sci. *4*, 23–23.

Nyyssola, A., Reinikainen, T., and Leisola, M. (2001). Characterization of glycine sarcosine*N*-methyltransferase and sarcosine dimethylglycine*N*-methyltransferase Appl. Environ. Microbiol. *67*, 2044–2050.

Onyenwoke, R.U., Brill, J.A., Farahi, K., and Wiegel, J. (2004). Sporulation genes in members of the low G+C Gram-type-positive phylogenetic branch (Firmicutes). Arch. Microbiol. *182*, 182–192.

Padan, E., Bibi, E., Masahiro, I., and Krulwich, T.A. (2005). Alkaline pH homeostasis in bacteria: New insights. Biochim. Biophys. Acta *1717*, 67–88.

Paidhungat, M., Ragkousi, K., and Setlow, P. (2001). Genetic requirements for induction of germination spores of *Bacillus subtilis* by Ca_2-dipicolinate. J. Bacteriol. *183*, 4886–4893.

Rahlfs, S., Aufurth, S., and Müller, V. (1999). The Na^+-F_1F_O-ATPase operon from *Acetobacterium woodii*. Operon structure and presence of multiple copies of *atpE* which encode proteolipids of 8- and 18-kDa. J. Biol. Chem. *274*, 33999–34004.

Roberts, M. (2005). Organic compatible solutes of halotolerant and halophilic microorganisms. Saline Syst. *1*, 5.

Roessler, M., and Müller, V.V. (1998). Quantitative and physiological analyses of chloride dependence of growth of *Halobacillus halophilus*. Appl. Environ. Microbiol. *64*, 3813–3817.

Roessler, M., and Müller, V. (2002). Chloride, a new environmental signal molecule involved in gene regulation in a moderately halophilic bacterium, *Halobacillus halophilus*. J. Bacteriol. *184*, 6207–6215.

Sandmann, G. (2001). Carotenoid biosynthesis and biotechnological application. Arch. Biochem. Biophys. *385*, 4–12.

Saum, S.H., and Müller, V. (2007). Salinity-dependent switching of osmolyte strategies in a moderately halophilic bacterium: glutamate induces proline biosynthesis in *Halobacillus halophilus*. J. Bacteriol. *189*, 6968–6975.

Saum, S.H., and Müller, V. (2008a). Growth phase-dependent switch in osmolyte strategy in a moderate halophile: ectoine is a minor osmolyte but major stationary phase solute in *Halobacillus halophilus*. Environ. Microbiol. *10*, 716–726.

Saum, S.H., and Müller, V. (2008b). Regulation of osmoadaptation in the moderate halophile *Halobacillus halophilus*: chloride, glutamate and switching osmolyte strategies. Saline Syst. *4*, 4.

Saum, S.H., Sydow, J.F., Palm, P., Pfeiffer, F., Oesterhelt, D., and Müller, V. (2006). Biochemical and molecular characterization of the biosynthesis of glutamine and glutamate, two major compatible solutes in the moderately halophilic bacterium *Halobacillus halophilus*. J. Bacteriol. *188*, 6808–6815.

Saum, S., Pfeiffer, F., Palm, P., Rampp, M., Schuster, S., Müller, V., and Oesterhelt, D. (2013). Chloride and organic osmolytes: a hybrid strategy to cope with elevated salinities by the moderately halophilic, chloride-dependent bacterium *Halobacillus halophilus*. Environ. Microbiol. *15*, 1619–1633.

Schmidt, S., Biegel, E., and Müller, V. (2009). The ins and outs of Na^+ bioenergetics in *Acetobacterium woodii*. Biochim. Biophys. Acta *1787*, 691–696.

Seaver, L.C., and Imlayt, J.A. (2001). Alkyl hydroperoxide reductase is the primary scavenger of endogenous hydrogen peroxide in *Escherichia coli*. J. Bacteriol. *183*, 7173–7181.

Shindo, K., Endo, M., Miyake, Y., Wakasugi, K., Morritt, D., Bramley, P.M., Fraser, P.D., Kasai, H., and Misawa, N. (2008). Methyl glucosyl-3,4-dehydro-apo-8′-lycopenoate, a novel antioxidative glyco-C(30)-carotenoic acid produced by a marine bacterium *Planococcus maritimus*. J. Antibiot. (Tokyo) *61*, 729–735.

Slonczewski, J.L., Fujisawa, M., Dopson, M., and Krulwich, T.A. (2009). Cytoplasmic pH measurement and homeostasis in bacteria and archaea. Adv. Microb. Physiol. *55*, 1–317.

Sorek, R., Kunin, V., and Hugenholtz, P. (2008). CRISPR – a widespread system that provides acquired resistance against phages in bacteria and archaea. Nat. Rev. Microbiol. *6*, 181–186.

Spring, S., Ludwig, W., Marquez, M.C., Ventosa, A., and Schleifer, K.-H. (1996). *Halobacillus* gen. nov., with descriptions of *Halobacillus litoralis* sp. nov. and *Halobacilus trueperi* sp. nov., and transfer of *Sporosarcina halophila* to *Halobacillus halophilus* comb. nov. Int. J. Syst. Bacteriol. *46*, 492–496.

Stragier, R., and Losick, R. (1996). Molecular genetics of sporuation in *Bacillus subtilis*. Annu. Rev. Genet. *30*, 297–341.

Tabor, C.W., and Tabor, H. (1985). Polyamines in microorganisms. Micrbiol. Rev. *49*, 81–99.

Taher, A.G. (1999). Inland saline lakes of Wadi El Natrun depression, Egypt. Int. J. Salt Lake Res. *8*, 149–170.

Takami, H., Nakasone, K., Takaki, Y., Maeno, G., Sasaki, R., Masui, N., Fuji, F., Hirama, C., Nakamura, Y., Ogasawara, N., *et al.* (2000). Complete genome sequence of the alkaliphilic bacterium *Bacillus halodurans* and genomic sequence comparison with *Bacillus subtilis*. Nucleic Acids Res. *28*, 4317–4331.

Takami, H., Takaki, Y., and Uchiyama, I. (2002). Genome sequence of *Oceanobacillus iheyensis* isolated from the Iheya Ridge and its unexpected adaptive capabilities to extreme environments. Nucleic Acids Res. *30*, 3927–3935.

Takami, H., Takaki, Y., Chee, G.-J., Nishi, S., Shimamura, S., Suzuki, H., Matsui, S., and Uchiyama, I. (2004). Thermoadaptation trait revealed by the genome sequence of thermophilic *Geobacillus kaustophilus*. Nucleic Acids Res. *32*, 6292–6303.

Tekaia, f., Yeramian, E., and Dujon, B. (2002). Amino acid composition of genomes, lifestyles of organisms and evolutionary trends: a global picture with correspondence analysis. Gene *297*, 51–60.

Terui, Y., Otnuma, M., Hiraga, K., Kawashima, E., and Oshima, T. (2005). Stabilization of nucleic acids by unusual polyamines produced by an extreme thermophile, *Thermus thermophilus*. Biochem. J. *388*, 427–433.

Turner, M., and Jervis, D.I. (1963). The distribution of pigmented *Bacillus* species in saltmarsh and other saline and non-saline soils. Nov. Hedw. *16*, 293–298.

Wagner, J., Etienne, H., Janel-Bintz, R., and Fuchs, R.P.P. (2002). Genetics of mutagenesis in *E. coli*: various combinations of translesion polymerases (Pol II, IV and V) deal with lesion/sequence context diversity. DNA Repair *1*, 159–167.

Wang, G., Oleczak, A., Walton, J.P., and Maier, R. (2005). Contribution of the *Helicobacter pylori* thiol peroxidase bacterioferritin comigratory protein to oxidative stress resistance and host colonization. Infect. Immun. 73, 378–384.

Zhao, B., Mesbah, N.M., Dalin, E., Goodwin, L., Nolan, M., Pitluck, S., Chertkov, O., Brettin, T.S., Han, J., Larimer, F.W., *et al.* (2011). Complete genome sequence of the anaerobic, halophilic alkalithermophile *Natranaerobius thermophilus* JW/NM-WN-LF^T. J. Bacteriol. *193*, 4023–4024.

Staying in Shape: The Haloarchaeal Cell Wall

6

Jerry Eichler, Adi Arbiv, Chen Cohen-Rosenzweig, Lina Kaminski, Lina Kandiba, Zvia Konrad and Shai Naparstek

Abstract

Haloarchaea are surrounded by different macromolecular structures comprising a cell wall. The composition of these cell walls is thought to contribute to the ability of these microorganisms to remain intact in the face of molar concentrations of salt. In many instances, the haloarchaeal cell is surrounded by a surface layer comprising a single protein component, the S-layer glycoprotein. Analysis of the S-layer glycoprotein has provided insight into the archaeal version of various post-translational modifications, such as glycosylation and lipid modification. In other cases, haloarchaeal cells are surrounded by glycan-based structures. In this chapter, selected aspects of haloarchaeal cell wall biology are considered.

Introduction

Today, the availability of a relatively large number of sequenced genomes provides the strongest support for the concept that organisms can be assigned to one of three distinct groups, namely the Eukarya, the Bacteria or the Archaea (Woese and Fox, 1977; Woese *et al.*, 1990; Graham *et al.*, 2000). In earlier times, however, when the sequencing of even single genes was considered a significant achievement, traits associated with the cell envelope helped distinguish Archaea from other life forms (Kandler and König, 1978, 1985, 1993; König, 2001).

Of the various archaeal groups considered in this context, it was the cell surface of the haloarchaea that drew the most interest, possibly due to the presence of bacteriorhodopsin, the light-driven proton pump, in the purple membrane of *Halobacterium halobium* (*salinarum*) (Oesterhelt and Stoeckenius, 1971; Henderson, 1975). Indeed, the cell envelopes of haloarchaea have proven to be rich sources of novel biological information. The unique ether-based membrane lipids of the archaeal membrane were first discerned in *Halobacterium cutirubrum* (*salinarum*) in 1962 (Sehgal *et al.*, 1962; Kates *et al.*, 1963), long before Archaea were recognized as a distinct life form. The surface (S)-layer glycoprotein of the *Halobacterium salinarum* cell envelope provided the first example of a non-eukaryal N-linked glycoprotein (Mescher and Strominger, 1976a,b) and also offered a relatively rare example of glycosylation of an Asn residue not found as part of the classic Asn-X-Ser/Thr sequon motif, where X is any residue but Pro (Gavel and von Heijne, 1980; Zeitler *et al.*, 1998). Halophilic Archaea have also provided examples of heterosaccharide-based structures not observed elsewhere (Niemetz *et al.*, 1997).

The present availability of complete genome sequences for haloarchaea isolated from a range of environmental conditions, along with the conclusions of previous biochemical and structural efforts, together provide considerable insight into the biogenesis, molecular composition and roles served by the cell envelopes surrounding different haloarchaea. In the following, chosen aspects of haloarchaeal cell envelope biology will be discussed.

Structure of haloarchaeal surface (S)-layers

Surface (S)-layers are two-dimensional crystalline arrays that comprise the sole layer beyond the plasma membrane in several haloarchaea. Indeed, of the various cell envelope types reported for haloarchaea, S-layers remain the best characterized. As early as 1956, electron microscopy revealed the *Hbt. halobium* (*salinarum*) surface as being covered by a hexagon-patterned monolayer comprising morphological units spaced approximately 17 nm apart (Houwink, 1956). Subsequent examination of thin-sectioned haloarchaeal cells revealed a 17-nm-thick cell wall beyond the plasma membrane (Stoeckenius and Rowen, 1967; Steensland and Larsen, 1969; Kirk and Ginzburg, 1972). X-ray diffraction was employed to demonstrate the presence of a protein layer some 8 nm from the haloarchaeal plasma membrane (Blaurock *et al.*, 1976). In these studies, morphological subunits presenting an inverted-parabola shape, thus creating a periplasmic-like space, were reported. Iodination of *Hbt. salinarum* surface proteins, together with proteolysis-based approaches, showed the S-layer to be the sole component of the S-layer glycoprotein (Mescher and Strominger, 1976a). Although the glycoprotein-based composition of the *Halobacterium* cell envelope had been previously suggested (Koncewicz, 1972; Mescher *et al.*, 1974), it was Mescher and Strominger (1976b) who isolated and described the *Hbt. salinarum* S-layer glycoprotein, thus providing the first example of non-eukaryal N-glycosylation.

Having been shown as being important for maintaining the proper shape of the cell (Mescher and Strominger, 1976a), a more refined view of the structure of the *Hbt. salinarum* S-layer was next sought. Relying on the *Hbt. salinarum* S-layer glycoprotein primary sequence (Lechner and Sumper, 1987), together with previously obtained X-ray diffraction data and electron microscopic images of negatively stained cell envelopes from *Haloferax* (then *Halobacterium*) *volcanii*, a three-dimensional reconstruction of the haloarchaeal S-layer glycoprotein and cell envelope was offered (Kessel *et al.*, 1988). Based on a 2 nm resolution, this model recruits six S-layer glycoproteins, organized into a 4.5 nm-thick dome-shaped pore around a narrow opening at the outermost face of the pore that widens as it approaches the membrane. The entire structure is thought to linked to the membrane by the predicted C-terminal trans-membrane domain of each S-layer glycoprotein. At the same time, an O-glycosylated region of the S-layer glycoprotein just upstream of the putative C-terminal trans-membrane domain is proposed to prop up the domed structure, acting as a spacer domain. While those agents responsible for maintaining the integrity of the S-layer structure are not known, divalent cations are somehow involved (Mescher and Strominger, 1976a; Kessel *et al.*, 1988). More recent tomography-based reconstruction of the *Hbt. salinarum* S-layer has offered a more realistic portrayal of the haloarchaeal cell wall (Trachtenberg *et al.*, 2000). Such studies, performed using intact cells maintained in their growth medium and rapidly frozen to maintain a high degree of sample preservation, revealed the *Hbt. salinarum* cell envelope as possessing the same basic architecture as the *Hfx. volcanii* S-layer. At 2 nm resolution, both strains display a matching hexagonal arrangement of the same basic, domed-shaped morphological unit, with each presenting identical 6-fold symmetry and identical 15 nm centre–centre spacing in each case. Nonetheless, despite these similarities in their S-layer architecture, *Hbt. salinarum* and *Hfx. volcanii* assume extremely different shapes; *Hbt. salinarum* appears as rods and *Hfx. volcanii* appears as indented disks. As such, it would seem that factors other than the S-layer determine cell shape.

Haloarchaeal S-layer glycoproteins

It is thought that haloarchaeal S-layers comprise a single glycoprotein species, the S-layer glycoprotein (Mescher and Strominger, 1976b; Sumper *et al.*, 1990). While genes encoding S-layer glycoproteins have been noted in available haloarchaeal genome sequences, only the S-layer glycoproteins from *Hbt. salinarum* and *Hfx. volcanii*, and to a some degree, that of *Haloarcula japonica* (Wakai *et al.*, 1997), have been examined in detail (Eichler, 2003).

The *Hbt. salinarum* S-layer glycoprotein is an 818 amino acid residue-long polypeptide, synthesized together with a 34 amino acid signal peptide (Lechner and Sumper, 1987). The *Hfx. volcanii* S-layer glycoprotein is a 794 amino acid residue-long polypeptide, also synthesized with a 34 amino acid signal peptide (Sumper *et al.*, 1990). While predicted to have molecular masses of 86.5 and 81.7 kDa, respectively, both S-layer glycoproteins migrate at the ~190 kDa position in SDS-PAGE due to an abundance of negatively charged residues, an adaptation designed to allow haloarchaeal proteins to fold properly in hypersaline conditions (Lanyi, 1984; Fukuchi *et al.*, 2003). Although the two polypeptides differ in terms of their glycosylation profile, each protein is modified by both N- and O-glycosylation. While the *Hbt. salinarum* S-layer glycoprotein includes 11 putative N-glycosylation sites, only three have been confirmed as being modified experimentally (Paul *et al.*, 1986; Lechner and Sumper, 1987). In contrast, Asn-17, a putative N-glycosylation site, is apparently not modified (Lechner and Sumper, 1987). Strikingly, the *Hbt. salinarum* S-layer glycoprotein is modified by two distinct N-linked glycans, each relying on a different linking sugar. Whereas Asn-2 is modified by a repeating sulfated pentasaccharide moiety linked through a *N*-acetylgalactosamine residue, other modified Asn residues present a glucose-linked, sulfated polysaccharide (Lechner and Wieland, 1989).

The *Hfx. volcanii* S-layer glycoprotein contains seven putative N-glycosylation sites, two of which were originally reported as containing a linear string of glucose subunits (Sumper *et al.*, 1990; Mengele and Sumper, 1992). More recent analysis has, however, led to a revision of the N-linked glycan composition (Abu-Qarn *et al.*, 2007; Magidovich *et al.*, 2010; Guan *et al.*, 2010; Calo *et al.*, 2011b). In these studies, Asn-13 and Asn-83 were shown as being decorated by a pentasaccharide comprising a hexose, two hexuronic acids, a methyl ester of hexuronic acid and a terminal mannose subunit. The composition of the glycan(s) identified at other *Hfx. volcanii* S-layer glycoprotein N-glycosylation sites were also partially characterized in the past, with glucose, galactose, mannose, and idose being reported (Sumper *et al.*, 1990; Mengele and Sumper, 1992). Thus, despite their similarities at the amino acid level (40.5% identity), the S-layer glycoproteins of the two species present very different glycosylation profiles. Accordingly, it had been proposed that the enhanced negative character and subsequent increase in surface charge density resulting from the sulfated nature of *Hbt. salinarum* S-layer glycoprotein N-linked glycans, relative to their non-sulfated counterparts in *Hfx. volcanii*, represents an adaptation to the enhanced saline surroundings encountered by the former species (Mengele and Sumper, 1992). Recently, this hypothesis has been called into question when it was shown that in *Hfx. volcanii* cells grown in the presence of 1.75 M NaCl, the S-layer glycoprotein is modified by a second glycan, this time comprising sulfated hexose, hexose, hexose, rhamnose, in addition to the pentasaccharide described above (Guan *et al.*, 2012).

Finally, both the *Hbt. salinarum* and *Hfx. volcanii* S-layer glycoproteins are O-glycosylated. In both proteins, glucose–galactose pairs decorate threonine residues clustered upstream of the proposed single membrane-spanning domain proceeding the C-terminus (Lechner and Sumper, 1987; Sumper *et al.*, 1990).

Glycosylation of haloarchaeal S-layer glycoproteins

Upon demonstration of the N-glycosylation of the *Hbt. salinarum* S-layer glycoprotein (Mescher and Strominger, 1976b), efforts to define the pathway responsible for the haloarchaeal version of this post-translational modification were initiated. Such studies revealed significant similarities between the archaeal process and the well-defined eukaryal N-glycosylation pathway, the only other version of N-glycosylation known at the time (Sumper, 1987; Lechner and Wieland, 1989).

In higher Eukarya, seven soluble nucleotide-activated sugar subunits are sequentially added to a dolichol pyrophosphate lipid carrier on the cytoplasmic face of the endoplasmic reticulum (ER) membrane. Following a 'flipping' of the lipid-linked heptasaccharide to face the ER lumen, seven additional sugar subunits are added, each derived from an individual dolichol phosphate carrier, previously charged in the cytoplasm and 'flipped' to face the ER lumen. Once the 14-member oligosaccharide has been assembled, the entire polymer is transferred, en bloc, to select Asn residues of nascent polypeptide targets translocating across the ER membrane via the Stt3 subunit of the oligosaccharide transferase complex (for reviews, see Burda and Aebi, 1999; Helenius and Aebi, 2004; Yan and Lennarz, 2005).

As in Eukarya, mono- and polysaccharide-charged phosphodolichol carriers have also been detected in *Hbt. salinarum* (Mescher *et al.*, 1976). The participation of sugar-charged dolichol pyrophosphate in *Hbt. salinarum* S-layer glycoprotein N-glycosylation was concluded based on the inhibition of this post-translational modification upon addition of bacitracin, an antibiotic that selectively interferes with the regeneration of pyrophosphate-containing dolichols (Mescher and Strominger, 1976a). Moreover, the fact that this membrane-impermeant antibiotic prevented glycosylation at the *Hbt. salinarum* S-layer glycoprotein Asn-2 position is indicative of a process implicating steps that occur on the outer surface of the plasma membrane, the topological homologue of the lumen-facing face of the ER membrane. The failure of bacitracin to hinder *Hfx. volcanii* S-layer glycoprotein N-glycosylation is in keeping with the detection of mono- and polysaccharide-charged dolichol phosphate, rather than dolichol pyrophosphate in this species, as the singly phosphorylated version of dolichol is not susceptible to bacitracin (Kuntz *et al.*, 1997; Eichler, 2001; Guan *et al.*, 2010). The modification of membrane-impermeant N-glycosylation site-containing peptides further supports the assignment of the outer membrane surface as the site of *Hbt. salinarum* S-layer glycoprotein N-glycosylation (Lechner *et al.*, 1985a). Subsequent studies involving the glucosyltransferase inhibitors, amphomycin, PP36 and PP55, also concluded that *Hfx. volcanii* S-layer glycoprotein glycosylation occurs on the outer cell surface (Zhu *et al.*, 1995). In contrast, the lipid-linked oligosaccharides that are ultimately delivered to the S-layer glycoprotein *Hbt. salinarum* and *Hfx. volcanii* are sulfated and methylated, respectively, in the cytoplasm (Lechner *et al.*, 1985a; Magidovich *et al.*, 2010). Likewise, the transient methylation experienced by the lipid-bound sulfated oligosaccharide in *Hbt. salinarum* is also predicted to transpire in the cytoplasm. While the role of this transient methylation remains unclear, its inhibition blocked *Hbt. salinarum* S-layer glycoprotein glycosylation (Lechner *et al.*, 1985b). Such processing of archaeal lipid-linked glycans prior to their transfer to the target protein seemingly does not occur in Eukarya (Varki, 1998). Furthermore, the ability of *Hbt. salinarum* to N-glycosylate the S-layer glycoprotein Asn-2 position upon replacement of the Ser-4 position by a Val, Leu or Asn residue (Zeitler *et al.*, 1998) suggests that glycosylation motifs apart from the classic Asn-X-Ser/Thr sequon, where X is any residue but proline (Gavel and von Heijne, 1990), are recognized in Archaea.

In recent years, the availability of an ever-growing number of archaeal genome sequences and the appearance of appropriate molecular tools for more and more species has made detailed dissection of archaeal N-glycosylation possible. *Hfx. volcanii* represents a leading model system in such efforts (Calo *et al.*, 2010a). Initially, Abu-Qarn and Eichler (2006) scanned the *Hfx. volcanii* genome for homologues of genes whose products are involved in N-glycosylation in Eukarya or Bacteria. In this manner, *Hfx. volcanii* homologues of genes encoding participants in both systems were identified. Support for these *Hfx. volcanii* ORFs as corresponding to true genes was provided by reverse-transcriptase polymerase chain reactions that confirmed the transcription of the vast majority of these sequences in cells grown to exponential phase in rich medium. To identify additional *Hfx. volcanii* N-glycosylation genes, sequences in the vicinity of the first set of sequences identified above were considered (Yurist-Doutsch and Eichler, 2009). Subsequent gene deletion, followed by mass spectrometric analysis of S-layer glycoprotein N-glycosylation sites, served to assign roles to the various gene products in the N-glycosylation process.

In this manner, a series of *agl* (*a*rchaeal *gl*ycosylation) genes encoding proteins involved in the assembly and attachment of a pentasaccharide to select Asn residues of the *Hfx. volcanii* S-layer glycoprotein was identified. Acting at the cytoplasmic face of the plasma membrane, AglJ, AglG, AglI and AglE sequentially add the first four pentasaccharide residues (i.e. a hexose, two hexuronic acids and the methyl ester of a hexuronic acid) onto a common dolichol phosphate carrier, while AglD adds the final pentasaccharide residue, mannose, to a distinct dolichol phosphate (Abu-Qarn *et al.*, 2007, 2008; Yurist-Doutsch *et al.*, 2008, 2010; Guan *et al.*, 2010; Kaminski *et al.*, 2010; Magidovich *et al.*, 2010). The use of dolichol phosphate by Archaea as the lipid carrier upon which the N-linked is assembled also holds true in eukaryal N-glycosylation (Burda and Aebi, 1999). In contrast, bacterial N-linked glycans are first assembled on a different isoprenoid, namely undecaprenol phosphate (Szymanski and Wren, 2005; Weerapana and Imperiali, 2006). N-glycosylation roles have been assigned to *Hfx. volcanii* AglF, a glucose-1-phosphate uridyltransferase (Yurist-Doutsch *et al.*, 2010), AglM, a UDP-glucose dehydrogenase (Yurist-Doutsch *et al.*, 2010) and AglP, a methyltransferase (Magidovich *et al.*, 2010). Indeed, AglF and AglM were shown to act in a sequential and coordinated manner *in vitro*, transforming glucose-1-phophosphate into UDP-glucuronic acid (Yurist-Doutsch *et al.*, 2010). In a reaction requiring the archaeal oligosaccharide transferase, AglB (Abu-Qarn and Eichler, 2006; Chaban *et al.*, 2006; Igura *et al.*, 2008), the lipid-linked tetrasaccharide and its precursors are delivered to select Asn residues of the S-layer glycoprotein. Finally, the terminal pentasaccharide residue, mannose, is transferred from its dolichol phosphate carrier to the protein-bound tetrasaccharide (Guan *et al.*, 2010). Current understanding of *Hfx. volcanii* N-glycosylation is depicted in Fig. 6.1.

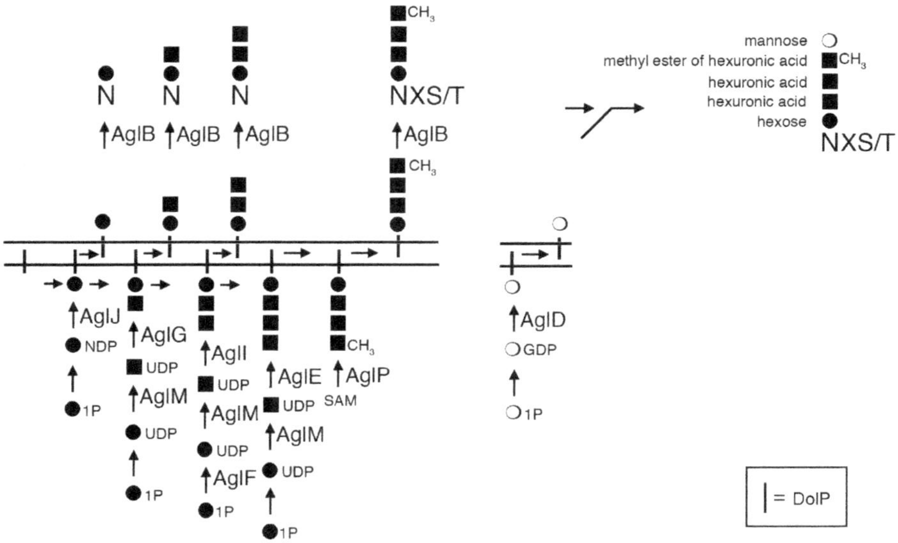

Figure 6.1 The current model of *Hfx. volcanii* N-glycosylation pathway. See text for details. The inset presents the symbol used to depict dolichol phosphate (DolP).

Finally, while the *Haloarcula marismortui* S-layer glycoprotein is modified by the same pentasaccharide as is N-linked to the corresponding *Hfx. volcanii* protein, the pathway of pentasaccharide assembly differs between the two species. In *Har. marismortui*, the pentasaccharide is first fully assembled onto a common dolichol phosphate carrier and only then delivered to select Asn residues of the S-layer glycoprotein, rather than relying on distinct dolichol phosphate carriers, as does *Hfx. volcanii*. Thus, although both originate from the Dead Sea, *Har. marismortui* and *Hfx. volcanii* have developed distinct N-glycosylation strategies (Calo *et al.*, 2011b).

Modulation of haloarchaeal S-layer glycoprotein N-glycosylation as an adaptive response

In *Hfx. volcanii*, cells lacking AglB (and hence unable to perform N-glycosylation) are viable (Abu-Qarn *et al.*, 2007). Still, absent or compromised N-glycosylation have detrimental effects on the ability of *Hfx. volcanii* to grow in high salt (Abu-Qarn *et al.*, 2007), on S-layer stability and architecture (Abu-Qarn *et al.*, 2007) and on the resistance of the S-layer to added protease (Yurist-Doutsch *et al.*, 2008, 2010; Kaminski *et al.*, 2010). Thus, while not essential for survival, it would seem that N-glycosylation is advantageous to *Hfx. volcanii* in certain situations. This, together with the fact that all but one of the known *agl* genes are clustered into a single gene island (Yurist-Doutsch and Eichler, 2009), raises the possibility that *Hfx. volcanii* modulates N-glycosylation of the S-layer glycoprotein in response to changing environmental conditions.

The first evidence supporting a role for modulated S-layer glycoprotein N-glycosylation in allowing *Hfx. volcanii* to cope with changing surroundings came from real-time PCR studies. Here, cells were raised in four arbitrarily selected conditions, namely growth in the presence of 4.8 M NaCl, growth in the presence of 1.75 M NaCl, growth to stationary phase and upon heat shock, and the levels of transcripts of a subset of adjacent *agl* genes were determined in each case. These levels were then compared with the amounts of the same transcripts obtained from cells grown to mid-exponential phase in rich medium. Such comparison revealed the coordinated up- or down-regulation of *agl* gene transcription as a function of growth condition (Yurist-Doutsch *et al.*, 2008, 2010).

The hypothesis that N-glycosylation of the S-layer glycoprotein can be modulated in response to changes in the environment was most recently substantiated when the N-linked glycan profile of the S-layer glycoprotein was compared in cells grown in medium containing 3.4 or 1.75 M NaCl (Guan *et al.*, 2012). At the higher salinity, S-layer glycoprotein Asn-13 and Asn-83 were shown to be modified by the pentasaccharide described above, while dolichol phosphate was shown to be modified by the tetrasaccharide comprising the first four pentasaccharide residues or by a single mannose residue that corresponds to the final pentasaccharide residue. In contrast, cells grown at low salinity contain dolichol phosphate modified by a distinct tetrasaccharide not seen in cells grown at high salinity, comprising a sulfated glucose, two glucoses and a rhamnose residue. The same tetrasaccharide was shown to modify S-layer glycoprotein Asn-498 in cells grown in low salt, at position not glycosylated in cells grown in the high salt medium. At the same time, Asn-13 and Asn-83 were modified by substantially less pentasaccharide in response to growth at the lower salinity (Fig. 6.2). Hence, in response to environmental salinity, *Hfx. volcanii* not only modulates the composition of the N-linked glycans decorating the S-layer glycoprotein but also which residues are subjected to this post-translational modification.

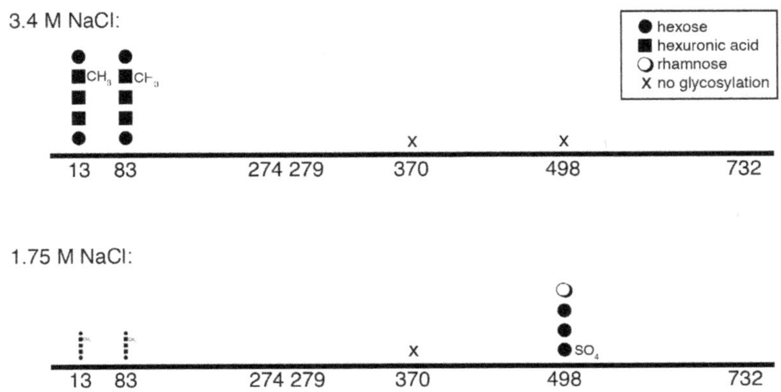

Figure 6.2 The *Hfx. volcanii* S-layer glycoprotein presents distinct N-linked glycan profiles in high and low salt surroundings. The N-linked glycan profile of the S-layer glycoprotein from *Hfx. volcanii* cells grown in either 3.4 or 1.75 M NaCl-containing medium was assessed by LC-ESI MS. In cells grown at high salinity, Asn-13 and Asn-83 are modified by a pentasaccharide comprising a hexose, two hexuronic acids, a methyl ester of a hexuronic acid and a mannose. In cells grown at low salinity, Asn-13 and Asn-83 are modified by the same pentasaccharide, albeit to a lesser extent, whereas Asn-498 is modified by a tetrasaccharide comprising a sulfated hexose, two hexoses and a rhamnose. Asn-370 is not modified in either growth condition. The inset presents the identities of sugars comprising the N-linked glycan.

Lipid-modification of haloarchaeal S-layer glycoproteins

In addition to glycosylation, haloarchaeal S-layer glycoproteins are also subject to post-translational lipid-modification. This was demonstrated upon incubation of *Hbt. salinarum* cells with the isoprene precursor, [^{3}H]-mevalonate, as well as with other tritiated archaeal lipid precursors, which led to selective incorporation of the radiolabel into the S-layer glycoprotein (Kikuchi *et al.*, 1999). Mass spectrometric analysis of the linked lipid moiety revealed it to be a novel diphytanylglycerol phosphate. While the manner by which the lipid is attached to the S-layer glycoprotein has yet to be described, it is thought to involve a phosphodiester-based linkage to either a Ser or Thr residue. Attempts to define the position of this modification in the protein assigned the added lipid to a 28kDa C-terminal tryptic fragment comprising residues 731–816. Moreover, upon considering the sequence similarity of the same C-terminal region in the *Hfx. volcanii* and *Haloarcula japonica* S-layer glycoproteins (Lechner and Sumper, 1987; Sumper *et al.*, 1990; Wakai *et al.*, 1997), it was predicted that here too a similar lipid modification transpires (Kikuchi *et al.*, 1999). In the case in *Hfx. volcanii*, lipid modification of the S-layer glycoprotein was indeed subsequently demonstrated (Eichler, 2001; Konrad and Eichler, 2002).

Studies addressing the biogenesis of the *Hfx. volcanii* S-layer glycoprotein provided further insight into the process of lipid modification of haloarchaeal S-layer glycoproteins. Through the use of [^{35}S] pulse-chase metabolic radiolabelling and the ribosomal-acting antibiotic, anisomycin, the *Hfx. volcanii* S-layer glycoprotein was shown to undergo a post-translational maturation step, reflected as an increase in the hydrophobicity and apparent molecular weight of the protein (Eichler, 2001). As growth in the presence of [^{3}H] mevalonic acid resulted in radiolabel being incorporated into the S-layer glycoprotein and given the ability of mevinolin, an inhibitor of 3-HMG-CoA reductase (involved in converting acetyl-CoA into mevalonic acid), to prevent S-layer glycoprotein maturation, it was concluded that the *Hfx. volcanii* S-layer glycoprotein is modified via lipid attachment, an event taking place late in the biogenesis of the protein, only after the protein has translocated across and inserted into the plasma membrane (Konrad and Eichler, 2002). Such lipid anchor attachment does not, however, occur in the absence of Mg^{2+} (Eichler, 2001), the presence of which is important for maintaining haloarchaeal S-layer integrity (Mescher and Strominger, 1976a; Kessel *et al.*, 1988). Finally, since the *Hbt. salinarum* S-layer glycoprotein undergoes a similar lipid-based maturation step as does its *Hfx. volcanii* counterpart, this event may represent a necessary step in S-layer glycoprotein biogenesis in other haloarchaea (Konrad and Eichler, 2002).

The cell envelope of the square haloarchaeon, *Haloquadratum walsbyi*

Most thalassic (seawater-derived) NaCl-saturated environments are dominated by *Haloquadratum walsbyi*, a square, non-motile pigmented haloarchaeal species (Oren *et al.*, 1996; Anton *et al.*, 1999; Benlloch *et al.*, 2001; Bolhuis *et al.*, 2004, 2006; Burns *et al.*, 2004). Able to tolerate up to 2 M $MgCl_2$, *Hqr. walsbyi* can survive conditions of extremely low water activity and, as such, is the last living organism remaining in crystallizer ponds before these become sterile, magnesium-saturated pools (Javor, 1984; Oren, 2002). The success of *Hqr. walsbyi* in such low water environments may be related to a cell envelope that is thought to include halomucin, a protein reminiscent of animal mucins in terms of amino acid sequence and domain organization (Bolhuis *et al.*, 2006). Halomucin may represent a specific adaptation to desiccation, acting much like the mucin-based cocoon of lungfish, creatures able to avoid dehydration and survive even after prolonged periods outside water (Chew *et al.*, 2004). In keeping with its sequence and possible functional similarities to mucins, halomucin contains numerous putative sites of N-glycosylation and sulfation. Such modifications would further contribute to the overall negative charge of the protein, contributing to the creation of an aqueous shield over the *Hqr. walsbyi* cell. A similar role has been attributed to the glycan moieties attached to the major glycosylated membrane-bound protein species coating *Thermoplasma acidophilum*, a cell-wall lacking thermoacidiphilic archaeon (Yang and Haug, 1979). As noted above, enhanced levels of negative charges on the glycan moieties decorating *Hbt. salinarum* S-layer glycoprotein have been cited as contributing to the ability of this species to survive in hypersaline surroundings (Mengele and Sumper, 1992). Of late, atomic force microscopy was employed to present a detailed analysis of the *Hqr. walsbyi* cell envelope (Saponetti *et al.*, 2011). Such efforts detected not only the presence of an S-layer, as seen in other haloarchaea (see above), but also a more external thin layer surrounding the cell, possibly composed of halomucin.

Animal mucins include glycans that are often capped by sialic acids, namely sugars possessing a nine-carbon backbone that serve to contribute rigidity to the protein-linked glycan (Sheehan *et al.*, 1991). Although widespread in Eukarya (Sheehan *et al.*, 1991; Sampathkumar *et al.*, 2006) and detected in some Bacteria (Vimr *et al.*, 2004), sialic acids have yet to be detected in Archaea, although a sialic-like sugar, 5-N-formyl-legionaminic acid, has been shown to be N-linked to a haloarchaeal viral protein generated in *Halorubrum* sp. strain PV6 (Kandiba *et al.*, 2012). Thus, while it remains to be determined whether halomucin contains sialic acids, it is of note that the *Hqr. walsbyi* genome includes homologues of the two essential sialic acid biosynthesis genes, *neuA* and *neuB* (Bolhuis *et al.*, 2006). The annotated *Hqr. walsbyi* genome also encodes two shorter homologues of halomucin and at least fourteen S-layer glycoprotein homologues that may contribute to the cell envelope of this species. Finally, *Hqr. walsbyi* may also be surrounded by a poly-γ-glutamate capsule, given the presence of genes homologous to bacterial sequences involved in the biosynthesis of this polymer (Ashiuchi and Misono, 2002).

Does the haloarchaeal cell envelope include a periplasmic space?

In addition to the N-glycosylation considered above, the S-layer glycoproteins of *Hbt. salinarum* and *Hfx. volcanii* also undergo O-glycosylation. In both cases, disaccharides comprising glucose and galactose are linked to threonine residues clustered upstream of the predicted C-terminal membrane-spanning domain (Wieland and Sumper, 1989; Sumper *et al.*, 1990). These O-linked glycans have been postulated as serving to 'prop up' the S-layer, thereby creating a periplasmic space between the plasma membrane and the inner surface of the S-layer surrounding the cell (Kessel *et al.*, 1988). The presence of such a structural compartment carries implications for a variety of cell processes. For instance, the presence of a periplasmic space would permit target proteins to undergo various post-translational modifications in a protected space. Such an arrangement would allow protein processing events to occur over a longer period that would be the case for a secreted substrate released directly to the extracellular world. Likewise, protein oligomerization events could exploit the existence of a physical barrier, i.e. the S-layer, to prevent the diffusion of any exported complex subunits until such time as oligomer assembly is complete. Although the presence of multimeric protein machines on the outer surface of the haloarchaeal plasma membrane has not yet been reported, such complexes have been reported in other archaeal species (Zolghadr *et al.*, 2011). Finally, the existence of a periplasmic space in haloarchaea could represent a compartment where elaborate nutrient capture, breakdown and uptake machines could work in tandem.

The heteropolysaccharide cell walls of *Halococcus morrhuae* and *Natronococcus occultus*

In addition to the relatively simple S-layer considered until now, certain haloarchaea rely on other cell envelope structures. Much less, however, is known of these assemblies. Still, what is currently known demonstrates the great variety in haloarchaeal cell envelope composition that exists. Two such examples are considered here.

Halococcus morrhuae is a coccoid haloarchaeon surrounded by a rigid cell wall sacculus appearing as a 50–60-nm wide electron-dense layer (Kandler and König, 1993). The realization made in the early 1970s that the cell wall of this species did not contain peptidoglycan, a major component of the bacterial cell wall, was a pillar upon which the assignment of the Archaea as a distinct phylogenetic kingdom rested (Wolfe, 2006). Since, the chemical composition of purified cell wall from *Hcc. morrhuae* stain CCM 859 has been determined and shown to contain glucose, mannose, galactose, glucuronic acid, galacturonic acid, glucosamine and gulosaminuronic acid, an aminouronic acid (Steber and Schleifer, 1975; Schleifer *et al.*, 1982). While the structure of this oligosaccharide has yet to be fully solved, it is thought to be arranged into three domains and to include sulfated subunits and N-acetylated amino sugars (Schleifer *et al.*, 1982; Kandler and König, 1998). In addition, the rigid *Hcc. morrhuae* cell wall also contains significant amounts of glycine, proposed to bridge the amino groups of glucosamines with the carboxyl groups of uronic or gulosaminuronic acid residues in the glycan strands (Steber and Schleifer, 1979). At present, nothing is known of the biosynthesis of this sulfated glycan.

Natronococcus occultus is a haloalkalophile that grows optimally in hypersaline environments with pH values between 9.5 and 10 (Tindall *et al.*, 1984) and is surrounded by a cell wall unique among the Archaea, comprising L-glutamate, N-acetyl-D-glucosamine, N-acetyl-D-galactosamine, D-galacturonic acid, D-glucuronic acid and D-glucose in a molar ratio of 5:7:1:8:0.5:0.3 (Niemetz *et al.*, 1997). The glycan comprising this cell wall polymer is thought to be formed from a 60 residue-long poly-(γ-L-glutamine) chain linked to two distinct heterosaccharides via N-amide linkages to the α-carboxylic group of the glutamine residues. Alternatively, two distinct poly-(γ-L-glutamine) chains may be linked to either heterosaccharide moiety. Here too, nothing is known concerning the biosynthesis of the cell wall polymer.

Future trends

The identification, isolation and sequencing of novel haloarchaeal species, together with the use of sophisticated tools for the genetic manipulation of haloarchaea and improved protocols for the purification of haloarchaeal proteins now available (Soppa *et al.*, 2006), will allow future studies on the haloarchaeal cell wall to answer questions not possible until now. Many such questions will focus on the S-layer glycoprotein. For instance, addressing the interplay between S-layer N- and O-glycosylation and conditions in the surroundings will provide insight into the adaptation of haloarchaeal to their environment. One could imagine future efforts at engineering such responses into host organisms to introduce enhanced halotolerance. Alternatively, further understanding of haloarchaeal S-layer glycoprotein glycosylation could serve as the basis for glyco-engineering efforts designed to introduce tailored glycans into proteins of interest. Indeed, such efforts have already begun, using *Hfx. volcanii* as a platform (Calo *et al.*, 2010b, 2011a).

Based on what is known to date, it would appear that in addition to the sole predicted membrane-spanning domain of the haloarchaeal S-layer glycoprotein located close to the C-terminus, a lipid-based moiety also anchors the protein to the membrane. Yet, paradoxically, the spheroplasts formed for transformation of *Hfx. volcanii* cells require EDTA-based treatment for release of the S-layer (Cline *et al.*, 1989). Such treatment would not release an integral membrane protein. As such, it is possible that two distinct populations of the S-layer glycoprotein exist, one anchored by a trans-membrane domain and one anchored by a lipid moiety. Future efforts will test this hypothesis and determine whether the former population serves as a precursor of the latter.

Finally, the application of modern imaging techniques, such as tomography and high-resolution structured illumination microscopy, to the study of haloarchaeal cell envelope architecture will provide novel structural insight into these assembly and composition of these entities. Such findings could form the basis for exploiting the two-dimensional lattice that is the haloarchaeal S-layer for a variety of applied uses.

Acknowledgements

Our work is supported by grants from the Israel Science Foundation (8/11) and the US Army Research Office (W911NF-11-1-520).

Web resources

Haloarchaeal genomes and related information is available at:

- http://www.halolex.mpg.de/public/
- http://edwards.sdsu.edu/halophiles/
- http://archaea.ucsc.edu/

An excellent source of technical tips for working with haloarchaea can be found at:

- http://www.haloarchaea.com/resources/halohandbook/

References

Abu-Qarn, M., and Eichler, J. (2006). Protein N-glycosylation in Archaea: Defining *Haloferax volcanii* genes involved in S-layer glycoprotein glycosylation. Mol. Microbiol. *61*, 511–525.

Abu-Qarn, M., Yurist-Doutsch, S., Giordano, A., Trauner, A., Morris, H.R., Hitchen, P., Medalia, O., Dell, A., and Eichler, J. (2007). *Haloferax volcanii* AglB and AglD are involved in N-glycosylation of the S-layer glycoprotein and proper assembly of the surface layer. J. Mol. Biol. *374*, 1224–1236.

Abu-Qarn, M., Giordano, A., Battaglia, F., Trauner, A., Morris, H.R., Hitchen, P., Dell, A., and Eichler, J. (2008). Identification of AglE, a second glycosyltransferase involved in N-glycosylation of the *Haloferax volcanii* S-layer glycoprotein. J. Bacteriol. *190*, 3140–3146.

Albers, S.V., and Driessen, A.J. (2005). Analysis of ATPases of putative secretion operons in the thermoacidophilic archaeon *Sulfolobus solfataricus*. Microbiology *151*, 763–773.

Allers, T., Ngo, H.P., Mevarech, M., and Lloyd, R.G. (2004). Development of additional selectable markers for the halophilic archaeon *Haloferax volcanii* based on the *leuB* and *trpA* genes. Appl. Environ. Microbiol. *70*, 943–953.

Antón, J., Llobet-Brossa, E., Rodríguez-Valera, F., and Amann, R. (1999). Fluorescence *in situ* hybridization analysis of the prokaryotic community inhabiting crystallizer ponds. Environ. Microbiol. *1*, 517–523.

Ashiuchi, M., and Misono, H. (2002). Biochemistry and molecular genetics of poly-γ-glutamate synthesis. Appl. Microbiol. Biotechnol. *59*, 9–14.

Baliga, N.S., Bonneau, R., Facciotti, M.T., Pan, M., Glusman, G., Deutsch, E.W., Shannon, P., Chiu, Y., Weng, R.S., Gan, R.R., *et al.* (2004) Genome sequence of *Haloarcula marismortui*: a halophilic archaeon from the Dead Sea. Genome Res. *14*, 2221–2234.

Benlloch, S., Acinas, S.G., Antón, J., López-López, A., Luz, S.P., and Rodríguez-Valera, F. (2001). Archaeal biodiversity in crystallizer ponds from a solar saltern: Culture versus PCR. Microb. Ecol. *41*, 12–19.

Blaurock, A.E., Stoeckenius, W., Oesterhelt, D., and Scherfhof, G.L. (1976). Structure of the cell envelope of Halobacterium halobium. J. Cell Biol. *71*, 1–22.

Bolhuis, H., te Poele, E., and Rodríguez-Valera, F. (2004). Isolation and cultivation of Walsby's square archaeon. Environ. Microbiol. *6*, 1287–1291.

Bolhuis, H., Palm, P., Wende, A., Falb, M., Rampp, M., Rodríguez-Valera, F., Pfeiffer, F., and Oesterhelt, D. (2006). The genome of the square archaeon Haloquadratum walsbyi: life at the limits of water activity. BMC Genomics 7, 169.

Burda, P., and Aebi, M. (1999). The dolichol pathway of N-linked glycosylation. Biochim. Biophys. Acta *1426*, 239–257.

Burns, D.G., Camakaris, H.M., Janssen, P.H., and Dyall-Smith, M.L. (2004). Cultivation of Walsby's square haloarchaeon. FEMS Microbiol. Lett. *238*, 469–473.

Calo, D., Kaminski, L., and Eichler, J. (2010a). Protein glycosylation in Archaea: sweet and extreme. Glycobiology *20*, 1065–1079.

Calo, D., Eilam, Y., Lichtenstein, R.G., and Eichler, J. (2010b). Towards glyco-engineering in Archaea: replacing *Haloferax volcanii* AglD with homologous glycosyltransferases from other halophilic archaea. Appl. Environ. Microbiol. *76*, 5684–5692.

Calo, D., Guan Z., and Eichler, J. (2011a). Glyco-engineering in Archaea: differential N-glycosylation of the S-layer glycoprotein in a transformed *Haloferax volcanii* strain. Microb. Biotechnol. *4*, 461–470.

Calo, D., Guan Z., Naparstek S., and Eichler, J. (2011b). Different routes to the same ending: comparing the N-glycosylation processes of *Haloferax volcanii* and *Haloarcula marismortui*, two halophilic archaea from the Dead Sea. Mol. Microbiol. *81*, 1166–1177.

Chaban, B., Voisin, S., Kelly, J., Logan, S.M., and Jarrell, K.F. (2006). Identification of genes involved in the biosynthesis and attachment of *Methanococcus voltae* N-linked glycans: insight into N-linked glycosylation pathways in Archaea. Mol. Microbiol. *61*, 259–268.

Chew, S.F., Chan, N.K., Loong, A.M., Hiong, K.C., Tam, W.L., and Ip, Y.K. (2004). Nitrogen metabolism in the African lungfish (*Protopterus dolloi*) aestivating in a mucus cocoon on land. J. Exp. Biol. *207*, 777–786.

Cline, S.W., Lam, W.L., Charlebois, R.L., Schalkwyk, L.C., and Doolittle, W.F. (1989). Transformation methods for halophilic archaebacteria. Can. J. Microbiol. *35*, 148–152.

Eichler, J. (2001). Post-translational modification unrelated to protein glycosylation follows translocation of the S-layer glycoprotein across the plasma membrane of the haloarchaeon *Haloferax volcanii*. Eur. J. Biochem. *268*, 4366–4373.

Eichler, J. (2003). Facing extremes: Archaeal surface-layer (glyco-) proteins. Microbiology *149*, 3347–3351.

Eichler, J., and Adams, M.W.W. (2005) Post-translational protein modification in Archaea. Microbiol. Mol. Biol. Rev. *69*, 393–425.

Falb, M., Pfeiffer, F., Palm, P., Rodewald, K., Hickmann, V., Tittor, J., and Oesterhelt, D. (2005). Living with two extremes: conclusions from the genome sequence of *Natronomonas pharaonis*. Genome Res. *15*, 1336–1343.

Fukuchi, S., Yoshimune, K., Wakayama, M., Moriguchi, M., and Nishikawa, K. (2003). Unique amino acid composition of proteins in halophilic bacteria. J. Mol. Biol. *327*, 347–357.

Gavel, Y., and von Heijne, G. (1990). Sequence differences between glycosylated and non-glycosylated Asn-X-Thr/Ser acceptor sites: implications for protein engineering. Protein Eng. *3*, 433–442.

Graham, D.E., Overbeek, R., Olsen, G.J., and Woese, C.R. (2000). An archaeal genomic signature. Proc. Natl. Acad. Sci. U.S.A. *97*, 3304–3308.

Guan, Z., Naparstek, S., Kaminski, L., Konrad, Z., and Eichler, J. (2010). Distinct glycan-charged phosphodolichol carriers are required for the assembly of the pentasaccharide N-linked to the *Haloferax volcanii* S-layer glycoprotein. Mol. Microbiol. *78*, 1294–1303.

Guan, Z., Naparstek S., Calo, D., and Eichler, J. (2012). Protein glycosylation as an adaptive response in Archaea: Growth at different salt concentrations leads to alterations in *Haloferax volcanii* S-layer glycoprotein N-glycosylation. Environ. Microbiol. *14*, 743–753.

Helenius, A., and Aebi, M. (2004). Roles of N-linked glycans in the endoplasmic reticulum. Annu. Rev. Biochem. *73*, 1019–1049.

Henderson, R. (1975). The structure of the purple membrane from *Halobacterium halobium*: analysis of the X-ray diffraction pattern. J. Mol. Biol. *93*, 123–128.

Houwink, A.L. (1956). Flagella, gas vacuoles and cell-wall structure in *Halobacterium halobium*; an electron microscope study. J. Gen. Microbiol. *15*, 146–150.

Igura, M., Maita, N., Kamishikiryo, J., Yamada, M., Obita, T., Maenaka, K., and Kohda, D. (2008). Structure-guided identification of a new catalytic motif of oligosaccharyltransferase. EMBO J. *27*, 234–243.

Javor, B. (1984). Growth potential of halophilic bacteria isolated from solar salt environments: carbon sources and salt requirements. Appl. Environ. Microbiol. *48*, 352–360.

Kaminski, L., Abu-Qarn, M., Guan, Z., Naparstek, S., Ventura, V.V., Raetz, C.R.H., Hitchen, P.G., Dell, A., and Eichler, J. (2010). AglJ participates in adding the linking saccharide in the *Haloferax volcanii* N-glycosylation pathway. J. Bacteriol. *192*, 5572–5579.

Kandiba, L., Aitio, O., Helin, J., Guan, Z., Permi, P., Bamford, D., Eichler, J., and Roine, E. (2012). Diversity in prokaryotic glycosylation: an archaeal-derived N-linked glycan contains legionaminic acid. Mol. Microbiol. *84*, 578–593.

Kandler, O., and König, H. (1978). Chemical composition of the peptidoglycan-free cell walls of methanogenic bacteria. Arch. Microbiol. *118*, 141–152.

Kandler, O., and König, H. (1985). Cell envelopes of archaebacteria. In The Bacteria. A Treatise on Structure and Function. Archaebacteria, Woese, C.R., and Wolfe, R.S., eds. (Academic Press, New York), vol. VIII, pp. 413–457.

Kandler, O., and König, H. (1993). Cell envelopes of archaea: structure and chemistry. In The Biochemistry of Archaea, Kates, M., Kusher, D.J., Matheson, A.T., eds. (Elsevier, Amsterdam), pp. 223–259.

Kandler, O., and König, H. (1998). Cell wall polymers in archaea (archaebacteria). Cell Mol. Life Sci. *54*, 305–308.

Kates, M., Sastry, P.S., and Yengoyan, L.S. (1963). Isolation and characterization of a diether analog of phosphatidyl glycerophosphate from *Halobacterium cutirubrum*. Biochim. Biophys. Acta *70*, 705–707.

Kessel, M., Wildhaber, I., Cohen, S., and Baumeister, W. (1988) Three-dimensional structure of the regular surface glycoprotein layer of *Halobacterium volcanii* from the Dead Sea. EMBO J. *7*, 1549–1554.

Kikuchi, A., Sagami, H., and Ogura, K. (1999). Evidence for covalent attachment of diphytanylglyceryl phosphate to the cell-surface glycoprotein of *Halobacterium halobium*. J. Biol. Chem. *274*, 18011–18016.

Kirk, R.G., and Ginzburg, M. (1972). Ultrastructure of two species of *Halobacterium*. J. Ultrastruct. Res. *41*, 80–94.

Koncewicz, M.A. (1972). Glycoproteins in the cell envelope of *Halobacterium halobium*. Biochem. J. *128*, 124P.

König, H. (2001). Archaeal cell walls. eLS DOI: 10.1038/npg.els.0000384

Konrad, Z., and Eichler, J. (2002). Lipid modification of proteins in Archaea: attachment of a mevalonic acid-based lipid moiety to the surface-layer glycoprotein of *Haloferax volcanii* follows protein translocation. Biochem. J. *366*, 959–964.

Kuntz, C., Sonnenbichler, J., Sonnenbichler, I., Sumper, M., and Zeitler, R. (1997). Isolation and characterization of dolichol-linked oligosaccharides from *Haloferax volcanii*. Glycobiology 7, 897–904.

Lanyi, J.K. (1974), Salt-dependent properties of proteins from extremely halophilic bacteria. Bacteriol. Rev. *38*, 272–290.

Lechner, J., and Sumper, M. (1987). The primary structure of a procaryotic glycoprotein. Cloning and sequencing of the cell surface glycoprotein gene of Halobacteria. J. Biol. Chem. *262*, 9724–9729.

Lechner, J., and Wieland, F. (1989). Structure and biosynthesis of prokaryotic glycoproteins. Annu. Rev. Biochem. *58*, 173–194.

Lechner, J., Wieland, F., and Sumper, M. (1985a) Biosynthesis of sulfated saccharides N-glycosidically linked to the protein via glucose. Purification and identification of sulfated dolichyl monophosphoryl tetrasaccharides from Halobacteria. J. Biol. Chem. *260*, 860–866.

Lechner, J., Wieland, F., and Sumper, M. (1985b). Transient methylation of dolichyl oligosaccharides is an obligatory step in Halobacterial sulfated glycoprotein biosynthesis. J. Biol. Chem. *260*, 8984–8989.

Magidovich, H., Yurist-Doutsch, S., Konrad, Z., Ventura, V.V., Hitchen, P.G., Dell, A., and Eichler, J. (2010). AglP is a S-adenosyl-L-methionine-dependent methyltransferase that participates in the N-glycosylation pathway of *Haloferax volcanii*. Mol. Microbiol. *76*, 190–199.

Mengele, R., and Sumper, M. (1992). Drastic differences in glycosylation of related S-layer glycoproteins from moderate and extreme halophiles. J. Biol. Chem. *267*, 8182–8185.

Mescher, M.F., and Strominger, J.L. (1976a). Structural (shape-maintaining) role of the cell surface glycoprotein of *Halobacterium salinarium*. Proc. Natl. Acad. Sci. U.S.A. *73*, 2687–2691.

Mescher, M.F., and Strominger, J.L. (1976b). Purification and characterization of a prokaryotic glycoprotein from the cell envelope of *Halobacterium salinarium*. J. Biol. Chem. *251*, 2005–2014.

Mescher, M.F., Strominger, J.L., and Watson, S.W. (1974). Protein and carbohydrate composition of the cell envelope of *Halobacterium salinarium*. J. Bacteriol. *120*, 945–954.

Mescher, M.F., Hansen, U., and Strominger, J.L. (1976). Formation of lipid-linked sugar compounds in *Halobacterium salinarium*. Presumed intermediates in glycoprotein synthesis. J. Biol. Chem. *251*, 7289–7294.

Niemetz, R., Karcher, U., Kandler, O., Tindall, B.J., and König, H. (1997). The cell wall polymer of the extremely halophilic archaeon *Natronococcus occultus*. Eur. J. Biochem. *249*, 905–911.

Ng, W.V., Kennedy, S.P., Mahairas, G.G., Berquist, B., Pan, M., Shukla, H.D., Lasky, S.R., Baliga, N.S., Thorsson, V., Sbrogna, J., *et al.* (2000). Genome sequence of *Halobacterium* species NRC-1. Proc. Natl. Acad. Sci. U.S.A. *97*, 12176–12181.

Oesterhelt, D., and Stoeckenius, W. (1971). Rhodopsin-like protein from the purple membrane of *Halobacterium halobium*. Nat. New Biol. *233*, 149–152.

Oren, A. (2002). Diversity of halophilic microorganisms: environments, phylogeny, physiology, and applications. J. Ind. Microbiol. Biotechnol. *28*, 56–63.

Oren, A., Duker, S., and Ritter, S. (1996). The polar lipid composition of Walsby's square bacterium. FEMS Microbiol. Lett. *138*, 135–140.

Paul, G.F., Lottspeich, F., and Wieland, F. (1986). Asparaginyl-N-acetylgalactosamine. Linkage unit of Halobacterial glycosaminoglycan. J. Biol. Chem. *261*, 1020–1024.

Sampathkumar, S.G., Li, A., and Yarema, K.J. (2006). Sialic acid and the central nervous system: perspectives on biological functions, detection, imaging methods and manipulation. CNS Neurol. Disord. Drug Targets *5*, 425–440.

Saponetti, M.S., Bobba, F., Salerno, G., Scarfato, A., Corcelli, A., and Cucolo, A. (2011). Morphological and structural aspects of the extremely halophilic archaeon *Haloquadratum walsbyi*. PLoS One *6*, e18653.

Schleifer, K.H., Steber, J., and Mayer, H. (1982). Chemical composition and structure of the cell wall of *Halococcus morrhuae*. Zentralbl. Bakt. Mikrobiol. Hyg. Ser. C3, 171–178.

Sehgal, S.N., Kates, M., and Gibbons, N.E. (1962). Lipids of *Halobacterium cutirubrum*. Can. J. Biochem. Physiol. *40*, 69–81.

Sheehan, J.K., Thornton, D.J., Somerville, M., and Carlstedt, I. (1991) Mucin structure. The structure and heterogeneity of respiratory mucus glycoproteins. Am. Rev. Respir. Dis. *144*, S4–S9.

Soppa, J. (2006). From genomes to function: haloarchaea as model organisms. Microbiology *152*, 585–590.

Steber, J., and Schleifer, K.H. (1975). *Halococcus morrhuae*: a sulfated heteropolysaccharide as the structural component of the bacterial cell wall. Arch. Microbiol. *105*, 173–177.

Steber, J., and Schleifer, K.H. (1979). N-glycyl-glucosamine: A novel constituent in the cell wall of *Halococcus morrhuae*. Arch. Microbiol. *123*, 209–212.

Steensland, H., and Larsen, H. (1969). A study of the cell envelope of the Halobacteria. J. Gen. Microbiol. *55*, 325–336.

Stoeckenius, W., and Rowen, R. (1967). A morphological study of *Halobacterium halobium* and its lysis in media of low salt concentration. J. Cell Biol. *34*, 365–393.

Sumper, M. (1987). Halobacterial glycoprotein biosynthesis. Biochim. Biophys. Acta *906*, 69–79.

Sumper, M., Berg, E., Mengele, R., and Strobel, I. (1990). Primary structure and glycosylation of the S-layer protein of *Haloferax volcanii*. J. Bacteriol. *172*, 7111–7118.

Szymanski, C.M., and Wren, B.W. (2005). Protein glycosylation in bacterial mucosal pathogens. Nat. Rev. Microbiol. *3*, 225–237.

Tindall, B.J., Ross, H.M.N., and Grant, W.D. (1984). *Natronobacterium* gen. nov. and *Natronococcus* gen. nov. – two genera of haloalkaliphilic archaebacteria. Syst. Appl. Microbiol. *5*, 41–57.

Trachtenberg, S., Pinnick, B., and Kessel, M. (2000). The cell surface glycoprotein layer of the extreme halophile *Halobacterium salinarum* and its relation to *Haloferax volcanii*: cryo-electron tomography of freeze-substituted cells and projection studies of negatively stained envelopes. J. Struct. Biol. *130*, 10–26.

Varki, A. (1998). Factors controlling the glycosylation potential of the Golgi apparatus. Trends Cell Biol. *8*, 34–40.

Vimr, E.R., Kalivoda, K.A., Deszo, E.L., and Steenbergen, S.M. (2004). Diversity of microbial sialic acid metabolism. Microbiol. Mol. Biol. Rev. *68*, 132–153.

Wakai, H., Nakamura, S., Kawasaki, H., Takada, K., Mizutani, S., Aono, R., and Horikoshi, K. (1997). Cloning and sequencing of the gene encoding the cell surface glycoprotein of *Haloarcula japonica* strain TR-1. Extremophiles *1*, 29–35.

Weerapana, E., and Imperiali, B. (2006). Asparagine-linked protein glycosylation: from eukaryotic to prokaryotic systems. Glycobiology *16*, 91R-101R.

Woese, C.R., and Fox, G.E. (1977). Phylogenetic structure of the prokaryotic domain: the primary kingdoms. Proc. Natl. Acad. Sci. U.S.A. *74*, 5088–5090.

Woese, C.R., Kandler, O., and Wheelis, M.L. (1990). Towards a natural system of organisms: proposal for the domains Archaea, Bacteria, and Eucarya. Proc. Natl. Acad. Sci. U.S.A. *87*, 4576–4579.

Wolfe, R.S. (2006). The archaea: a personal overview of the formative years. In The Prokaryotes. Archaea. Bacteria: Firmicutes, Actinomycetes, Falkow, S., Rosenberg, E., Schleifer K.H., and Stackebrandt, E., eds. (Springer, Berlin), vol. 3, pp. 3–9.

Yan, A., and Lennarz, W.J. (2005). Unraveling the mechanism of protein N-glycosylation. J. Biol. Chem. *280*, 3121–3124.

Yang, L.L., and Haug, A. (1979). Purification and partial characterization of a procaryotic glycoprotein from the plasma membrane of *Thermoplasma acidophilum*. Biochim. Biophys. Acta *556*, 265–277.

Yurist-Doutsch, S., and Eichler, J. (2009). Manual annotation, transcriptional analysis and protein expression studies reveal novel genes in the *agl* cluster responsible for N-glycosylation in the halophilic archaeon *Haloferax volcanii*. J. Bacteriol. *191*, 3068–3075.

Yurist-Doutsch, S., Abu-Qarn, M., Battaglia, F., Morris, H.R., Hitchen, P.G., Dell, A., and Eichler, J. (2008). *aglF*, *aglG* and *aglI*, novel members of a gene cluster involved in the N-glycosylation of the *Haloferax volcanii* S-layer glycoprotein. Mol. Microbiol. *69*, 1234–1245.

Yurist-Doutsch, S., Magidovich, H., Ventura, V.V., Hitchen, P.G., Dell, A., and Eichler, J. (2010). N-glycosylation in Archaea: on the coordinated actions of *Haloferax volcanii* AglF and AglM. Mol. Microbiol. *75*, 1047–1058.

Zeitler, R., Hochmuth, E., Deutzmann, R., and Sumper, M. (1998). Exchange of Ser-4 for Val, Leu or Asn in the sequon Asn-Ala-Ser does not prevent N-glycosylation of the cell surface glycoprotein from *Halobacterium halobium*. Glycobiology *8*, 1157–1164.

Zhu, C.R., Drake, R.R., Schweingruber, H., and Laine, R.A. (1995) Inhibition of glycosylation by amphomycin and sugar nucleotide analogs PP36 and PP55 indicates that *Haloferax volcanii* β-glycosylates both glycoproteins and glycolipids through lipid-linked sugar intermediates. Arch. Biochem. Biophys. *319*, 355–364.

Zolghadr, B., Klingl, A., Rachel, R., Driessen, A.J., and Albers, S.V. (2011). The bindosome is a structural component of the *Sulfolobus solfataricus* cell envelope. Extremophiles *15*, 235–244.

Cell Cycle and Polyploidy in Haloarchaea

7

Karolin Zerulla, Anke Baumann and Jörg Soppa

Abstract

Halobacterium salinarum has a very strict cell cycle control and stops cell division when replication is inhibited. Synchronized cultures were used to quantify cell cycle-specific changes of the transcriptome and the proteome. The number of cycling transcripts and proteins is much smaller than in other species. An explanation might be the surprising finding that *Hbt. salinarum* does not have an S-phase but replicates constitutively. It is the first species found to have replication uncoupled from other cell cycle processes like intracellular DNA transport and cell division. *Hbt. salinarum*, *Haloferax volcanii* and *Haloferax mediterranei* are all polyploid with 15 to 25 copies of their major chromosome. It is tempting to speculate that polyploidy might be typical for haloarchaea. The copy numbers of the minor chromosomes and plasmids differ from those of the major chromosome and from one another, thus the dosage of haloarchaeal genes depends on their localization on different replicons and can differ by as much as a factor of five. Nine different possible evolutionary advantages of polyploidy for haloarchaea are discussed, including a low mutation rate, high desiccation/X-ray irradiation resistance, survival over geological times or at extraterrestrial places, gene redundancy enabling the existence of heterozygous cells, relaxation of replication control, global gene dosage control, statistical instead of stochastical regulation of gene expression, and the usage of DNA as phosphate storage polymer. The majority of these evolutionary advantages require the presence of intermolecular gene conversion, which could indeed be shown to exist in haloarchaea.

Introduction

The cell cycle has to be tightly regulated to ensure that the cells are constant in size and genetic material over many generations. Specifically it has to be guaranteed that the cell mass doubles after the birth of cells until the next cell division, that the entire DNA is replicated exactly once, that DNA segregation has occurred with high fidelity so that the two daughter cells get identical genetic material, and that a septum is formed at the right time at mid-cell. Cell cycle regulation has been extensively studied in eukaryotes and bacteria, and a few reviews summarize the state of the art (eukaryotes: Cai and Tu, 2012; Cross *et al.*, 2011; David *et al.*, 2012; Uhlmann *et al.*, 2011; bacteria: Errington, 2010; Jenal, 2009; Thanbichler, 2010; Touzain *et al.*, 2011; Zaritsky *et al.*, 2011). In Archaea, cell cycle research has been restricted to the crenarchaeon *Sulfolobus acidocaldarius* and the euryarchaeon *Halobacterium salinarum*. Recent reviews summarize results obtained with *Sulfolobus* (reviews:

Samson and Bell, 2011; Bernander, 2007), while the results obtained with *Hbt. salinarum* are summarized in this contribution.

Cell cycle-specific intracellular DNA transport has led to the hypothesis that *Hbt. salinarum* only has a single copy of its chromosome and is thus monoploid. Therefore it was a surprise when quantification of the chromosome copy number using two different methods led to the discovery that *Hbt. salinarum* and *Haloferax volcanii* are both polyploid (Breuert *et al.*, 2006). Quantification of the chromosome copy number of seven species of Euryarchaeota of six genera revealed that none of them is monoploid and thus oligoploidy or polyploidy seems to be widespread and typical for this kingdom of Archaea (Soppa, 2011). In contrast, all analysed Crenarchaeota (seven species of four genera) were found to be monoploid and thus there is a stark dichotomy in the archaeal phylogenetic lineages. This is different in bacteria: monoploid, oligoploid and polyploid species are intermixed in various phylogenetic groups. However, only a minor fraction of bacterial species is monoploid and thus oligoploidy/polyploidy is typical for bacteria. This is true for different groups of bacteria, e.g. Proteobacteria, Cyanobacteria, and Gram-positive bacteria (Griese *et al.*, 2011; Pecararo *et al.*, 2011; Böttinger, B., Zerulla, K., and Soppa, J., unpublished data). Therefore, it seems that polyploidy evolved several times independently in various groups of prokaryotes. In this contribution we focus on haloarchaea and summarize recent results about the regulated ploidy of various replicons of *Hbt. salinarum* and two species of *Haloferax*. In addition, nine possible evolutionary advantages of polyploidy for haloarchaea are discussed. Last but not least, it is shown that haloarchaea possess an effective system for intermolecular gene conversion, which is a prerequisite for the majority of the discussed evolutionary advantages. Finally, an outlook is given and future trends are discussed.

The cell cycle of *Halobacterium salinarum*

Synchronization of cell cultures

A method for the synchronization of *Hbt. salinarum* cultures has been established, which allowed to characterize cell cycle-specific changes of gene expression as well as additional cell cycle-related phenomena (Herrmann and Soppa, 2002; Baumann and Soppa, 2007; A. Baumann and J. Soppa, unpublished data). The method makes use of aphidicolin, an inhibitor of eukaryotic replication that was shown to block DNA synthesis also in *Hbt. salinarum* (Forterre *et al.*, 1984). *Hbt. salinarum* has a strictly controlled cell cycle, thus the replication block results in a total inhibition of cell division. Other cellular processes remain undisturbed, and therefore incubation of a haloarchaeal cell culture with aphidicolin leads to a steady increase in the average cell length. When the inhibitor is removed after an optimized time of incubation, all cells of the culture divide synchronously. Fig. 7.1 shows the average cell length as well as the cell density during the incubation with aphidicolin during two subsequent cell cycles. As is typical for whole culture synchronization methods, the degree of synchronization is much lower during the second cell cycle and synchronization is totally lost soon thereafter. However, the degree of synchronization is about 100% during the first cell cycle, and thus the first cycle was used to obtain the results described in the following paragraphs. The high degree of synchrony was also verified by a third criterion, i.e. microscopic identification of cells in the process of septum formation, which was exclusively found in the boxed time windows of Fig. 7.1.

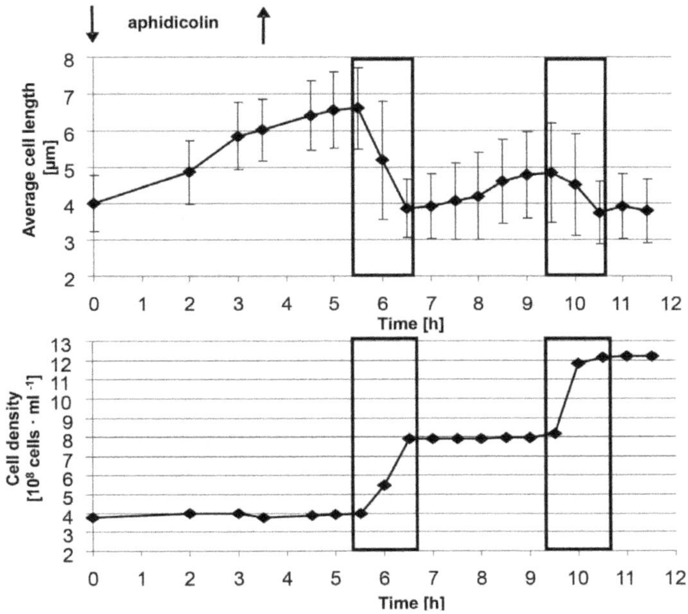

Figure 7.1 Synchronization of *Hbt. salinarum* cultures. *Hbt. salinarum* cultures can be synchronized by treating the cells for an optimized time with the replication inhibitor aphidicolin (see arrows). Synchronization is analysed by quantification of the average cell length (top panel) and the cell density (bottom panel). Cells with visible constrictions, which are in the process of septum formation, are exclusively observed during the boxed time periods. The first two cell cycles after inhibitor removal are shown.

Notably *Hbt. salinarum* is one of only very few prokaryotic species for which an efficient synchronization procedure has been established. Well-studied bacterial species for which a synchronization method is available include *Escherichia coli* which can be synchronized by isolating newborn cells with a 'baby machine', *Caulobacter crescentus* which can be synchronized by isolating swarmer cells, *Bacillus subtilis*, which can be synchronized making use of the outgrowth of spores, and *Synechococcus* which can be synchronized by daily cycles of light and darkness. The only additional archaeal species for which a synchronization method has been established is *Sulfolobus acidocaldarius*. Initially dilution of stationary phase cultures was used to generate partially synchronized cultures (Hjort and Bernander, 1999), but recently usage of a 'baby machine' has been introduced to generate highly synchronized cultures (Duggin *et al.*, 2008). It should be noted that *S. solfataricus* is a representative of the kingdom of Crenarchaeota, while *Hbt. salinarum* belongs to the kingdom of Euryarchaeota. The availability of synchronization procedures for one species of each kingdom has revealed that an 'archaeal cell cycle' does not exist, but that many species- or kingdom-specific differences are found. Some differences will be mentioned below; a detailed discussion about the cell cycle of *Sulfolobus* is given in several recent reviews (Bernander, 2007; Samson *et al.*, 2011).

Cell cycle specific changes of the transcriptome and the proteome

To get a global overview of cell cycle-dependent transcript level regulation, synchronized cultures were used to analyse transcriptome changes using a DNA microarray (Baumann *et al.*, 2007). After removal of aphidicolin, aliquots were taken every 30 minutes for five hours, so that the first synchronized cycle was completely covered. At all eleven time points the transcriptomes of the synchronized culture were compared to a control unsynchronized culture. Three biological replicates were performed and average values were calculated. Transcripts of nearly 2500 genes could be analysed, which represent 88% of all ~2800 genes of *Hbt. salinarum*. Unexpectedly, only 87 of the transcript levels were cell cycle-regulated. These 87 genes could be clustered into seven groups of kinetically co-regulated genes. The regulatory profiles of the first four clusters are shown in Fig. 7.2. Often several or all members of gene clusters or operons were grouped into the same cluster, e.g. all five genes of the ABC transporter operon OE4300–OE4304. Cluster 1 contains all three subunits of the glycerol-3-phosphate dehydrogenase (OE3763–OE3765). Cluster 3 is comprised exclusively of genes encoding enzymes involved in purine biosynthesis. Counterintuitively, the genes for pyrimidine biosynthesis enzymes are not cell cycle-regulated, but constitutively expressed.

Overall the number of cell cycle-regulated transcripts is only 3% smaller than in other species, e.g. the archaeon *Sulfolobus acidocaldarius* (18%), the bacterium *Caulobacter crescentus* (19%), or the eukaryotes *Schizosaccharomyces pombe* (10%) and *Homo sapiens*

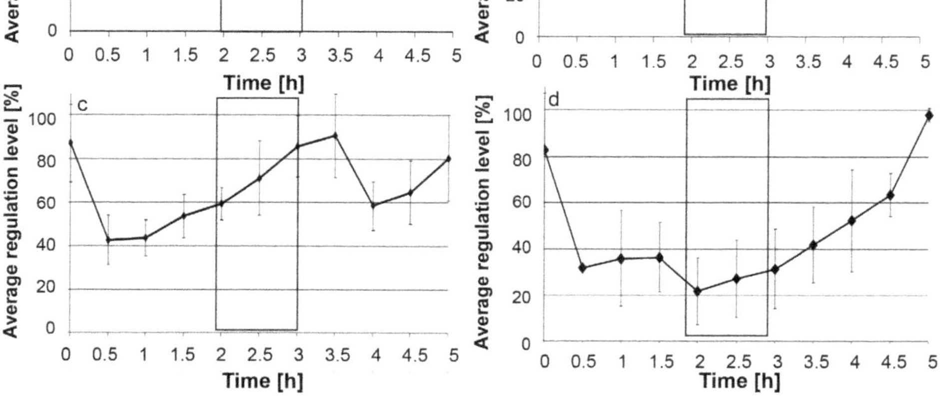

Figure 7.2 Cell cycle-specific transcriptome changes. *Hbt. salinarum* cultures were synchronized and at the indicated time points aliquots were removed. RNA was isolated and compared to the RNA of a mock-treated cultures using DNA microarrays. Transcript levels were analysed and 87 transcripts were found to have cell cycle-specific concentration changes. These transcripts could be sorted into seven groups of kinetically co-regulated genes. Average values of four of these groups are shown (modified after Baumann *et al.*, 2007).

(18–28%) (Cho *et al.*, 2001; Laub *et al.*, 2000; Lundgren and Bernander, 2007; Marguerat *et al.*, 2006; Whitfield *et al.*, 2002). Noteworthy, the levels of many transcripts that are cell cycle-regulated in other species are constitutively expressed in *Hbt. salinarum*. For example, eight out of nine transcripts encoding Cdc6 homologues have a constant level throughout the cell cycle. The same is true for histones and DNA polymerases. This might be correlated with the lack of a clear S-phase in *Hbt. salinarum*, as described below (see 'Relaxation of replication control' below).

In addition to the characterization of transcriptome changes cell cycle-specific changes of the proteome were analysed (Baumann, A., and Soppa, J., unpublished data). Again, cultures were synchronized and 14 aliquots were removed that represented one complete cell cycle. 2D differential gel electrophoresis (DIGE) was used to quantify protein level changes. A mixed sample comprised of equal aliquots of all time points was used as internal control and labelled with one fluorescent dye. Two other fluorescent dyes were used to label the extracts of two different time points. All three samples were mixed, separated by 2D gel electrophoresis, and scanned at three different excitation wavelengths. The values of the two time points were normalized using the internal control. As the experiment was performed in triplicates, in total 42 samples were analysed on 21 2D gels. More than 600 protein spots could be quantified at each time point in each of the three replicates. The results of the three biological replicates were averaged and the time series was examined for proteins with cyclic concentration changes. Unexpectedly, less than 30 proteins out of the more than 600 analysed exhibited cell cycle-dependent cyclic concentration changes. Several of them were analysed by peptide mass fingerprinting, but none had a cell cycle-related function. Taken together, similar to the transcriptome analyses also the proteome analyses indicated that *Hbt. salinarum* differs from other prokaryotes and eukaryotes as the fraction of cell cycle-specific concentration changes of its constituents is much smaller.

Cell cycle specific DNA transport

Synchronized cultures were also used to analyse the intracellular localization of the DNA. The localization was not constant but that the DNA was dynamically transported within the cell (Fig. 7.3; Herrmann and Soppa, 2002). First, the DNA was evenly distributed throughout the cell. Later, it was concentrated in the middle of the cell, and subsequently it was transported towards the cell poles, so that the middle of the cell was devoid of DNA. After that, a septum formed at mid-cell giving rise to two newborn daughter cells of equal size and DNA content At first these results were thought to be very similar to the 'factory model of replication' that has been described for *B. subtilis* (Lemon and Grossman, 1998). In this view the replicating DNA polymerases are localized exclusively at or near mid-cell. The un-replicated DNA is transported to mid-cell, is replicated there and the replicated chromosomes are transported to the cell poles, starting with the origin region and ending with the terminus region. However, subsequently it was revealed that *Hbt. salinarum* does not have an S-phase, but replication takes place at all time points shown in Fig. 7.3 (see 'Relaxation of replication control' below). In addition, the chromosome of *Hbt. salinarum* has more than one replication origin (Coker *et al.*, 2009), and thus the simple model that is true for a bacterium with a single replication origin like *B. subtilis* cannot apply to *Hbt. salinarum*. Taken together, the results indicate that *Hbt. salinarum* has totally uncoupled replication from segregation, and that segregation requires concentration of all DNA at mid-cell before equal amounts are transported to the two cell poles.

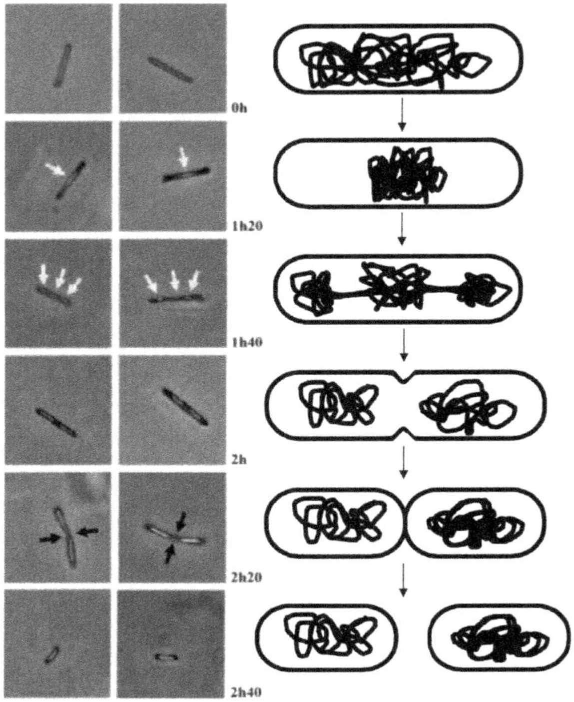

Figure 7.3 Cell cycle-specific intracellular DNA transport. *Hbt. salinarum* cultures were synchronized and aliquots were removed at different time points throughout the cell cycle. The cells were fixed, DNA was stained and the cells were observed with a fluorescence microscope. Left panel: Two typical cells are shown for selected time points (overlay of microscopic and fluorescence pictures). At two time points fluorescent foci are indicated by white arrows, at one time point the DNA-free area at mid-cell is indicated with black arrows. Right panel: A schematic overview of cell cycle-specific intracellular DNA transport.

Cell cycle specific oscillation of the cAMP concentration

In addition to transcript and protein levels also the level of cAMP was quantified throughout the cell cycle making use of synchronized cultures of *Hbt. salinarum*. Interestingly it was revealed that it was constant throughout most of the cell cycle, but that there were two very short peaks of cAMP concentration increase at the beginning and at the end of the period in which all cells divided (see figure 7 in Baumann *et al.*, 2007). In both peaks the increase was about a factor of two compared with the basal cAMP level. The results indicate that cAMP is used as a signal important for cell cycle progression at two times in the cell cycle. Whether cAMP is also a starvation signal in haloarchaea, like in *E. coli* and other bacteria, is currently unknown. If it would be so, cAMP could integrate carbon source supply with cell cycle progression. Cell cycle-dependent cAMP fluctuations have also been observed in *Saccharomyces cerevisiae* and in *Tetrahymena pyriformis*, and an integrative function of cAMP has also been hypothesized for these eukaryotes (Dickinson *et al.*, 1977; Müller *et al.*, 2003; Smith *et al.*, 1990; Watson and Berry, 1977).

Regulated polyploidy in haloarchaea

Growth phase dependent ploidy of *Halobacterium salinarum* and *Haloferax* species

The cell cycle-dependent intracellular DNA transport (see above) had led to the hypothesis that *Hbt. salinarum* is monoploid. Therefore, it was surprising that quantification of the genome copy number revealed that *Hbt. salinarum* is polyploid and contains about 25 genome copies in exponential phase and 15 genome copies in stationary phase (Breuert *et al.*, 2006). The results were obtained using two independent methods, i.e. quantitative Southern blotting and a newly developed Real Time PCR method. Fig. 7.4 shows the quantification of five different sites of the chromosome using Real Time PCR. The results are virtually identical irrespective of the localization near or far from an origin, in contrast to fast growing *E. coli* cells that are mero-oligoploid and harbour more origins than termini. Quantification of the copy numbers of the three megaplasmids of *Hbt. salinarum* revealed that the growth phase-dependent differential copy number regulation of pHs1 is even larger than that of the chromosome, while the copy number of pHs2 and pHs3 is about five and is totally independent of the growth phase (compare Fig. 7.5). Therefore, the gene dosage of *Hbt. salinarum* genes depends on the replicon and the growth phase and spans the numbers from 5 to 25.

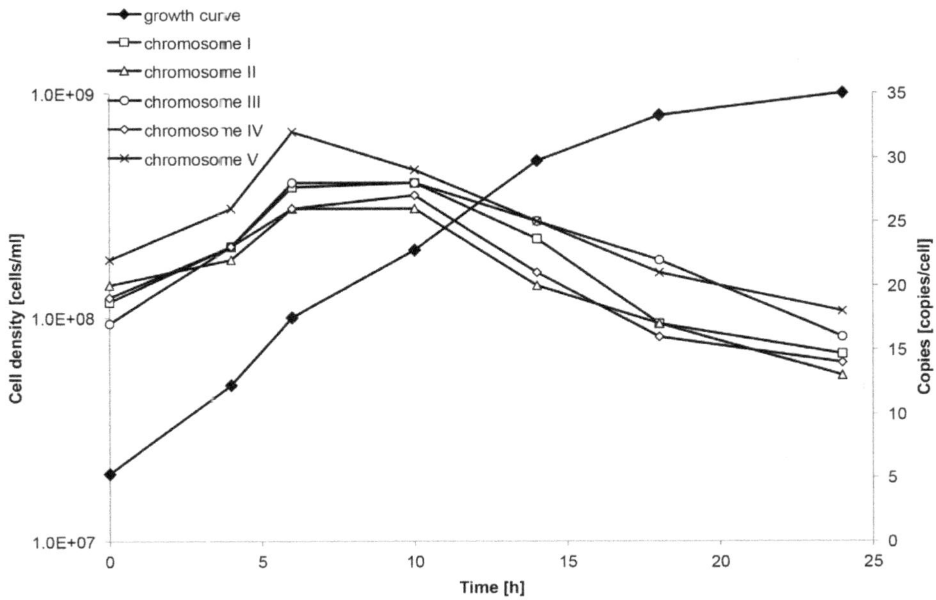

Figure 7.4 Quantification of the copy number of five sites distributed over the major chromosome of *Hbt. salinarum*. A growth curve of a *Hbt. salinarum* cultures is shown (left Y-axis). At five sites the copy number of the major chromosome was quantified using the Real Time PCR method described by Breuert *et al.* (2006), and the copy number at different time points is plotted versus time (right Y-axis).

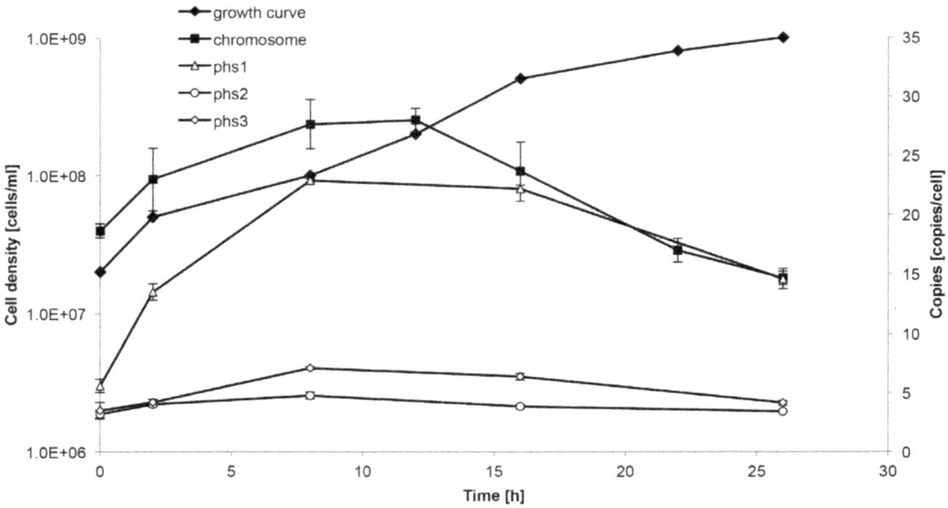

Figure 7.5 Copy number of *Hbt. salinarum* replicons. A growth curve of a *Hbt. salinarum* cultures is shown (left Y-axis). The copy numbers of the four different replicons of *Hbt. salinarum* were quantified at different times of the growth curve and the copies numbers are plotted versus time (right Y-axis) (reprinted from Breuert *et al.*, 2006).

Analysis of *Hfx. volcanii* replicons revealed that the copy numbers of the major chromosome as well as of the minichromosomes pHV1, pHV3 and pHV4 are growth phase-regulated (Zerulla, K., and Soppa, J., unpublished data). The numbers of all four replicons are down-regulated in stationary phase. Again, the dosage of *Hfx. volcanii* genes depends on the growth phase and the respective replicon and can be as low as eight (pHV4 in stationary phase) and as high as 50 (pHV3 in exponential phase). A third analysed species is *Haloferax mediterranei*. It is also polyploid and the copy number of the chromosome is also growth phase-regulated (Zerulla, K., and Soppa, J., unpublished results). Table 7.1 summarizes results obtained with the three species of two haloarchaeal genera. As all of them are polyploid and show growth phase-dependent copy number regulation, it is tempting to speculate that this might well be true for other species and might be characteristic for haloarchaea.

Single cell analysis of the ploidy level

Quantification methods like quantitative Southern blotting, real-time PCR or spectroscopic assays reveal average copy numbers of cell populations and cannot tell anything about the differences between single cells. Therefore, ten randomly selected colonies of *Hfx. volcanii* were used to inoculate cultures and to analyse the DNA amounts in single cells of these cultures using a Fluorescent Activated Cell Sorter (FACS). It turned out that the distribution of the DNA content was identical in all ten independent clones (Zerulla, K., and Soppa, J., unpublished results). Therefore, the copy number is highly regulated at the single cell level and the variance in a population of *Hfx. volcanii* cells is very small.

Table 7.1 Genome copy numbers and replicons of three haloarchaeal species

		Growth phase	
Organism	Location	Exponential	Stationary
Hbt. salinarum	CHR	28.0 ± 1.0	15.0 ± 0.6
	phs1	23.0 ± 0.1	14.0 ± 0.9
	phs2	4.8 ± 0.3	3.5 ± 0.1
	phs3	7.1 ± 0.1	4.2 ± 0.2
Hfx. volcanii	CHR	39.9 ± 5.3	27.9 ± 5.4
	pHV1	25.7 ± 3.4	10.3 ± 4.3
	pHV3	35.9 ± 5.5	28.2 ± 6.4
	pHV4	15.0 ± 2.1	7.8 ± 2.5
Hfx. mediterranei	CHR	20.3 ± 6.1	10.6 ± 3.8

Influence of external parameters on the ploidy level

Characterization of the chromosome copy number of the bacterium *Azotobacter vinelandii* had shown that only fast growing cells are polyploid, in contrast to slow-growing cells which are monoploid (Maldonado *et al.*, 1994). Therefore, the authors concluded that polyploidy of *A. vinelandii* might not exist outside of the laboratory.

To investigate whether the ploidy level of haloarchaea is also growth rate-dependent, *Hbt. salinarum* was grown at three different conditions. A fast-growing culture incubated aerobically under optimal conditions at 42°C was compared with two slowly growing cultures, one culture incubated at 30°C and one culture grown anaerobically by arginine fermentation. The chromosome copy numbers of all three cultures were identical in all growth phases, showing that for *Hbt. salinarum* the ploidy level is independent of growth rate (Breuert *et al.*, 2006). The same results were obtained with *Hfx. volcanii* that was grown with two different carbon sources (Zerulla, K., and Soppa, J., unpublished results). Again, the ploidy levels of fast-growing and slow-growing cells were identical underscoring that in haloarchaea the growth rate does not influence the chromosome copy number.

Evolutionary advantages of haloarchaeal polyploidy

In the following paragraphs nine possible evolutionary advantages are discussed that might have driven the development of polyploidy in haloarchaea. For all but one of these advantages experimental evidence at least in the form of model experiments exists. It should be noted that not all advantages apply to all species. For example, *Hbt. salinarum* and *Hfx. volcanii* are both polyploid and have similar numbers of chromosomes, but only the former is highly resistant to X-irradiation. Further evolutionary advantages exist that do not apply to haloarchaea and several of those are mentioned in the last paragraph.

Low mutation rates

When a mutation occurs in one copy of the chromosome of a polyploid species, another copy with the wild-type information can be used as a template for DNA repair. Species that make use of such a repair mechanism would have a lower apparent mutation rate than species that lack this pathway or monoploid species. For *Hfx. volcanii* it has in fact been observed that the mutation rate is much lower than in other species that have been analysed with the same genetic assay (Mackwan *et al.*, 2007). Spontaneous mutations in the *pyrE2* gene were selected based on the resistance of the resultant mutants against 5-fluoroorotic acid. A rate of about 2×10^{-8} mutations per cell division was observed for *pyrE2*, which was used to calculate a genomic mutation rate of 4.5×10^{-4}. This value is about one order of magnitude lower than the genomic mutation rates of $2–5 \times 10^{-3}$ determined for other species (Mackwan *et al.*, 2007).

Resistance against desiccation and irradiation

Double strand breaks (DSBs) are a much more severe kind of mutation compared to point mutations (mismatch mutations, alkylations, oxidations etc.). In the laboratory X-irradiation is commonly used to induce DSBs. However, DSBs are also induced by desiccation and thus are a common threat for terrestrial microorganisms. In addition, when replication forks pass single strand breaks, a double-strand break is produced in the daughter genome and therefore DSBs are very common in all organisms. It is long known that the bacterium *Deinococcus radiodurans*, which is oligoploid and contains about eight copies of the chromosome, is highly resistant to X-irradiation as well as desiccation (Hansen, 1978). Within hours it can restore complete chromosomes out of small overlapping fragments of totally shattered chromosomes. It could be shown that this is a two-stage process that includes DNA synthesis followed by homologous recombination (Zahradka *et al.*, 2006).

Hbt. salinarum was also shown to be highly resistant to ^{60}Co gamma irradiation as well as desiccation and high vacuum (Kottemann *et al.*, 2005). More than 20% of the cells survived 20 days of desiccation or incubation under high vacuum. The radiation resistance of a LD_{10} of 5 kGy is 20-fold higher than that of *E. coli*. The resistance of *Hbt. salinarum* is higher in the exponential phase than in stationary phase cells, which fits to the growth phase dependent regulation of the ploidy level described above.

Four cycles of gamma irradiation followed by regrowth led to the selection of two independent *Hbt. salinarum* mutants that had an increased resistance ($LD_{50} > 11$ kGy) and were described to have a higher resistance than any known natural isolate (Deveaux *et al.*, 2007). Both mutants had an elevated level of a single strand binding protein, in agreement with the involvement of homologous recombination in the repair of shattered chromosomes. Similar to *Deinococcus* also *Hbt. salinarum* can restore complete chromosomes out of overlapping fragments, albeit somewhat slower (Kottemann *et al.*, 2005).

While the ability to restore complete chromosomes from small fragments require overlaps and is thus possible only in oligoploid and polyploid species, it should be noted that there is no direct correlation between the number of chromosomes and the degree of resistance against DSBs. *Hbt. salinarum* harbours 3-fold more chromosomes than *D. radiodurans*, but is slightly less resistant to DSBs. The polyploid *Hfx. volcanii* is not at all resistant against gamma irradiation, underscoring that not all possible evolutionary advantages apply to all species that could theoretically make use of them. In contrast to its low resistance to X-irradiation, *Hfx. volcanii* was found to be very resistant to desiccation (Zerulla, K., and Soppa, J., unpublished results). Furthermore, the survival rate depended on the number of chromosomes present in the cell. Cells with only two chromosomes (generated by phosphate starvation, see 'DNA as a phosphate-storage polymer' below) survived a 12-day desiccation period much less efficiently than cells with the normal number of 20 chromosomes.

Survival over geological times

Several research groups have repetitively reported the isolation of haloarchaea from ancient halites of different ages, one as old as about 250 million years (Fendrihan *et al.*, 2006; McGenity *et al.*, 2000; Sankaranarayanan *et al.*, 2011; Schubert *et al.*, 2010; Stan-Lotter *et al.*, 2002; Vreeland *et al.*, 2007; and references therein). The earliest reports date back exactly 50 years (Dombrowski, 1963; Tasch, 1963). In addition, the isolation and characterization of DNA from salt of different geological ages has been reported, the oldest sample was claimed to have an age of 429 million years (Park *et al.*, 2009). However, the matter always has been and still is heavily debated. One major counterargument is that DNA simply does not have the chemical stability to survive extended times without destruction (Dose *et al.*, 1991; Heebsgaard *et al.*, 2005; Nickle *et al.*, 2002; review: Grant, 2004). Another major counterargument is that the 16S rRNA sequences of the isolated 'ancient' haloarchaea are extremely similar to modern haloarchaea and that they do not branch off very early in evolutionary trees. Therefore, applying normal evolution rates results in the conclusion that they cannot be as old as anticipated; on the contrary, they cannot be old at all (references as above). In summary, whatever the artefact may be, the reports of isolation of living haloarchaea from ancient halites must be false in spite of the extremely great care that was taken in external sterilization of salt samples prior to cultivation. In the extreme case the surfaces of salt samples were deliberately contaminated prior to sterilization to unambiguously proof that the sterilization protocol is 100% effective (Sankaranarayanan *et al.*, 2011).

However, the now-known fact that haloarchaea are polyploid might resolve the chemical stability contradiction. If double-strand breaks can easily be repaired using additional copies of the chromosome, a non-limited number of DSBs can occur without compromising the viability of salt-embedded haloarchaea. In addition, gene conversion (see below) might explain the similarity between ancient haloarchaea and recent species.

As the halites are not solid rocks but include brine-filled inclusions, the energy for DNA repair might stem from several sources. It has been proposed that haloarchaea might use glycerol and other resources that leak from co-included dead *Dunaliella* algae (Schubert *et al.*, 2009). Ultimately haloarchaea might cannibalize dead individuals of their own species. The energy needed for survival in brine-filled inclusions probably is extremely very low as growth does not occur and the chemical and physical conditions do not favour DNA damage, e.g. absence of UV light, reactive oxygen species, or alkylating chemicals from active metabolism. It has been observed that haloarchaea can form several small spherical particles from rod-shaped cells that have been proposed to represent dormant states (Fendrihan *et al.*, 2012). Noteworthy, such particles have also been observed in ancient halite. The controversy might never become resolved. However, polyploidy certainly adds another aspect to the topic of living haloarchaea from ancient halites.

Survival in extraterrestrial environments

For more than a decade it has been discussed whether haloarchaea might be able to grow or survive on other planets, e.g. on Mars, or might survive interplanetary transport, e.g. on meteorites. The discussion has been stimulated by the discoveries that Mars may still contain ground water and that long ago the conditions on Mars were much more similar to the current conditions on Earth than today. A boost experiment was performed in 1994 as a *Synechococcus* isolate and an archaeon originally designated *Haloarcula* sp. but now described as *Halorubrum chaoviator* (Mancinelli *et al.*, 2009) were exposed to a space environment for 15 days and survival rates were determined afterwards (Mancinelli *et al.*, 1998). For *Hrr. chaoviator* the survival rate was 10^{-7} for cells exposed to vacuum alone and 2×10^{-8} for cells exposed to vacuum and UV-irradiation. Of course, these rates are low, nevertheless they document that haloarchaea can survive real space conditions. It was also shown that several species of haloarchaea including *Halococcus dombrowskii* survive simulated Mars conditions (Fendrihan *et al.*, 2009). Similar results have also been reported for methanogenic Archaea (Kendrick and Kral, 2006; Morozova *et al.*, 2007), opening the fascinating possibility that microorganisms might exist on other planets and might have even survived interplanetary transport on meteorites (Dornmayer-Pfaffenhuemer *et al.*, 2011), e.g. from Mars to Earth. Of course, survival under these extreme conditions would probably depend on a very high capacity of repairing broken chromosomes, which is enhanced in polyploid species (see above).

Gene redundancy: a population of genomes instead a population of cells

In polyploid cells carrying several or many copies of each gene, the theoretical possibility exists that some copies can be mutated without losing the wild-type information on the remaining copies. This could be important under conditions when the cells are unable to grow and favourable mutations might restore the growth capacity. In growing cells the genomes could be equalized again by gene conversion, which is discussed below. While this is a theoretical consideration, at least the existence of heterozygous cells under specific selection conditions in the laboratory has been proven for haloarchaea as well as for other groups of polyploid prokaryotes. Selection in the absence of leucine and tryptophan led to a heterozygous *Hfx. volcanii* culture that contained two different types of genomes in every cell, one with the *leuB* gene at the *leuB* locus and the other with the *trpA* gene at the *leuB* locus (Lange *et al.*, 2011). Often heterozygous cells are the result of an attempt to replace a gene with a selection marker. If the respective gene turns out to be essential, heterozygous cells are selected that contain genomes with the selection marker as well as genomes with the essential gene. This has been observed for methanogenic Archaea (Hildenbrand *et al.*, 2011; Stock *et al.*, 2010) as well as for several species of Cyanobacteria (Labarre *et al.*, 1989; Nodop *et al.*, 2008; Spence *et al.*, 2004; Takahama *et al.*, 2004). The ease of selecting heterozygous polyploid cells in the laboratory in diverse species of prokaryotes indicates that heterozygous cells also arise in natural populations of polyploid prokaryotes. However, experimental evidence is still missing.

Relaxation of replication control

The cell cycle is divided into several phases, notably containing a so-called S-phase (synthesis phase) that is the time of replication. For monoploid cells with one replication origin it is important that replication is initiated faithfully once and only once every cell cycle. Only a tight replication control in combination with a tight segregation control guarantee that both daughter cells get a chromosome and that the number of chromosomes is constant over many generations. Polyploid cells theoretically do not have these necessities; both segregation control as well as replication control could be relaxed. And indeed two polyploid species have been described that have a relaxed segregation control. In *Synechocystis* PCC 6803 the two daughter cells do not get exactly the same amount of DNA, on average one daughter cell gets about 5% more than the other, but the difference can be much bigger and exceed the factor of two (Schneider *et al.*, 2007). The loss of control is even bigger in *Methanocaldococcus jannaschii*, which divides into cells of unequal size that can get very different amounts of DNA (Malandrin *et al.*, 1999). This is not the case for *Hbt. salinarum*, which forms a septum exactly at mid-cell and that distributes the DNA equally to both daughter cells (Breuert, S., and Soppa, J., unpublished results).

However, *Hbt. salinarum* has lost a strict cell-cycle-specific control of replication and does not have a S-phase. Eight synchronized cultures were incubated for 30 minutes with 5-bromo-2′-deoxyuridine (BrdU) and the amount of BrdU that was incorporated into genomic DNA during replication was quantified with an anti-BrdU antiserum. As shown in Fig. 7.6, no BrdU was incorporated in the first 30 minutes after removal of the inhibitor aphidicolin that was used for culture synchronization, but identical amounts of BrdU were incorporated in the seven 30-minute intervals thereafter. Notably, DNA synthesis took place before cell division, during cell division, and after cell division (Fig. 7.6). Equal DNA synthesis during the whole cell cycle of four hours offers the obvious advantage that the enzymatic capacity needed to produce 25 copies of the chromosome is much smaller than in the case that replication would be confined to a short S-phase. To our knowledge such a total lack of an S-phase has never been described before for any prokaryotic species. Only for the polyploid cyanobacterium *Synechococcus elongatus* PCC 7942 it has been described that replication initiation is not totally synchronized (Watanabe *et al.*, 2012). However, a peak time of DNA synthesis prior to cell division was observed, and replication was totally blocked in the dark. The lack of further examples is most probably due to the facts that only a small number of species can be synchronized and that a lack of an S-phase has not been assumed. We predict that analysis of DNA synthesis of further polyploid species will lead to the discovery of more examples of species with a relaxation or a loss of a tight control of replication, segregation and/or septum formation.

Global gene dosage control

The expression level of a gene is often correlated to the gene dosage. Therefore, regulation of the degree of polyploidy in response to growth phases or environmental conditions offers a way to concomitantly regulate the dosage of all genes and thus to globally regulate gene expression. If species contain several replicons, the dosage of genes can be replicon-specific. This effect is very pronounced for *Hbt. salinarum*, i.e. the main chromosome has a copy number of up to 25, while two other replicons have only a copy number of five (Fig. 7.5). In addition, the copy number of two replicons is growth-phase regulated, while the copy number of the other two replicons is independent of the growth phase. Therefore, the ratio of the genes on different replicons is not constant, but variable, e.g. the ratio between pHS1 and pHS2 varies between 1 and 4. It has not been experimentally tested whether the differential replicon-specific gene dosages really result in differences in gene expression, but this will be performed making use of a strain with a deletion of the *dhfr* (dihydrofolate reductase) gene, which can be used to place the *dhfr* gene on the different replicons and quantify the amount of the expression product DHFR. Differential regulation of the ploidy level has also been observed in other species. For example, fast growing *Methanococcus acetivorans* cells

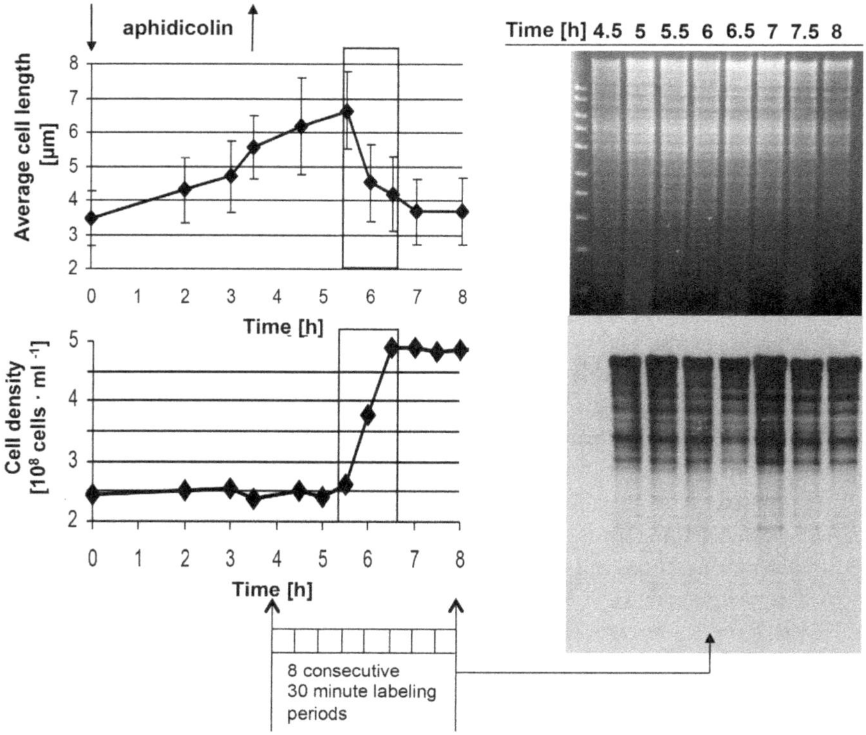

Figure 7.6 *Hbt. salinarum* lacks an S-phase. Eight *Hbt. salinarum* cultures were synchronized as described by Baumann *et al.* (2007) (left panels). For the four hours following removal of the inhibitor aphidicolin, each of the eight cultures was treated with 5-bromo-2′-deoxyuridine (BrdU) for 30 minutes and harvested directly afterwards. Genomic DNA was isolated from the eight cultures, digested with a restriction enzyme and run on an agarose gel (panel top right). The DNA was blotted to a nylon membrane and incorporated BrdU was visualized using an anti-BrdU antibody coupled with an enzymatic reaction with a chemiluminescence substrate (panel bottom right). No BrdU incorporation was observed in the first 30 minutes following inhibitor removal, while identical amounts were incorporated during the next seven 30 minute time intervals, which represent times before, during, and after cell division.

contain nearly 20 copies of the chromosome, in stark contrast to slow growing cells, which contain only 2–3 copies of the chromosome (Hildenbrand *et al.*, 2011). Another example is *A. vinelandii*, which contains 40 genome copies during fast growth and only one genome copy during slow growth (Maldonado *et al.*, 1994). Also in these cases it is tempting to speculate that the very high changes in gene dosage could be correlated with expression levels, but also in these cases experimental verifications have not been performed. Nevertheless, global regulation of the gene dosage in polyploid species in response to growth phase and/or external conditions does not seem to be exceptional, but rather the rule.

Shifting regulation from stochastics to statistics

Monoploid species contain just one copy of each gene, including the regulatory sequences upstream of the coding region. If the intracellular concentrations of transcription factors are low, which is typically the case, then gene expression becomes stochastic in two different aspects. First, the average concentration of transcription factors can be around or even lower than one molecule per cell and thus a fraction of the cells does not contain a single molecule. A quantitative determination of 1000 different YFP tagged proteins in single cells of *E. coli* revealed that the average intracellular protein amount ranges from 0.1 molecule per cell to 10,000 molecules per cell (Taniguchi *et al.*, 2010). Second, the interaction between transcription factors and their respective recognition motif in the genome is a binding equilibrium determined by association and dissociation rates. At the low concentrations of both binding partners present in monoploid species there are phases when the transcription factor is bound and transcription is blocked and phases when it is dissociated from the DNA and transcription happens. Thus gene expression happens as a probabilistic burst of the synthesis of mRNA and concomitantly protein. The development in recent years of single cell analysis methods that are sensitive enough to detect single molecules of mRNAs and proteins have allowed the experimental verification that gene expression in *E. coli* and other bacteria really happens in this stochastic fashion (Tanigushi *et al.*, 2010; review: Broude, 2011). One consequence is that genetically identical individual bacteria in a population are heterogeneous concerning their intracellular concentrations of mRNAs and proteins and can be heterogeneous concerning their phenotypes (de Jong *et al.*, 2012).

This situation totally changes when the concentrations of the two binding partners, DNA recognition motifs as well as transcription factors, are considerably higher. In haloarchaea the number of genome copies is 20–25 and thus more than one order of magnitude higher than in monoploid species. The intracellular concentration of transcription factors is not known, but it seems likely that the number of molecules is at least as great as the number of DNA binding sites, otherwise a complete inhibition of gene expression would not be possible. However, a complete lack of gene expression has been observed for a variety of haloarchaeal genes in the absence of a cognate inducer. Therefore, also the intracellular concentration of transcription factors in haloarchaea might be one order of magnitude higher than in monoploid species. If that were true, then the enhanced concentration of both binding partners would results in a different mode of gene regulation than in monoploid species, it would not be the stochastic on/off switching described above for *E. coli* but a statistical smooth regulation of the strength of gene expression. In addition, the heterogeneity of cellular composition and phenotypes of a genetically identical population of haloarchaea would be much smaller than a population of monoploid species. Experimental verification of this potential evolutionary advantage of polyploidy in haloarchaea would not only require the quantification of transcription factors, but also the development of single cell techniques with single molecule sensitivity, which is complicated by the high intracellular salt concentration.

DNA as a phosphate-storage polymer

Polyploid cells contain much more phosphate, carbon and nitrogen bound in their DNA as monoploid cells. It could be shown that *Hfx. volcanii* can grow for several generations in synthetic medium in the absence of phosphate. Under these conditions *Hfx. volcanii* sacrifices all genetic advantages of polyploidy discussed above and reduces the number of chromosomes from 30 to only 2 (Zerulla, K., Chimileski, S., Gophna, U., Papke, T., and Soppa, J., in revision). Upon re-addition of phosphate to phosphate-starved cells the number of chromosomes increases very fast and reaches the original level within a few hours. These results reveal that *Hfx. volcanii* and probably also other polyploid haloarchaea use their genomic DNA not only as genetic material, but also as a phosphate-storage polymer. It is currently believed that DNA was added to the biology of cells because it is much more stable than RNA and therefore replaced the original RNA genomes. Based on the results obtained with *Hfx. volcanii* an alterative view seems possible. DNA might have been added to the biology of cells early in evolution as a phosphate-storage polymer because it is much more stable than the alternative phosphate-storage polymer polyphosphate and might have evolved into its role as genetic material only later.

Further advantages not applying to haloarchaea

The relaxation of the DNA segregation control observed in *M. jannaschii* and *S. elongatus* has already been discussed above. Another advantage is the possibility to enlarge the size of cells. An extreme example is *Epulopiscium* sp. that can reach cell lengths of longer than 600 μm and cell volumes five orders of magnitude larger than *E. coli*. The genome copy number correlates with the cell volume and is extremely high, 'small' cells contain about 50,000 genome copies and large cells about 200,000 genome copies (Mendell *et al.*, 2008). Polyploidy is probably essential for very large prokaryotic cells because diffusion would not allow an equal distribution of mRNAs and proteins in the cell if it would contain only a single chromosome. In line with this reasoning the chromosomes of *Epulopiscium* sp. are arranged at the cell periphery all around the cell. The evolutionary advantage of large prokaryotic cells is that they escape predators of normal-sized prokaryotes.

Another putative evolutionary advantage is the possibility to perform several cell divisions without the need to replicate the DNA between them. This has been discussed as an advantage for pathogens soon after entering a host. Using this strategy a single cell could multiply very fast into several cells and enhance the chance that one or more of the descendants will escape the immune system of the host and firmly establish itself in the host.

Gene conversion and escape from 'Muller's ratchet'

The possible evolutionary advantages of polyploidy described in paragraphs 4.1 to 4.5 all require that the information of one copy of the chromosome can be transported to another copy of the chromosome, a process called 'gene conversion'. Gene conversion is defined as the non-reciprocal transfer of information between two homologous sequences. Gene conversion has mostly been studied in eukaryotes, somewhat in bacteria, and hardly at all in Archaea (Lawson *et al.*, 2009). Gene conversion has been detected as a process occurring during meiosis in fungi about 50 years ago (Holliday, 1964). In bacteria gene conversion is nearly exclusively studied as an intramolecular interaction between two sites of the same chromosome, leading to the concerted evolution of ribosomal RNA operons and gene families as well as being one of two molecular mechanisms underlying antigenic or phase variation (Santoyo and Romero, 2005). However, the evolutionary advantages discussed above would require gene conversion between the same sites of two different copies of the chromosome (intermolecular gene conversion). Only two studies about intermolecular gene conversion with prokaryotes have been performed, one of those was using *Hfx. volcanii*.

The study made use of a double deletion mutant that contained a genome lacking both the *trpA* gene encoding an enzyme involved in tryptophan biosynthesis and the *pyrE* gene encoding an enzyme involved in pyrimidine biosynthesis (Allers *et al.*, 2004). Using a triple selection scheme a heterozygous mutant was generated that contained two different types of genomes, one with the *leuB* gene encoding an enzyme involved in leucine biosynthesis at the *leuB* locus and the other with the *trpA* gene at the *leuB* locus (Lange *et al.*, 2011). In the absence of both amino acids the mutant was forced to retain both types of genomes simultaneously in the cell. This strain was subsequently cultivated under three different conditions: (1) in the presence of only leucine, selecting only for the presence of the *trpA* gene, (2) in the presence of tryptophan, selecting only for the presence of the *leuB* gene, and (3) in the presence of both amino acids, i.e. in the absence of any selection (compare Fig. 7.7). At different times during growth aliquots were taken and the copy numbers of both types of genomes were quantified. It was revealed that under all three conditions gene conversion led to an equalization of genome copies. Conditions (1) and (3) led to a loss of the *trpA*-type genome within a very short time period of only several days. Condition (2) led to a loss of the *leuB*-type genome, but it took several weeks. These results were in line with the design of the experiment, because the two directions were not symmetrical. Gene conversion from the *trpA*-type to the *leuB*-type genome required DNA synthesis of only less than 50 nt, while gene conversion from the *leuB*-type to the *trpA*-type genome required the synthesis of nearly 1 kbp (Lange *et al.*, 2011). The most important result of this study was that genome equalization via gene conversion occurred in the total absence of selection.

To our knowledge only one additional study about intermolecular gene conversion in prokaryotes exists. It made use of a heterozygous strain of *Methanococcus maripaludis* that was a result of the attempt to replace an essential gene with a resistance gene (Stock *et al.*, 2010). This strain contained a low copy number of the genome with the essential gene (1–2) and a high copy number of the genome with the resistance gene (about 15). It was incubated in the presence of three different concentrations of the antibiotic, thereby varying the strength of the selection, and in the absence of the antibiotic equalling the absence of any selection (Hildenbrand *et al.*, 2011). Again, the most important result was that even in the absence of selection gene conversion led to an equalization of the genome copies, in this case in the direction of the essential gene that could not be totally lost, while in the absence of selection the resistance gene was not needed. As expected, under full selection pressure the situation remained unchanged. Interestingly, low selection pressures led to intermediate results. Thus it seemed that *M. maripaludis* retained only so many copies of the genome with the resistant gene as necessary for growth and 'tried' to enhance the number of genomes with the essential gene, albeit the gene product of the essential gene was needed only in very low amounts and the growth rates of cells containing one copy of the essential gene versus 20 copies of the essential gene were identical.

A third example of intermolecular gene conversion has been reported in chloroplasts, which are the descendants of endosymbiotic Cyanobacteria. It could be shown that in a mixture of mutated and wild type chloroplast genomes deleterious mutations were eliminated by gene conversion (Khakhlova and Bock, 2006). The sparseness of results about intermolecular gene conversion in prokaryotes is probably due to the lack of studies and it can be predicted that it occurs in many more polyploid species.

The occurrence of intermolecular gene conversion solves several puzzles and is an escape from 'Muller's ratchet'. 'Muller's ratchet' is a theoretical prediction that asexual polyploid should be unable to exist (Muller, 1964). The polyploidy of haloarchaea has often been questioned because it is readily possible to isolate point mutants or to construct designed changes of the chromosome. It has been reasoned that this would be impossible for a

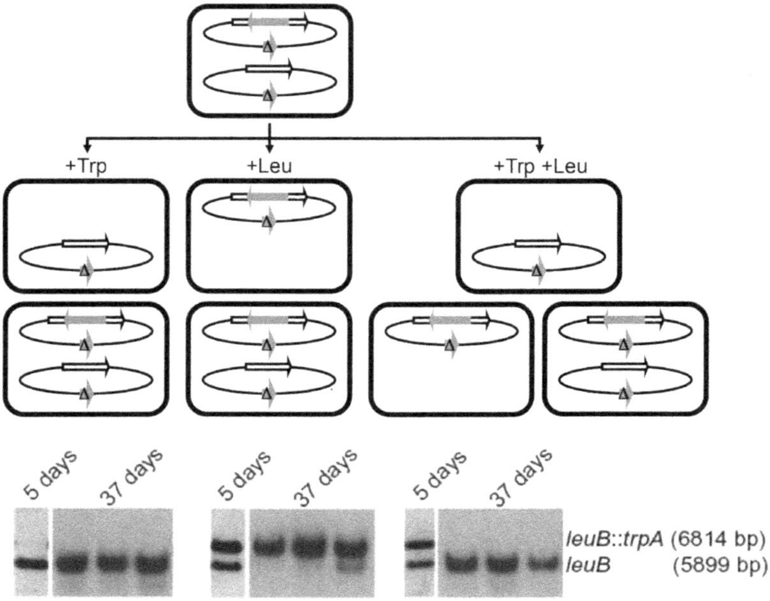

Figure 7.7 Gene conversion leads to the equalization of *Hfx. volcanii* chromosomes in the absence of selection. Using a triple selection scheme a heterozygous *Hfx. volcanii* mutant was generated that contained two types of chromosomes simultaneously, which carried either the *leuB* gene or the *trpA* gene at the *leuB* locus (schematically shown on top). In synthetic medium in the absence of leucine and tryptophan the mutant is forced to retain both type of chromosomes. In the presence of only tryptophan or only leucine, there is selection for only one type of chromosome, while in the presence of both tryptophan and leucine there is no selection at all. The mutant culture was used to inoculate cultures that were grown with the additions of the two amino acids as indicated (each condition in triplicates). The theoretically possible outcomes are shown schematically in the middle of the figure. After 5 days and 37 days, respectively, aliquots were removed and the *leuB* locus was analysed by Southern blotting. A probe was used that hybridized identically to the *leuB* gene and the *leuB::trpA* gene. As can be seen, under all three conditions the genomes were equalized by gene conversion, even in the total absence of selection (taken from Lange *et al.*, 2011).

polyploid species because the mutations arise in a single copy of the chromosome and for statistical reasons it cannot spread to all other e.g. 20 chromosomes. This counterargument became invalid as it was shown that genomes are equalized by intermolecular gene conversion (Lange *et al.*, 2011). The reason is that inevitably deleterious mutations in essential genes must accumulate up to a point that two daughter cells inherit chromosomes that do not have the native version of an essential gene on any of their chromosomes. Again, intermolecular gene conversion solves this problem and is thus an escape from 'Muller's ratchet'.

Conclusions and outlook

The three analysed species of haloarchaea are all polyploid and thus polyploidy might be widespread in and typical for haloarchaea. The copy number is tightly regulated and influenced by growth phase as well as environmental conditions. There are various evolutionary advantages of polyploidy for haloarchaea, and experimental evidence or at least indications exist for nearly all of the nine discussed evolutionary advantages. Polyploidy might explain why haloarchaea can survive in halites over geological times, what has often been questioned based on the lack of stability of DNA. The ease of selection of heterozygous cells in haloarchaea (and other polyploid species of Archaea and bacteria) in model experiments indicate that under specific environmental conditions heterozygous cells arise also in nature, and these have advantages compared to homozygous wild-type cells. Intermolecular gene conversion has been shown to exist in haloarchaea and to be an effective mechanism for the equalization of genomes in growing cells. While recent years have led to many novel and surprising results in this area of research, many challenges remain to be tackled in the future, and a few questions will be shortly discussed below.

It has been shown that the copy number of haloarchaeal replicons is tightly regulated, but the molecular mechanism is totally unknown. We could show that a genomic region of only very few kbp around the origin of replication transfers the ability of autonomous replication to a suicide plasmid. This was attempted for two replication origins of the major chromosome and it was successful in both cases (Zerulla, K., and Soppa, J., unpublished data). These haloarchaeal artificial chromosomes (HACs) can easily be manipulated and will allow unravelling the molecular details of copy number regulation. The molecular details of gene conversion are also not known. It can be expected that the RadA protein is involved, but other participating components are unknown. In addition, nothing is known about the regulation of gene conversion. A system is needed that enables an easy, fast, sensitive and quantitative analysis of gene conversion in the absence of selection.

Nine putative evolutionary advantages of polyploidy in haloarchaea have been discussed. It remains to be elucidated how many and which of these advantages apply to a given species e.g. *Hfx. volcanii,* under various conditions. It would also be interesting to clarify whether species newly isolated from ancient halites are really polyploid. Another question is whether all 25 copies of the chromosome are transcribed to the same extent, or whether 'working copies' and 'genetic storage copies' of the chromosome exist. While this is theoretically possible to unravel, it will be extremely demanding because it requires establishing single cell methods with single molecule sensitivity in the high salt cytoplasm of haloarchaea. During recent years haloarchaea have been used as model organisms for other species of polyploid prokaryotes. For example, intermolecular gene conversion has not been studied in any species of bacteria. Hopefully research with haloarchaea will continue to supply exciting results and insights also in the future.

References

Allers, T., Ngo, H.P., Mevarech, M., and Lloyd, R.G. (2004). Development of additional selectable markers for the halophilic archaeon *Haloferax volcanii* based on the *leuB* and *trpA* genes. Appl. Environ. Microbiol. *70*, 943–953.

Baumann, A., Lange, C., and Soppa, J. (2007). Transcriptome changes and cAMP oscillations in an archaeal cell cycle. BMC Cell Biol. *8*, 21.

Bernander, R. (2007). The cell cycle of *Sulfolobus*. Mol. Microbiol. *66*, 557–562.

Breuert, S., Allers, T., Spohn, G., and Soppa, J. (2006). Regulated polyploidy in halophilic archaea. PLoS One *1*, e92.

Broude, N.E. (2011). Analysis of RNA localization and metabolism in single living bacterial cells: achievements and challenges. Mol. Microbiol. *80*, 1137–1147.

Cai, L., and Tu, B.P. (2012). Driving the cell cycle through metabolism. Annu. Rev. Cell Dev. Biol. *28*, 59–87.

Cho, R.J., Huang, M., Campbell, M.J., Dong, H., Steinmetz, L., Sapinoso, L., Hampton, G., Elledge, S.J., Davis, R.W., and Lockhard, D.J. (2001). Transcriptional regulation and function during the human cell cycle. Nat. Genet. *27*, 48–54.

Coker, J.A. DasSarma, P., Capes, M., Wallace, T., McGarrity, K., Gessler, R., Liu, J., Xiang, H., Tatusov, R., Berquist, B.R., *et al.* (2009). Multiple replication origins of *Halobacterium* sp. strain NRC-1. J. Bacteriol. *191*, 5253–5261.

Cross, F.R., Buchler, N.E., and Skotheim, J.M. (2011). Evolution of networks and sequences in eukaryotic cell cycle control. Philos. Trans. R. Soc. Lond. B Biol. Sci. *366*, 3532–3544.

David, M., Petit, D., and Bertoglio, J. (2012). Cell cycle regulation of Rho signaling pathways. Cell Cycle *11*, 3003–3010.

DeVeaux, L.C., Muller, J.A., Smith, J., Petrisko, J., Wells, D.P., and DasSarma, S. (2007). Extremely radiation-resistant mutants of a halophilic archaeon with increased single-stranded DNA-binding protein (RPA) gene expression. Radiat. Res. *168*, 507–514.

Dickinson, J.R., Graves, M.G., and Swoboda, B.E.P. (1977). Cyclic AMP metabolism in the cell cycle of *Tetrahymena pyriformis*. FEBS Lett. *62*, 152–154.

Dombrowski, H. (1963). Bacteria from the Paleozoic salt deposits. Ann. N. Y. Acad. Sci. *108*, 453–460.

Dornmayer-Pfaffenhuemer, M., Legat, A., Schwimberski, K., Fendrihan, S., and Stan-Lotter, H. (2011). Responses of haloarchaea to simulated microgravity. Astrobiology *11*, 199–205.

Dose, K., Bieger-Dose, A., Kerz, O., and Gill, M. (1991). DNA-strand breaks limit survival in extreme dryness. Orig. Life Evol. Biosph. *21*, 177–187.

Duggin, I.G., McCallum, S.A., and Bell, S.D. (2008). Chromosome replication dynamics in the archaeon *Sulfolobus acidocaldarius*. Proc. Natl. Acad. Sci. U.S.A. *105*, 16737–16742.

Errington, J. (2010). From spores to antibiotics via the cell cycle. Microbiology *156*, 1–13.

Fendrihan, S., Legat, A., Pfaffenhuemer, M., Gruber, C., Weidler, G., Gerbl, F., and Stan-Lotter, H. (2006). Extremely halophilic archaea and the issue of long-term microbial survival. Rev. Environ. Sci. Biotechnol. *5*, 203–218.

Fendrihan, S., Berces, A., Lammer, H., Musso, M., Ronto, G., Polacsek, T.K., Holzinger, A., Kolb, C., and Stan-Lotter, H. (2009). Investigating the effects of simulated Martian ultraviolet radiation on *Halococcus dombrowskii* and other extremely halophilic archaebacteria. Astrobiology *9*, 104–112.

Fendrihan, S., Dornmayr-Pfaffenhuemer, M., Gerbl, F.W., Holzinger, A., Grosbacher, M., Briza, P., Erler, A., Gruber, C., Platzer, K., and Stan-Lotter, H. (2012). Spherical particles of halophilic archaea correlate with exposure to low water activity – implications for microbial survival in fluid inclusions of ancient halite. Geobiology *10*, 424–433.

Forterre, P., Elie, C., and Kohiyama, M. (1984). Aphidicolin inhibits growth and DNA synthesis in halophilic archaebacteria. J. Bacteriol. *159*, 800–802.

Grant, W.D. (2004). Life at low water activity. Philos. Trans. R. Soc. Lond. B. Biol. Sci. *359*, 1249–1266; discussion 1266–1247.

Griese, M., Lange, C., and Soppa, J. (2011). Ploidy in Cyanobacteria. FEMS Microbiol. Lett. *323*, 124–131.

Hansen, M.T. (1978). Multiplicity of genome equivalents in the radiation-resistant bacterium *Micrococcus radiodurans*. J. Bacteriol. *134*, 71–75.

Hebsgaard, M.B., Phillips, M.J., and Willerslev, E. (2005). Geologically ancient DNA: fact or artefact? Trends Microbiol. *13*, 212–220.

Herrmann, U., and Soppa, J. (2002). Cell cycle-dependent expression of an essential SMC-like protein and dynamic chromosome localization in the archaeon *Halobacterium salinarum*. Mol. Microbiol. *46*, 395–409.

Hildenbrand, C., Stock, T., Lange, C., Rother, M., and Soppa, J. (2011). Genome copy numbers and gene conversion in methanogenic archaea. J. Bacteriol. *193*, 734–743.

Hjort, K., and Bernander, R. (1999). Changes in cell size and DNA content in *Sulfolobus* cultures during dilution and temperature shift experiments. J. Bacteriol. *181*, 5669–5675.

Holliday, R. (1964). A mechanism for gene conversion in fungi. Genet. Res. *5*, 282–304.

Jenal, U. (2009). The role of proteolysis in the *Caulobacter crescentus* cell cycle and development. Res. Microbiol. *160*, 687–695.

de Jong, I.G., Veening, J.W., and Kuipers, O.P. (2012). Single cell analysis of gene expression patters during carbon starvation in *Bacillus subtilis* reveals large phenotypic variation. Environ. Microbiol. *14*, 3110–3121.

Kendrick, M.G., and Kral, T.A. (2006). Survival of methanogens during desiccation: implications for life on Mars. Astrobiology *6*, 546–551.

Khakhlova, O., and Bock, R. (2006). Elimination of deleterious mutations in plastid genomes by gene conversion. Plant J. *46*, 85–94.

Kottemann, M., Kish, A., Iloanusi, C., Bjork, S., and DiRuggiero, J. (2005). Physiological responses of the halophilic archaeon *Halobacterium* sp. strain NRC1 to desiccation and gamma irradiation. Extremophiles *9*, 219–227.

Labarre, J., Chauvat, F., and Thuriaux, P. (1989) Insertional mutagenesis by random cloning of antibiotic resistance genes into the genome of the cyanobacterium *Synechocystis* strain PCC 6803. J. Bacteriol. *171*, 3449–3457.

Lange, C., Zerulla, K., Breuert, S., and Soppa, J. (2011). Gene conversion results in the equalization of genome copies in the polyploid haloarchaeon *Haloferax volcanii*. Mol. Microbiol. *80*, 666–677.

Laub, M.T., McAdams, H.H., Feldblyum, T., Fraser, C.M., and Shapiro, L. (2000). Global analysis of the genetic network controlling a bacterial cell cycle. Science *290*, 2144–2148.

Lawson, M.J., Jiao, J., Fan, W., and Zhang, L. (2009). A pattern analysis of gene conversion literature. Comp. Funct. Genomics 76512.

Lemon, K.P., and Grossman, A.D. (1998). Localization of bacterial DNA polymerase: evidence for a factory model of replication. Science *282*, 1516–1519.

Lundgren, M., and Bernander, R. (2007). Genome-wide transcription map of an archaeal cell cycle. Proc. Natl. Acad. Sci. U.S.A. *104*, 2939–2944.

McGenity, T.J., Gemmell, R.T., Grant, W.D., and Stan-Lotter, H. (2000). Origins of halophilic microorganisms in ancient salt deposits. Environ. Microbiol. *2*, 243–250.

Mackwan, R.R., Carver, G.T., Drake, J.W., and Grogan, D.W. (2007). An unusual pattern of spontaneous mutations recovered in the halophilic archaeon *Haloferax volcanii*. Genetics *176*, 697–702.

Malandrin, L., Huber, H., and Bernander, R. (1999). Nucleoid structure and partition in *Methanococcus jannaschii*: an archaeon with multiple copies of the chromosome. Genetics *152*, 1315–1323.

Maldonado, R., Jimenez, J., and Casadesus, J. (1994). Changes of ploidy during the *Azotobacter vinelandii* growth cycle. J. Bacteriol. *176*, 3911–3919.

Mancinelli, R., White, M., and Rothschild, L. (1998). Biopan-survival I: Exposure of the osmophiles *Synechococcus* sp. (Nägeli) and *Haloarcula* sp. to the space environment. Life Sciences Exobiology *22*, 327–334.

Mancinelli, R.L., Landheim, R., Sánchez-Porro, C., Dornmayr-Pfaffenhuemer, M., Gruber, C., Legat, A., Ventosa, A., Radax, C., Ihara, K., White, M.R., *et al.* (2009). *Halorubrum chaoviator* sp. nov., a haloarchaeon isolated from sea salt in Baja California, Mexico, Western Australia and Naxos, Greece. Int. J. Syst. Evol. Microbiol. *59*, 1908–1913.

Marguerat, S., Jensen, T.S., de Lichtenber, U., Wilhelm, B.T., Jensen, L.J., and Bahler, J. (2006). The more the merrier: comparative analysis of microarray studies on cell cycle-regulated genes in fission yeast. Yeast *23*, 262–277.

Mendell, J.E., Clements, K.D., Choat, J.H., and Angert, E.R. (2008) Proc. Natl. Acad. Sci. U.S.A. *105*, 6730–6734.

Morozova, D., Mohlmann, D., and Wagner, D. (2007). Survival of methanogenic archaea from Siberian permafrost under simulated Martian thermal conditions. Orig. Life Evol. Biosph. *37*, 189–200.

Müller, D., Exler, S., Aguilera-Vazquez, L., Guerrere-Martin, E., and Reuss, M. (2003). Cyclic AMP mediates the cell cycle dynamics of energy metabolism in *Saccharomyces cerevisiae*. Yeast *20*, 351–367.

Nickle, D.C., Learn, G.H., Rain, M.W., Mullins, J.I., and Mittler, J.E. (2002). Curiously modern DNA for a "250 million-year-old" bacterium. J. Mol. Evol. *54*, 134–137.

Nodop, A., Pietsch, D., Hocker, R., Becker, A., Pistorius, E.K., Forchhammer, K., and Michel, K.P. (2008) Transcript profiling reveals new insights into the acclimation of the mesophilic fresh-water cyanobacterium *Synechococcus elongatus* PCC 7942 to iron starvation. Plant Physiol. *147*, 747–763.

Park, J.S., Vreeland, R.H., Cho, B.C., Lowenstein, T.K., Timofeeff, M.N., and Rosenzweig, W.D. (2009). Haloarchaeal diversity in 23, 121 and 419 MYA salts. Geobiology *7*, 515–523.

Pecoraro, V., Zerulla, K., Lange, C., and Soppa, J. (2011). Quantification of ploidy in proteobacteria revealed the existence of monoploid, (mero-)oligoploid and polyploid species. PLoS One *6*, e16392.

Samson, R.Y., and Bell, S.D. (2011). Cell cycles and cell division in the archaea. Curr. Opin. Microbiol. *14*, 350–356.

Sankaranarayanan, K., Timofeeff, M.N., Spathis, R., Lowenstein, T.K., and Lum, J.K. (2011). Ancient microbes from halite fluid inclusions: optimized surface sterilization and DNA extraction. PLoS One *6*, e20683.

Santoyo, G., and Romero, D. (2005). Gene conversion and concerted evolution in bacterial genomes. FEMS Microbiol. Rev. *29*, 169–183.

Schneider, D., Fuhrmann, E., Scholz, I., Hess, W.R., and Graumann, P.L. (2007) Fluorescence staining of live cyanobacterial cells suggest non-stringent chromosome segregation and absence of a connection between cytoplasmic and thylakoid membranes. BMC Cell Biol. *8*, 39.

Schubert, B., Lowenstein, T., Timofeeff, M., and Parker, M. (2009). How do prokaryotes survive in fluid inclusions in halite for 30 k.y.? Geology *37*, 1059–1062.

Schubert, B.A., Lowenstein, T.K., Timofeeff, M.N., and Parker, M.A. (2010). Halophilic Archaea cultured from ancient halite, Death Valley, California. Environ. Microbiol. *12*, 440–454.

Smith, M.E., Dickinson, J.R., and Wheals, A.E. (1990). Intracellular and extracellular levels of cyclic AMP during the cell cycle of *Saccharomyces cerevisiae*. Yeast *6*, 53–60.

Soppa, J. (2011). Ploidy and gene conversion in Archaea. Biochem. Soc. Trans. *39*, 150–154.

Spence, E., Bailey, S., Nenninger, A., Moller, S.B., and Robinson, C. (2004) A homolog of Albino3/OxaI is essential for thylakoid biogenesis in the cyanobacterium *Synechocystis* sp. PCC6803. J. Biol. Chem. *279*, 55792–55800.

Stan-Lotter, H., Pfaffenhuemer, M., Legat, A., Busse, H.J., Radax, C., and Gruber, C. (2002). *Halococcus dombrowskii* sp. nov., an archaeal isolate from a Permian alpine salt deposit. Int. J. Syst. Evol. Microbiol. *52*, 1807–1814.

Stock, T., Selzer, M., and Rother, M. (2010). *In vivo* requirement of selenophosphate for selenoprotein synthesis in archaea. Mol. Microbiol. *75*, 149–160.

Takahama, K., Matsuoki, M., Nagahama, K., and Ogawa, T. (2004) High-frequency gene replacement in Cyanobacteria using a heterologous rps12 gene. Plant Cell Physiol. *45*, 333–339.

Taniguchi, Y., Choi, P.J., Li, G.W., Chen, H., Babu, M., Hearm, J., Emili, A., and Xie, X.S. (2010). Quantifying *E. coli* proteome and transcriptome with single-molecule sensitivity in single cells. Science *329*, 533–538.

Tasch, P. (1963). Dead and viable fossil salt bacteria. Univ. Wichita Bull. *39*, 2–7.

Thanbichler, M. (2010). Synchronization of chromosome dynamics and cell division in bacteria. Cold Spring Harbor Perspect. Biol. *2*, a000331.

Touzain, F., Petit, M.A., Schbath, S., and El Karoui, M. (2011). DNA motifs that sculpt the bacterial chromosome. Nat. Rev. Microbiol. *9*, 15–26.

Uhlmann, F., Bouchoux, C., and Lopez-Aviles, S. (2011). A quantitative model for cyclin-dependent kinase control of the cell cycle: revisited. Philos. Trans. R. Soc. Lond. B Biol. Sci. *366*, 3572–3583.

Vreeland, R., Jones, J., Monson, A., Rosenzweig, W., Lowenstein, T., Timofeeff, M., Satterfield, C., Cho, B., Park, J., Wallace, A., *et al.* (2007). Isolation of live Cretaceous (121–112 million years old) halophilic Archaea from primary salt crystals. Geomicrobiol. J. *24*, 275–282.

Watanabe, S., Ohbayashi, R., Shiwa, Y., Noda, A., Kanesaki, Y., Chibazakura, T., and Yoshikawa, H. (2012). Light-dependent and asynchronous replication of cyanobacterial multi-copy chromosomes. Mol. Microbiol. *83*, 856–865.

Watson, C.D., and Berry, D.R. (1977). Fluctuations in cAMP levels during the cell cycle of *Saccharomyces cerevisiae*. FEMS Microbiol. Lett. *1*, 175–178.

Whitfield, M.L., Sherlock, G., Saldanha, A.J., Murray, J.I., Ball, C.A., Alexander, K.E., Matese, J.C., Perou, C.M., Hurt, M.M., Brown, P.O., *et al.* (2002). Identification of genes periodically expressed in the human cell cycle and their expression in tumors. Mol. Biol. Cell *13*, 1977–2000.

Zahradka, K., Slade, D., Bailone, A., Sommer, S., Averbeck, D., Petranovic, M., Lindner, A.B., and Radman, M. (2006). Reassembly of shattered chromosomes in *Deinococcus radiodurans*. Nature *443*, 569–573.

Zaritsky, A., Wang, P., and Vischer, N.O. (2011). Instructive simulation of the bacterial cell division cycle. Microbiol. *157*, 1876–1885.

Cell Regulation by Proteolytic Systems and Protein Conjugation

8

Julie Maupin-Furlow

Abstract

Proteases and protein conjugation systems are important in regulating cell function. Archaea synthesize (or are predicted to encode) numerous types of regulatory proteases including proteasomes, Lon protease, intramembrane cleaving proteases and others. Of these, proteasomes are demonstrated to be important in stress responses and essential to the growth of halophilic Archaea. Halophilic Archaea also synthesize a protein conjugation system termed sampylation in which different ubiquitin-like SAMPs are conjugated to protein targets through isopeptide bonds. While as yet unknown, sampylation is thought to regulate cellular functions by targeting proteins for proteasome-mediated degradation and for other non-proteolytic modifications. This chapter is focused on how Archaea may regulate cell function through proteolysis and protein conjugation with emphasis on the halophilic Archaea.

Introduction

Proteases not only serve to remove unassembled, misfolded or foreign proteins but also to regulate important cellular processes. Enzyme-mediated protein cleavage is used by cells to control metabolic and environmental adaptations, cell differentiation, cell-cycle progression, signal transduction, transcription and other important biological functions (McAdams and Shapiro, 2009; Ciechanover, 2010; Gur *et al.*, 2011). Proteolysis is advantageous over other control mechanisms in that it is rapid and irreversible, thus, enabling the cell to efficiently reduce the levels of distinct components, guarantee loss of protein function for regulation, and even active proteins by clearing polypeptide precursors. Complete turnover of a protein comes at a high energy cost considering the amount of energy the cell devotes to the synthesis and productive folding of proteins. To avoid uncontrolled degradation of proteins, proteolysis is highly selective at the initial stages of protein substrate cleavage.

Archaea synthesize a wide variety of proteases, typically linked to cell regulation, including intramembrane-cleaving proteases, Lon and proteasomes (Maupin-Furlow *et al.*, 2005). Energy-dependent proteases, such as Lon, and proteasomes, typically have associated mechanisms to recognize dysfunctional or regulatory proteins for processive degradation to small peptides. Contrastingly, intramembrane-cleaving proteases often cleave and release membrane-anchored proteins or peptides for cellular signalling events, appendage assembly on the cell surface, or other biological functions.

An archaeal ubiquitin-like system of protein conjugation (termed sampylation) was recently discovered (Humbard *et al.*, 2010a). Similar to ubiquitylation, sumolyation and other protein conjugation systems of eukaryotes, the sampylation system of Archaea involves the covalent attachment of small archaeal ubiquitin-like modifier proteins or SAMPs through an isopeptide bond to target proteins. While the biological role(s) of sampylation is yet unknown, proteins modified by sampylation are thought to be degraded by proteasomes and/or be altered in their activity, location and/or ability to form multisubunit interactions with other protein partners.

This chapter is focused on how proteolytic systems and protein conjugation may be used by halophilic Archaea to regulate cell function. Molecular mechanisms of regulation will be integrated into the discussion when available. Studies of other organisms will also be incorporated as needed to provide a broad perspective on the topic. Proteolytic systems will include intramembrane-cleaving proteases, LonB-type proteases, and proteasomes. The recently discovered ubiquitin-like system of protein conjugation (termed sampylation) will also be discussed.

Intramembrane proteolysis

Intramembrane proteolysis is thought to be a very ancient mechanism used by cells to control stress responses, metabolism, differentiation, development and other biological processes (Koonin *et al.*, 2003; Urban and Shi, 2008). Intramembrane proteolysis is used by cells to cleave (pre)protein targets within their single-pass transmembrane domain, resulting in the liberation of soluble proteins or peptide fragments that can act as molecular effectors (e.g. transcription factor, quorum sensing peptide) or components of external structures (e.g. type IV pili homologues such as the archaeal flagella). Intramembrane proteolysis is also used to degrade membrane-retained protein fragments that are no longer needed (e.g. remnant signal peptides after their production by signal peptidase).

Intramembrane cleaving proteases

Proteases that catalyse intramembrane proteolysis are termed intramembrane-cleaving proteases or I-CLiPs. I-CLiPs have the common ability to hydrolyse peptide bonds within lipid bilayers and can be divided into three major groups (metallo-, serine and aspartyl proteases) based on the proteolytic active sites discovered so far (Wolfe, 2009; Erez *et al.*, 2009). Members within each I-CLiP group can be quite diverse in primary sequence, but all have multipass transmembrane segments and conserved active site motifs likely derived through convergent evolution (Koonin *et al.*, 2003). Site-2 proteases (S2Ps) and rhomboids (RHOs) are of the metalloprotease and serine protease group, respectively. I-CLiPs of the aspartyl protease group (termed GXGD-type) include the prokaryotic type IV prepilin peptidase (TFPP), the eukaryotic signal peptide peptidase (SPP), and presenilin (the catalytic subunit of the γ-secretase complex). Depending on the orientation of the active site within the membrane, I-CLiPs cleave either type 1 (N_{out}–C_{in}) or type 2 (N_{in}–C_{out}) membrane proteins (Ng *et al.*, 2007; Golde *et al.*, 2009; Wolfe, 2010). For clarification, both prokaryotic signal peptide peptidase A (SppA) (found in bacteria, Archaea and plant chloroplasts) and eukaryotic SPP degrade remnant signal peptides. However, in contrast to the eukaryotic SPP, which is an I-CLiP, the prokaryotic SppA is quite distinct and is instead composed of a soluble dome-shaped catalytic domain (with proposed Ser as the nucleophile and Lys as the general base) anchored to the membrane by a single N-terminal transmembrane segment (Kim *et al.*, 2008).

Archaeal I-CLiPs

I-CLiP homologues (of the aspartyl, metallo- and serine type) are widely distributed in Archaea (Koonin *et al.*, 2003; Maupin-Furlow *et al.*, 2005; Kinch *et al.*, 2006; Torres-Arancivia *et al.*, 2010). In fact, most halophilic Archaea encode two to five members of each I-CLiP group with some of these intramembrane protease domains fused to large hydrophilic N- or C-terminal domains of separate function. For example, rhomboid homologues closely related to *Hfx. volcanii* HVO_0727 are common among halophilic Archaea and include an extensive N-terminal AN1-type Zn finger domain (known to bind DNA, RNA, protein and/or lipid substrates in other proteins) (Klug, 1999; Matthews and Sunde, 2002; Hall, 2005; Brown, 2005; Gamsjaeger *et al.*, 2007). In analogy to eukaryotes and bacteria, archaeal I-CLiPs are presumed to regulate cell function. The discussion below is focused on archaeal I-CLiPs that have been biochemically or genetically characterized including the TFPP- and presenilin-like aspartyl proteases as well as the intramembrane metalloproteases.

TFPP-like aspartyl proteases

TFPP/TFPP-like proteases use a diaspartyl active site (coupling an Asp residue with a distant GXGD motif) to cleave type 2 membrane proteins (Fig. 8.1). Often these proteases

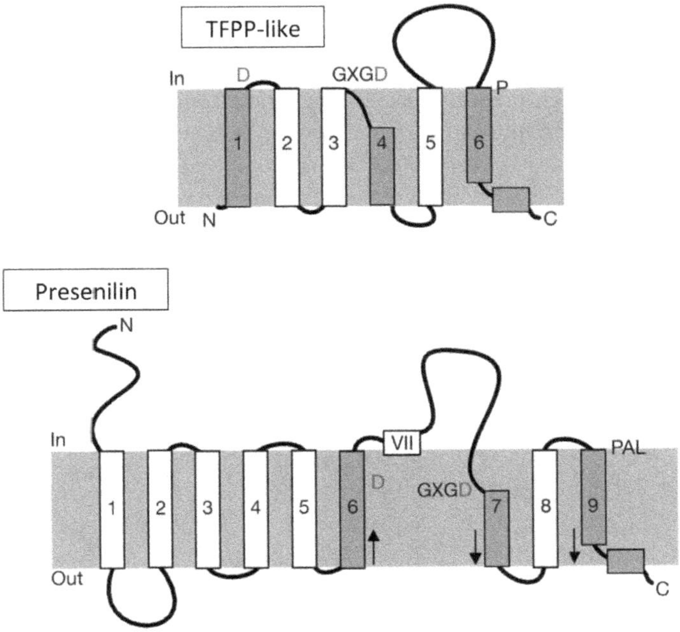

Figure 8.1 TFPP-like (FlaK) and presenilin aspartyl proteases. Membrane topology diagrams of the two diaspartyl intramembrane proteases are depicted with conserved active site residues (Asp of FlaK TM1 and presenilin TM6, GXGD motif of FlaK TM4 and presenilin TM7) and C-terminal proline (including P or PAL). [Figure modified from (Hu *et al.*, 2011) with permission.]

cleave N-terminal leader peptides from precursor proteins resulting in their secretion and assembly into type IV pilin-like structures on the cell surface. Archaeal TFPP-like aspartyl proteases that have been characterized include *Methanococcus* spp. FlaK, *Sulfolobus solfataricus* PibD, and more recently *Haloferax volcanii* PibD (Ng *et al.*, 2007; Tripepi *et al.*, 2010; Hu *et al.*, 2011; Pohlschröder *et al.*, 2011). The most extensively studied of these is FlaK, which has recent crystal structure data to guide understanding TFPP-like protease function (Hu *et al.*, 2011). In relation to biological function, the FlaK and PibD proteases cleave the N-terminal signal peptide from preflagellin prior to its incorporation into the growing flagellar filament, a step essential to flagellar assembly and motility (Bardy and Jarrell, 2002, 2003; Albers *et al.*, 2003). While FlaK appears somewhat narrow in its substrate preferences, PibD has a relatively broad specificity for substrates including the ability to cleave signal peptides as short as three amino acids (Ng *et al.*, 2006, 2009). Thus, PibD is thought to cleave not only preflagellin but also preproteins that assemble into non-flagellar type IV pilus-like structures used for surface adhesion (with biochemical and genetic evidence to support this function in *S. solfataricus* and *Hfx. volcanii*) (Albers *et al.*, 2003; Szabo *et al.*, 2006; Tripepi *et al.*, 2010; Zolghadr *et al.*, 2011).

Presenilin-like aspartyl proteases

Only recently have presenilin-like aspartyl proteases been identified and characterized in Archaea. Presenilin proteases are related to the TFPP-like proteases in their use of two aspartate residues in the active site (Fig. 8.1), but differ in membrane topology and their ability to cleave type 1 (not type 2) membrane proteins (Wolfe, 2010). Recently, homologues of presenilin were identified in Archaea by scanning genome sequences for the diaspartyl signature (YD and GXGD) and C-terminal PAL motifs common to presenilins (Torres-Arancivia *et al.*, 2010). DUF1119/COG3389 proteins were identified in Korarchaeota and Euryarchaeota as candidate presenilin-like I-CLiPs and *Methanoculleus marisnigri* MCMJR1 was selected from this group for further purification based on its high level production in recombinant *Escherichia coli* (Torres-Arancivia *et al.*, 2010). MCMJR1 was purified and found to cleave the artificial substrate β-amyloid precursor protein at several transmembrane sites without the need for other protein co-factors (in contrast to eukaryotic presenilin, which is proteolytically active only when associated with other γ-secretase subunits). While it is likely that the archaeal presenilin-like proteases regulate and maintain cellular homeostasis, physiological substrates are yet to be discovered.

Intramembrane metalloproteases

Site-2 metalloprotease (S2P) family members are widespread in all domains of life and are characterized by a conserved HESSHXnDG metal-binding motif and at least four transmembrane helices (Lewis and Thomas, 1999; Rudner *et al.*, 1999; Brown *et al.*, 2000; Kinch *et al.*, 2006). Most halophilic Archaea encode at least two S2P homologues with conserved active site residues. The S2P family derives its name from the mammalian S2P enzyme which cleaves sterol-regulatory element-binding protein (SREBP) within a transmembrane segment at site-2 (Rawson *et al.*, 1997). This cleavage is required for release of the cytoplasmic domain of SREBP from the endoplasmic reticulum (ER) membrane, which can then function as a transcription factor in the activation of genes regulating cholesterol and fatty acid metabolism (Brown *et al.*, 2000). In bacteria, S2P-mediated untethering of soluble protein domains by cleavage from transmembrane anchors contributes to pheromone production (An *et al.*, 1999), polarity determination (Chen *et al.*, 2006), sporulation (Yu and Kroos, 2000), and stress responses (Alba *et al.*, 2002). Similarities of present-day bacterial and eukaryal S2P family members in their ability to respond to unfolded protein stress (Ye *et al.*, 2000; Alba *et al.*, 2002) combined with comparative genomics suggest intramembrane metalloproteases were used in ancient signalling cascades to report extracytoplasmic stresses (Kinch *et al.*, 2006).

Crystal structures of an S2P homologue from the archaeon *Methanocaldococcus jannaschii* (mjS2P) have provided the first insight into S2P structure (Fig. 8.2) (Feng *et al.*, 2007). Among 40 S2P homologues from various bacteria and Archaea, the transmembrane core domain of mjS2P (residues 1–224) was the most amendable to crystallography (Feng *et al.*, 2007). Like soluble metalloproteases, the catalytic active site of mjS2P is formed by three residues (His54, His58 and Asp148) that coordinate zinc, a glutamate residue (Glu55) likely involved in water deprotonation and an asparagine residue (Asn140) positioned to interact with the substrate carbonyl. Open and closed conformations are detected in mjS2P crystal structures suggesting the route for substrates access to the active site is controlled

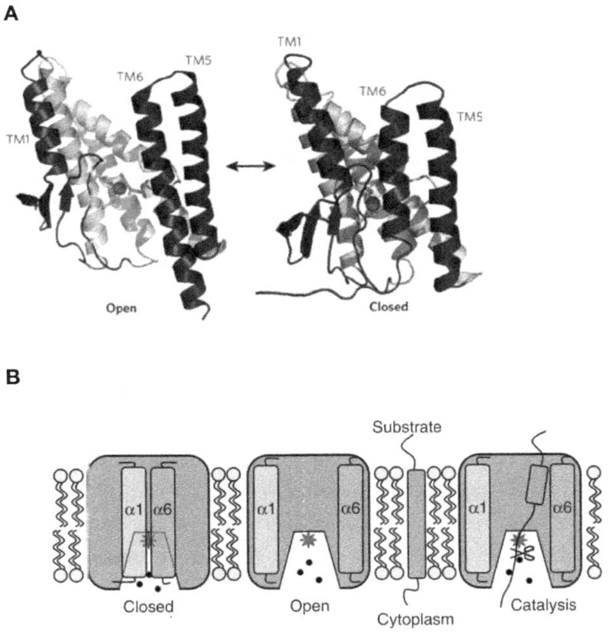

Figure 8.2 Intramembrane metalloprotease (mjS2P). (A) Crystal structures suggest mjS2P adopts an open and closed conformation that is used to gate substrate access to the active site. Transmembrane helices mobile between these two conformations (TM1 and TM5-TM6 in dark blue), catalytic zinc ion (purple ball), and side chains of residues coordinating zinc (cyan sticks) are indicated. (B) Cartoon of S2P mechanism based on crystal structure with water molecules (black dots) thought to access the catalytic active site (red asterisk) through a channel from the cytosol. [Figure modified from (Feng *et al.*, 2007; Erez *et al.*, 2009) with permission]. A colour version of this figure is located in the plate section at the back of the book.

(Fig. 8.2). An mjS2P channel from the cytosolic side of the membrane to the catalytic active site (buried within transmembrane domains) is proposed to allow passage of water molecules and, thus, enable peptide bond hydrolysis (Fig. 8.2).

The biological substrates of archaeal S2Ps are as yet unknown, but (in analogy to eukaryotes and bacteria) these substrates are likely tethered to membranes in an inactive form and activated upon release of a soluble domain by the S2P. Consistent with this model, mjS2P can cleave sites within the segments of a protein substrate that would typically span a membrane as demonstrated using the artificial substrate protein *Caenorhabditis elegans* CED-9 in detergent micelles (Feng *et al.*, 2007). Interestingly, many S2P gene homologues are linked in apparent operons with *traB*-like genes in halophilic Archaea (e.g. *Hfx. volcanii* HVO_1555 and HVO_1554, respectively). In the bacterium *Enterococcus faecalis*, TraB is a transmembrane protein that functions in association with Eep (an S2P homologue) in mediating pheromone response (Clewell, 2011). Eep is required for processing the precursors of both the cAD1 peptide pheromone and iAD1 inhibitor peptide, while TraB influences the amount of mature cAD1 peptide pheromone that is secreted (Clewell, 2011). Although yet to be examined experimentally, it is reasonable to speculate that the haloarchaeal S2P and TraB homologues are functionally linked in cell regulation.

Energy-dependent proteases

Energy-dependent proteases are involved in general and regulatory proteolysis. Energy-dependent proteases degrade not only dysfunctional proteins, but also proteins that must be removed at specific times and/or cellular locations for proper control of cell function (Gottesman and Maurizi, 1992; Gottesman, 1999; Gur *et al.*, 2011; Maupin-Furlow, 2012). Post-translational mechanisms are often used to target substrate proteins to energy-dependent proteases (e.g. binding to an adaptor protein, covalent attachment of molecular groups such as ubiquitin) (Kirstein *et al.*, 2009; Maupin-Furlow, 2012). The advantage of linking proteolysis to energy dependence is that the cell can maintain checkpoints to insure that the proper protein is selected for degradation. As the name implies, the protease itself requires energy (ATP binding/hydrolysis) to unfold the protein and enable this substrate to contact the proteolytic active sites. Mechanisms that target the substrate protein to the energy-dependent protease can even require energy such as the ATP-dependent activation (adenylation) of ubiquitin prior to protein conjugation (Hochstrasser, 2009).

Four major groups of energy-dependent proteases are currently known including Lon, FtsH, Clp, and HslUV/proteasome (Lupas *et al.*, 1997). Only two of these groups are universally conserved in Archaea, the proteasomes and Lon proteases (Maupin-Furlow *et al.*, 2005). Interestingly, while highly diverse in amino acid sequence, these four groups of proteases share a number of common structural features. The proteases are all self-compartmentalized with their (relatively non-specific) proteolytic active sites sequestered within an (often gated) chamber that protects the cell from random proteolysis (Maupin-Furlow, 2012). Proteins with folded structure are prevented from entering the proteolytic chamber unless assisted by separate proteins or protein domains of the AAA+ superfamily (ATPases associated with diverse cellular activities) (Ogura and Wilkinson, 2001; Dougan *et al.*, 2002a; Tucker and Sallai, 2007). The AAA+ component of energy-dependent proteolysis uses ATP binding/hydrolysis to unfold the substrate protein, open the protease gates, and translocate the protein into the protease chamber for destruction (Bar-Nun and Glickman,

2012). Typically, energy-dependent proteolysis is processive. Once a substrate undergoes hydrolytic attack by energy-dependent proteases (or the associated proteolytic chamber), the protein is completely degraded to small peptides without the release of large intermediates (Thompson *et al.*, 1994; Akopian *et al.*, 1997; Kisselev *et al.*, 1998, 1999; Jennings *et al.*, 2008; Licht and Lee, 2008; Hsieh *et al.*, 2011). Processive degradation is thought to be important in preventing the release of protein fragments that might interfere with cell function. For instance, protein fragments could promote the formation of protein aggregates or catalyse functions that are no longer regulated due to degradation of regulatory domains. The non-processive hydrolysis of proteins (with release of large fragments) is reported for energy-dependent proteases but appears less common than processive degradation (Wang *et al.*, 1999; Ferrington *et al.*, 2001; Cardozo and Michaud, 2002; Denny, 2004).

Lon proteases

Lon proteases are highly conserved among all organisms and considered one of the most streamlined operational units to mediate energy-dependent proteolysis. In bacteria and eukaryotic organelles, Lon proteases have a prominent role in degrading denatured and misfolded proteins (Kowit and Goldberg, 1977; Rigas *et al.*, 2012). In *E. coli*, Lon protease is also demonstrated to maintain the quality control of presecretory proteins (Sakr *et al.*, 2010) and regulate the half-life of native proteins (e.g. cell division inhibitor SulA and UV mutagenesis protein UmuD) (Sonezaki *et al.*, 1995; Gonzalez *et al.*, 1998; Ishii *et al.*, 2000). As with other energy-dependent proteases, Lon proteases are also implicated in protein folding (Suzuki *et al.*, 1997).

Lon proteases are tandem fusions of AAA+ and Ser-Lys catalytic dyad protease domains and can be divided into two major subfamilies, A-Lon and B-Lon (Rotanova *et al.*, 2004, 2006). A distinguishing feature of these two groups is the presence of a multilobed A-Lon N-terminal (LAN) domain that is absent from the B-Lons. A-Lons are soluble and common to the bacterial cytoscl and eukaryotic organelles (i.e. mitochondrial matrix, chloroplast stroma) (Swamy and Goldberg, 1982; Adam *et al.*, 2001; Venkatesh *et al.*, 2012). B-Lon homologues are widespread among Archaea and often include transmembrane domains emerging from their AAA+ domain. While certain Gram-positive bacteria (e.g. *Bacillus subtilis*) harbour A- and B-Lons, both of these bacterial Lons appear soluble and devoid of membrane anchors (Serrano *et al.*, 2001).

Recent biochemical and structural studies provide insights into the molecular mechanism of archaeal B-Lon function. Similarly to A-Lons (Stahlberg *et al.*, 1999; Park *et al.*, 2006), the archaeal B-Lons are purified from recombinant *E. coli* as ring-like particles (of six identical subunits) that are functional in energy-dependent proteolysis autonomous of other AAA+ complexes (Fukui *et al.*, 2002; Besche and Zwickl, 2004; Botos *et al.*, 2005; Cha *et al.*, 2010). However, unlike the A-Lons, archaeal B-Lons are anchored to cell membranes (Fukui *et al.*, 2002; Besche *et al.*, 2004).

A number of crystal structures of A-/B-Lons are available that provide insight into Lon function. In particular, structures of AAA+ and protease domains of A-/B-Lons have been pieced together (Rotanova *et al.*, 2006), and structures of an intact B-Lon (*Ton* B-Lon of the hyperthermophilic archaeon *Thermococcus onnurineus*) are also available (Cha *et al.*, 2010). For *Ton* B-Lon, a soluble, stable form of the particle (with both AAA+ and protease domains) was generated for crystallography by deletion of the membrane anchoring region and site-directed mutagenesis of the Ser-Lys catalytic dyad (to prevent its self-degrading

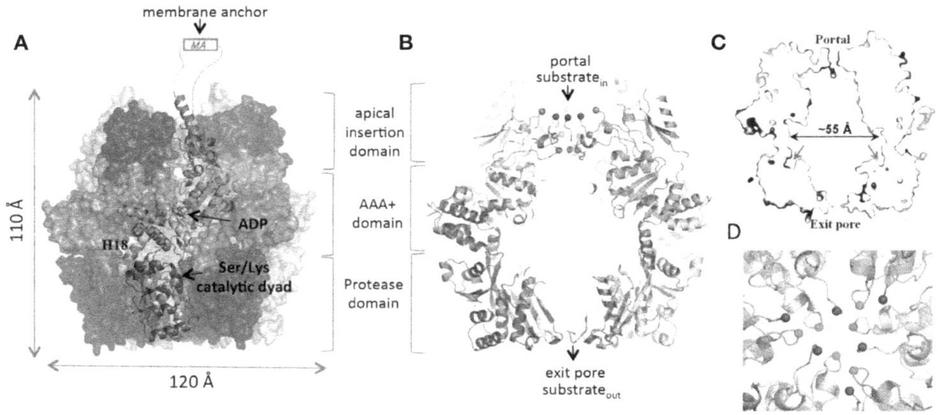

Figure 8.3 Archaeal B-Lon protease. (A) Crystal structure of *Thermococcus onnurineus* B-Lon (*Ton*LonB) hexamer with five subunits in surface representation and one monomer as a ribbon diagram. Insert 1 (which is attached to the deleted membrane anchor) is depicted as an orange ribbon. ADP binding site is indicated as stick figure and lavender balls indicate positions where Ser-Lys catalytic dyad for protein degradation likely residues. (B) Ribbon diagram of *Ton*LonB with a vertical section through the centre. Substrate entry portal with residues likely gating the opening indicated (Cα atoms of F216 and M275 in red and green balls, respectively). Substrate exit pore is also indicated. (C) Surface rendering of *Ton*LonB vertical slab with red arrows indicating substrate binding grooves. (D) Top view of TonLonB residues gating the entry portal indicated as above. [Figure modified from (Cha *et al.*, 2010) with permission]. A colour version of this figure is located in the plate section at the back of the book.

activity) (Cha *et al.*, 2010). From a 2.0 Å structure of *Ton* B-Lon, a three-tiered hexagonal cylinder with a large sequestered proteolytic chamber is evident (Fig. 8.3). Much like FtsH, the proteolytic active sites of *Ton* B-Lon are sequestered within the same chamber as the AAA+ molecular chaperone used for unfolding protein substrates during protein degradation. This streamlined architecture is in contrast to the Clp and HslUV/proteasome groups, which have AAA+ and proteolytic components separated by a relatively narrow opening likely used for the passage of proteins during their destruction. The general structure of *Ton* B-Lon, thus, suggests that the protein unfolding and degradation activities are intimately linked. In addition, proteolysis is unidirectional with substrates entering via an entry portal adjacent to the cell membrane and small peptide products exiting from a relatively narrow exit pore that opens to the cytosol (or soluble protein component).

Like other energy-dependent proteases, *Ton* B-Lon appears gated with an aromatic-hydrophobic (Ar-φ) loop at its presumed entry portal (Fig. 8.3) (Cha *et al.*, 2010). Ar-φ loops are often crucial for the unfolding and degradation of proteins by energy-dependent proteases and are thought to grip and pull on substrate proteins (via hydrophobic interactions), with cycles of ATP hydrolysis driving this motion (Martin *et al.*, 2008b). Consistent with this function, a site directed variant (F216A) in the Ar-φ loop renders *Ton* B-Lon relatively inactive in ATP-dependent proteolysis (Cha *et al.*, 2010). Likewise, structural differences between free and nucleotide (ADP) bound states of *Ton* B-Lon suggest nucleotide exchange causes the diameter and shape of the substrate entry portal to fluctuate and bring the Ar-φ loops into contact with different regions of the substrate (Cha *et al.*, 2010).

Additional flexible loops within the entry portal (termed Ins2 and 3) are also displaced by nucleotide binding and are thought to facilitate the active translocation of bound proteins towards the proteolytic sites (Cha *et al.*, 2010).

In contrast to the A-Lons, less is known regarding the substrates and biological roles of the B-Lons that are common to Archaea. Detailed analysis of archaeal B-Lon substrate specificity has not been conducted and physiological substrates remain to be determined. Archaeal B-Lons can hydrolyse folded proteins in an ATP-dependent manner, but the protein substrates used in these *in vitro* assays are from non-archaeal sources (e.g. bovine α-casein, *E. coli* UmuD, synthetic fusion proteins) (Fukui *et al.*, 2002; Besche *et al.*, 2004; Cha *et al.*, 2010).

In analogy to the A-Lons, the B-Lons of Archaea are thought to be important regulators of cell function. Bolstering this idea is the finding that some archaeal B-Lon genes (*Pyrococcus* spp.) contain inteins (Pietrokovski, 2001). Genes containing inteins are often vital for host survival (Pietrokovski, 2001). Inteins are selfish DNA elements inserted in-frame and translated together with their host proteins. The intein-containing precursor protein undergoes an autocatalytic protein splicing reaction resulting in two products: the host protein and the intein-derived protein fragment. Consistent with this functional model, the intein of *Pyrococcus abyssi* B-Lon mediates efficient protein self-splicing when produced in recombinant *E. coli* (O'Brien *et al.*, 2010).

Study has been initiated on understanding the function of B-Lons in halophilic Archaea. In the genomes of halophilic Archaea, the coding sequences for the B-Lons are often in close proximity to *cct3* (a heat-shock inducible gene encoding a subunit of the thermosome in *Hfx. volcanii*) (Large *et al.*, 2002) and genes which may encode membrane-associated proteases (Sastre *et al.*, 2011). Indeed, preliminary analysis of *Natrialba magadii* transcripts by RT-PCR suggests B-Lon is co-transcribed with its transmembrane protease gene neighbour (Sastre *et al.*, 2011). Interestingly, transcription of Lon protease is heat shock inducible in bacteria (Chin *et al.*, 1988) and FtsH (the energy-dependent protease associated in bacterial cell membranes) is functionally linked to other transmembrane proteases (e.g. HtpX) (Sakoh *et al.*, 2005). While the substrates and catalytic activities are as yet unknown, *Nab. magadii* (*Nm*) B-Lon has been purified from recombinant *E. coli* and used to generate polyclonal antibodies for immunoblot analysis (Sastre *et al.*, 2011). Based on immunoblot, *Nm* B-Lon was detected in membrane (vs. cytosolic) fractions at relatively constant levels throughout all phases of growth (Sastre *et al.*, 2011). In contrast, the ATP-dependent caseinolytic activity of the membrane-associated fractions was only detected in stationary phase cells, prompting the researchers to propose that intracellular signals activated the B-Lon of *Nab. magadii* (Sastre *et al.*, 2011). While preliminary evidence also suggests *Nab. magadii* B-Lon binds DNA, it is unclear whether this DNA binding activity is specific and/or used to regulate function similarly to the A-Lons of bacteria and eukaryotes (Fu *et al.*, 1997; Fu and Markovitz, 1998; Lu *et al.*, 2003; Liu *et al.*, 2004).

Proteasomes

Proteasomes are energy-dependent proteases that are widespread in eukaryotes, Archaea and some bacteria (i.e. actinobacteria) (Maupin-Furlow, 2012). Proteasomes are composed of a cylindrical 20S core particle (CP) of four-stacked heptameric rings (Fig. 8.4) that harbour 6–14 proteolytic active sites within the interior. Subunits of the CP are of the related

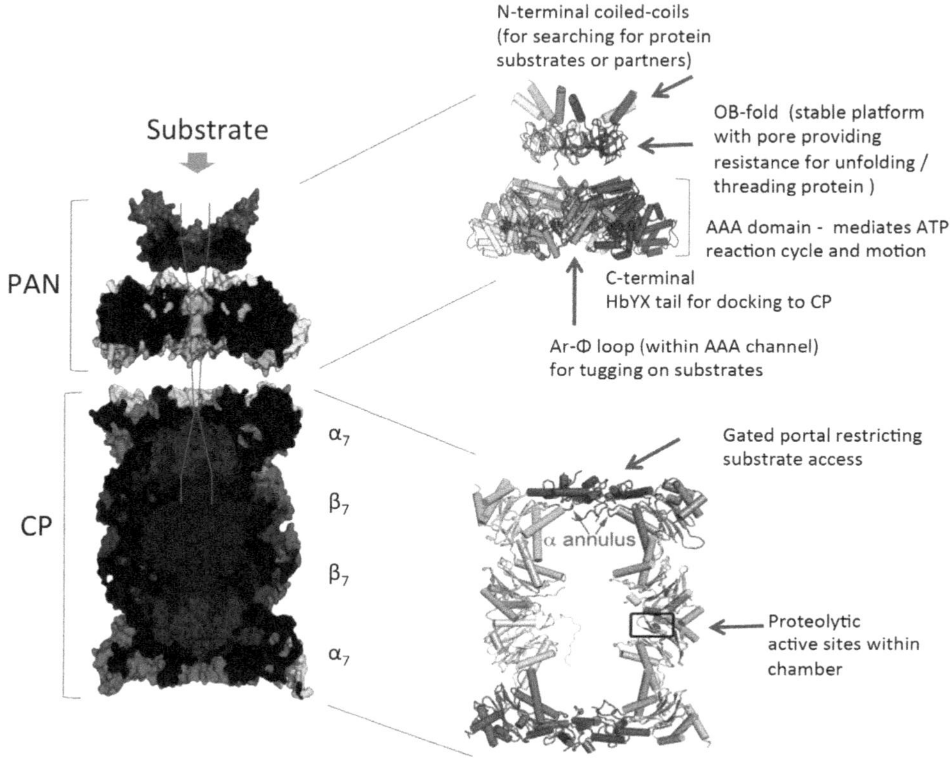

Figure 8.4 Archaeal proteasomes. Proteasomes are composed of a catalytic core particle (CP) that catalyses the hydrolysis of peptide bonds. CPs are barrel-like structures composed of four stacked heptameric rings of α- and β-type subunits (indicated by α_7 and β_7, respectively). CPs have a central channel with narrow openings on each end that are gated by α subunit N-termini. The proteolytic active sites formed by the N-terminal Thr residues of β-type subunits (indicated in red) line the central cavity of the CP. CPs associate with a variety of regulators including the archaeal PAN. PAN is a hexameric ATPase that docks (via its C-terminal HbYX motif) to the ends of CPs. PAN has coiled-coil, OB-fold, Ar-Φ loop and AAA motor components that coordinate to unfold proteins and facilitate their passage into the central chamber of the CP for destruction. [Figure modified from (Zhang *et al.*, 2009a; Stadtmueller and Hill, 2011) with permission]. A colour version of this figure is located in the plate section at the back of the book.

α- and β-type subfamilies with the α-type forming the outer two rings and β-type forming the inner two rings that harbour the proteolytic active sites. CPs have openings on each end of the cylinder that serve as substrate entry portals, which are gated by the N-termini of the α-type subunits. While small peptides can enter through this portal, folded proteins are too bulky. To degrade folded proteins, the CPs associate with separate AAA+s that require ATP to unfold proteins, open the CP gates, and enable proteins to be translocated to the proteolytic interior of the CP for degradation. Proteins are typically degraded by a processive mechanism, with small peptides released as products and little evidence of release of large fragments, which (if uncontrolled) would compromise cell function.

Mix and match protease domains with regulators/chaperones

Proteasomes and Clp proteases are similar in their ability to mix and match the association of protease domains with AAA+ or other regulatory/chaperone complexes. This flexibility is in contrast to Lon and FtsH proteases, which are synthesized as tandem protease–AAA+ fusions from a single gene. In the Clp system (of bacteria and eukaryotic organelles), the protease domain (ClpP) associates with different AAA+ complexes (e.g. ClpX, ClpA) to form energy-dependent proteases with distinct but overlapping substrate specificities (Gottesman *et al.*, 1998; Flynn *et al.*, 2003). In *E. coli,* a ClpS adaptor can deliver N-end rule substrates and aggregated model substrates to ClpAP for degradation (Dougan *et al.*, 2002b; Erbse *et al.*, 2006; Roman-Hernandez *et al.*, 2011), while SspB delivers ssrA-tagged cargo for destruction by ClpXP (Levchenko *et al.*, 2000; Flynn *et al.*, 2001). Proteasomes, likewise, are not a uniform population of complexes and, instead, include CPs interacting with distinct regulators. For instance, the 26S proteasomes of eukaryotes are complexes of CPs with 19S caps or regulatory particles (RPs) that harbour regulatory particle triple-A (Rpt, AAA+) subunits (Finley *et al.*, 1998; Voges *et al.*, 1999). The ATP-driven chaperone valosin-containing protein (VCP)/p97/Cdc48 is also linked to the activity of CPs in eukaryotes (Meyer *et al.*, 2012). CPs can even interact with non-AAA+ complexes such as the 11S proteasome activator (PA28), which increases the ability of CPs to degrade small peptides (Stadtmueller and Hill, 2011).

Proteasome-activating nucleotidases

In Archaea, the most-studied AAA+ complexes thought to interact with CPs are the proteasome-activating nucleotidases or PANs (Fig. 8.4). Archaeal PANs and CPs have been instrumental in the quest to understand the molecular details of proteasome structure and function. PANs were first identified based on their close homology to the regulatory particle triple-A (Rpt, AAA+) subunits of eukaryotic 26S proteasomes and are now known to interact with and stimulate ATP-dependent proteolysis by CPs (Zwickl *et al.*, 1999; Wilson *et al.*, 2000; Smith *et al.*, 2005, 2011). Crystal structures of CPs and subdomains of PANs have been pieced together, and a model of how proteasomes work has emerged (Lowe *et al.*, 1995; Groll *et al.*, 2003; Djuranovic *et al.*, 2009; Zhang *et al.*, 2009a,b; Yu *et al.*, 2010). Chemical cross-linking mass spectrometry (CXMS) coupled with available structural data has also been used to develop models of halophilic CPs, less amendable to crystallography (Karadzic *et al.*, 2012).

PANs interact with CPs and together this complex can mediate the energy-dependent degradation of proteins (Fig. 8.4). Alone, PANs associate into a doughnut-like hexameric ring (Smith *et al.*, 2005; Medalia *et al.*, 2009; Zhang *et al.*, 2009a). PAN can unfold proteins by energy-dependent transfer of substrate through its hexameric ring and can thread the unfolded protein into the CP channel for proteolytic destruction (Smith *et al.*, 2005; Zhang *et al.*, 2009b). In addition, PAN has a C-terminal HbYX motif (where Hb is a hydrophobic amino acid and X is any amino acid) that is important for docking and opening the gates on each end of the CP cylinder. Thus, in current proteasome models, the C-terminal domain of PAN is in contact with the outer rings of the CP, and the central channels of PAN and CP are positioned coaxial, forming a long central tunnel for the passage of substrate proteins (Fig. 8.4). Coiled-coil (CC) tentacle-like structures within the N-termini of PAN are thought to stretch out from the ends of the proteasome complex searching for substrates (or protein partners). Once substrates are recognized, Ar-Φ (pore-1) loops that line the central passage of the AAA+ domain of PAN are thought to grip and tug portions of the substrate protein into the PAN channel by ATP fuelled motions (similarly to other energy-dependent proteases) (Wang *et al.*, 2001; Hinnerwisch *et al.*, 2005; Inobe *et al.*, 2008; Martin *et al.*, 2008a,b; Zhang *et al.*, 2009a,b). PAN also has an oligonucleotide/oligosaccharide binding (OB)-fold domain that forms a narrow pore (16 Å diameter) immediately adjacent to the coiled-coil domains (Fig. 8.4). OB-folds are generally stable (Alexandrescu *et al.*, 1995, 1999). Thus, the OB domain of PAN is thought to provide a rigid platform that blocks the movement of folded proteins through its narrow pore as the protein substrate is being tugged and pulled by the Ar-Φ loop (Zhang *et al.*, 2009a). Passive resistance of the OB-fold domain would promote protein unfolding. Ultimately, the energy-dependent translocation of a protein through PAN results in its unfolding and passage into the central chamber of the CP for destruction.

The stoichiometry of ATP binding to PAN and functional effects of this binding imply an ordered reaction cycle occurs during energy-dependent proteolysis (Smith *et al.*, 2011). Within the hexameric ring of PAN, subunits that are opposite each other appear to partner-up and mediate an ordered reaction cycle of ATP binding, hydrolysis and release (Smith *et al.*, 2011). This repetitive reaction cycle is thought to coordinate and drive conformational changes in the ring (in particular, tugging/pulling motions of the Ar-Φ loops within the channel) that are critical for protein unfolding.

Additional regulators likely to bind archaeal CPs

One caveat to study of PAN is that its direct association with CPs has not been demonstrated in an archaeal cell. Thus, formation of the PAN and CP complex is speculated to be transient with additional regulatory components or chaperones likely to bind CPs in the cell. Consistent with this proposal, PAN does not bind CPs with high affinity, and PAN binding is mediated through a C-terminal HbYX motif that is common to other archaeal components (e.g. PbaA) (Kusmierczyk *et al.*, 2011). Other points that suggest CPs interact with more than just PAN are that many archaeal genomes that encode CPs do not have PAN homologues (Maupin-Furlow *et al.*, 2004), and CPs are demonstrated to be essential for growth while PANs are dispensable (at least in *Hfx. volcanii*) (Zhou *et al.*, 2008). These observations, combined with the inability of CPs to degrade folded proteins, have prompted speculation that the VAT/p97/Cdc48 family of AAA+ proteins may function in energy-dependent proteolysis by CPs in Archaea (Maupin-Furlow, 2012). ATPases of the VAT/p97/Cdc48 family are closely linked to proteasome function in eukaryotes (Meyer *et al.*, 2012) and can unfold proteins in Archaea (based on *in vitro* analysis of VAT from *Thermoplasma acidophilum*) (Gerega *et al.*, 2005). Recent studies now provide evidence for Cdc48 association and function with CPs in archaea (Barthelme and Sauer, 2012, 2013).

Proteasomes in halophilic Archaea

One advantage of using halophilic Archaea as model systems is that numerous genetic tools are available that can be coupled with protein biochemistry to examine proteasome function in an archaeal cell. Proteasomes and associated regulatory complexes have been studied in a variety of halophilic Archaea including *Hfx. volcanii, Halobacterium salinarum* R1 and *Haloarcula marismortui.* For review and reference, a timeline of studies on haloarchaeal regulatory proteases and protein conjugation systems is outlined in Table 8.1.

Table 8.1 Timeline of regulatory proteases and protein conjugation systems studied in halophilic Archaea

System	Important finding(s)/achievement(s)	Reference
Hfx. volcanii proteasome	First report that haloarchaea encode/synthesize CPs. CP subtypes purified, sequenced and demonstrated to catalyse the hydrolysis of peptides and unfolded proteins. First description of prokaryotic with CP subtypes. CP genes (two α-type, one β-type) cloned by reverse genetics. Large propeptide (49 aa) of β precursor demonstrated to be removed in CPs to expose N-terminal Thr	Wilson *et al.* (1999)
Har. marismortui proteasome	Active CP of α and β subunits purified and used for 3D reconstruction from negatively stained electron micrographs	Franzetti *et al.* (2002)
Hfx. volcanii proteasome	Subunit topology of two CP subtypes ($\alpha1_7\beta_7\beta_7\alpha1_7$ and $\alpha1_7\beta_7\beta_7\alpha2_7$) determined by chemical cross-linking	Kaczowka and Maupin-Furlow (2003)
	Proteasome protein levels monitored in cell culture. PAN-A, α1 and β relatively constant throughout growth. PAN-B and α2 dramatically increase during stationary phase	Reuter *et al.* (2004)
	GFP reporter developed and used to demo that proteins with hydrophobic C-terminal tails are at low levels, but increase upon addition of proteasome inhibitor	Reuter and Maupin-Furlow (2004)
	CP proteins are modified post-translationally (Nα-acetylation of α1/2, phosphorylation of β)	Humbard *et al.* (2006)
	Proteasomal genes are cotranscribed with gene neighbours (RNase P, MOSC domain, SAM-methyltransferase homologues)	Gil *et al.* (2007)
	Proteasome inhibitor has global impact on cellular proteome (e.g. increase in proteins related to sulfur mobilization, translation, metabolism and cell division)	Kirkland *et al.* (2007)
	PAN-A deletion has global impact on proteome including increase in phosphoproteins. MS-based proteomics used to identify nine Ser/Thr/Tyr phosphosites including two that are unique to PAN-A mutant strain. Additional proteins altered in PAN-A mutant strain were identified	Kirkland *et al.* (2008)
Hbt. salinarum proteasome	Immunoanalysis/RNA mapping provide preliminary evidence that PAN-A and -B do not associate, PAN-B and CP interact in ATP-dependent manner and PAN N-termini are heterogeneous. Heterogeneity proposed to have biological role in proteolysis	Chamieh *et al.* (2008)
Hfx. volcanii proteasome	Systematic single and double deletion of CP and PAN genes (including conditional lethal and trans-complement strains) used to demonstrate: CPs are essential; PAN-A/B are dispensable; α1 and PAN-A are required for optimal growth at low salt and in the presence of L-canavanine (stimulates protein unfolding); α1 is required for optimal growth after exposure to thermal stress; α2 and PAN-B do not complement α1 and PAN-A mutations, respectively, suggesting distinct roles for these homologues in the cell	Zhou *et al.* (2008)
	N-terminal penultimate residue of α1 influences the Nα-acetylation and protein levels α1 as well as the growth rate, and stress responses of the cell	Humbard *et al.* (2009)
	PAN-A mutation increases the half-life of PCNA (archaeal DNA-sliding clamp protein)	Kirkland and Maupin-Furlow (2009)

Table 8.1 continued

System	Important finding(s)/achievement(s)	Reference
	Proteins with hydrophobic C-terminal residues appear unstable (may be linked to proteasome function)	Reuter *et al.* (2010)
	Proteasome proteins are phosphorylated (α1, α2, PAN-A) and *O*-methylated (α1). Mutations in α1 phosphosites display dominant negative phenotypes for cell viability and pigmentation. Rio1 Ser/Thr kinase phosphorylates α1 (but not phosphosite variants). Phosphorylated PAN-A is found associated with PAN-B	Humbard *et al.* (2010b)
	Third CP subtype ($\alpha2_7\ \beta_7\beta_7\alpha2_7$) purified from *Hfx. volcanii*. Two symmetrical CPs ($\alpha1_7\beta_7\beta_7\alpha1_7$, $\alpha2_7\beta_7\beta_7\alpha2_7$) analysed by chemical cross-linking, MS and *in silico* modelling	Karadzic *et al.* (2012)
Hfx. volcanii sampylation	First report that ubiquitin-like proteins (SAMP1/2) form protein conjugates in Archaea	Humbard *et al.* (2010a)
	E1- and ubiquitin-like proteins demonstrated to mediate both protein conjugation and sulfur transfer in Archaea	Miranda *et al.* (2011)
Nab. magadii Lon	Lon membrane-associated and transcriptionally linked to protease gene homologue. Lon may bind DNA	Sastre *et al.* (2011)
Hbt. salinarum proteasome	Stress proposed to regulate PAN-proteasome system in *Halobacterium*. Increased peptidase activity detected in CP-associated fractions after thermal and low salt stresses. CP and PAN transcripts may be linked to stress. PAN-A and PAN-B thought associate after exposure to stress	Chamieh *et al.* (2012)

From work using *Hfx. volcanii* as a model system, a picture of archaeal proteasome function is revealed. The results of these studies demonstrate that halophilic Archaea can synthesize a population of proteasomal subtypes including PANs of PAN-A and PAN-B homo- and hetero-oligomeric composition (Humbard *et al.*, 2010b) and CPs composed of β subunits that associate with outer rings of different α subunit composition (symmetrical CPs in $\alpha1_7\beta_7\beta_7\alpha1_7$ and $\alpha2_7\beta_7\beta_7\alpha2_7$ configuration as well as asymmetrical CPs of $\alpha1_7\beta_7\beta_7\alpha2_7$ stoichiometry) (Wilson *et al.*, 1999; Kaczowka and Maupin-Furlow, 2003; Karadzic *et al.*, 2012). Similarly to eukaryotes, proteasomal proteins of *Hfx. volcanii* are modified by Nα-acetylation (acetylation of the α-amino group of N-terminal Met), phosphorylation of Ser/Thr residues, *O*-methylesterification of Asp/Glu residues and cleavage of a large propeptide (49 aa) to expose the active site N-terminal Thr of the CP β-type subunit (Wilson *et al.*, 1999; Humbard *et al.*, 2006, 2010b). Study of *Hfx. volcanii* proteasomes also reveals that the levels of proteasomal proteins are modulated with growth phase: PAN-B and α2 proteins at highest levels in stationary versus log phase, while PAN-A, α1 and β protein levels are relatively constant throughout the different phases of growth (Reuter *et al.*, 2004). Interestingly, like eukaryotes, CPs are essential: a lethal phenotype is observed after conditional depletion of either the single β gene or both α-type genes (Zhou *et al.*, 2008) suggesting that CPs are important in regulating cell division, protein quality control, or other important functions in *Hfx. volcanii*. In contrast, the viability of *panA panB* double mutants indicates PANs are dispensable (Zhou *et al.*, 2008) signifying CPs have function(s) independent of PANs and may use alternative AAA+ ATPases (e.g. Cdc48/p97/VAT homologues) for some proteolytic functions. Consistent with the thought that archaeal proteasomes have a role in protein quality control, PAN-A and α1 are needed for *Hfx. volcanii* to overcome stresses such as exposure to L-canavanine (an amino acid analogue that causes protein unfolding) and growth at suboptimal salt conditions (Zhou *et al.*, 2008). The α1 protein is also needed for cells to overcome thermal stress (Zhou *et al.*, 2008).

Targeting proteins for proteolysis

To avoid the widespread and uncontrolled breakdown of proteins (synthesized at high energy cost), cells have mechanisms to select proteins from their milieu for destruction by proteases. The mechanisms used by cells to target proteins for degradation vary but often involve changes in protein structure. In this section, mechanisms thought to be used by Archaea for selective proteolysis are discussed including sampylation (an archaeal form of ubiquitylation), the N-end rule pathway (including Nα-acetylation of N-terminal residues), and phosphorylation. Other mechanisms not discussed, but which may also control proteolysis in Archaea, include N ε-acetylation of lysine residues, glycosylation and methylation. This latter suggestion is based on the demonstration of these types of post-translational modifications in Archaea (Soppa, 2010; Calo *et al.*, 2010; Botting *et al.*, 2010; Streif *et al.*, 2010; Humbard *et al.*, 2010b) and their known association with proteolysis in eukaryotes (Ravid and Hochstrasser, 2008; Pradhan *et al.*, 2009; Bertaggia *et al.*, 2012).

Ubiquitin-like protein conjugation

Small archaeal ubiquitin-like modifier proteins (SAMPs) form isopeptide bonds to lysine residues of proteins by a ubiquitylation-like process termed sampylation (Humbard *et al.*, 2010a) (Fig. 8.5). This mechanism of post-translational modification is predicted for all Archaea and demonstrated for *Hfx. volcanii* (Humbard *et al.*, 2010a). Sampylation appears to be an ancient form of protein modification based on its conservation with the TtuBC protein conjugation system recently demonstrated for the hyperthermophilic bacterium *Thermus thermophilus* (Shigi, 2012).

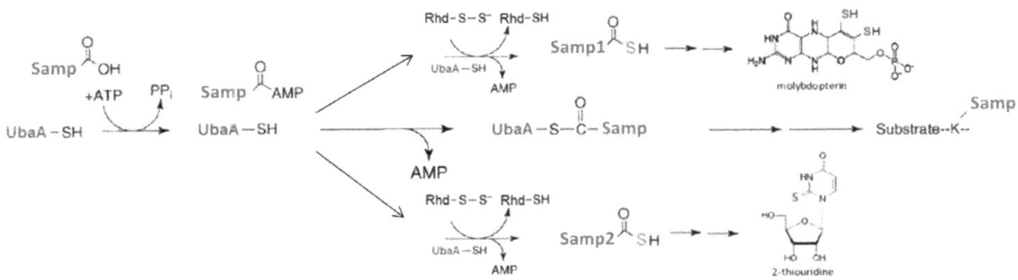

Figure 8.5 Sampylation, an archaeal form of ubiquitylation. In *Hfx. volcanii*, small archaeal ubiquitin-like modifier proteins (SAMPs) can form isopeptide bonds to proteins by a mechanism termed sampylation that is reminiscent of ubiquitylation in its dependence upon the E1 homologue UbaA for function. However, unlike ubiquitylation, E2 and E3 homologues are not predicted for sampylation. UbaA and SAMP1/2 also appear to be linked to sulfur mobilization in the biosynthesis of MoCo and thiolated tRNA, where rhodanese homologues (Rhd) may provide the sulfur needed for SAMP thiocarboxylation.

SAMPs are members of the β-grasp fold superfamily found in all domains of life. Small proteins with β-grasp folds include the ubiquitin and ubiquitin-like proteins that form protein conjugates in eukaryotes (Burroughs *et al.*, 2007). MoaD/ThiS-type proteins, which are required for sulfur transfer to form biomolecules (e.g. molybdenum cofactor, thiamine, and thiolated tRNA), are also members of the β-grasp fold superfamily (Burroughs *et al.*, 2007). A diglycine motif, on the extreme C-terminus of the β-grasp fold proteins, is typically required for function (i.e. protein conjugation, sulfur transfer). Most Archaea encode more than one small protein with a C-terminal diglycine motif that is predicted to form a β-grasp fold structure. In *Hfx. volcanii*, two of these ubiquitin-like proteins (SAMP1/2) are demonstrated to form protein conjugates (Humbard *et al.*, 2010a). The C-terminal diglycine motif of SAMP1/2 is required for function (Humbard *et al.*, 2010a). SAMP1and its methanogen homologue have a β-grasp fold structure based on solution NMR and X-ray crystallography (Ranjan *et al.*, 2010; Jeong *et al.*, 2011), and SAMP2 is predicted to have a related structural fold based on modelling (Jeong *et al.*, 2011).

The standard ubiquitin modification occurs through an isopeptide linkage between the α-carboxyl of the C-terminal glycine of ubiquitin and the ε-amino group of a lysine residue on the target protein (Hochstrasser, 2009). A three-step cascade catalyses the formation of this isopeptide bond. The E1 or ubiquitin-activating enzyme forms an ATP-dependent thioester linkage between ubiquitin and the cysteine residue of the E1 active site. Next, an E2, or conjugating enzyme, binds E1 and transfers the ubiquitin to E2. Finally, the E2 binds an E3, or ubiquitin ligase, which has bound the protein target and catalyses the formation of the isopeptide bond to the target protein.

Sampylation has many aspects that are analogous to ubiquitylation. Like ubiquitin, SAMP is conjugated to proteins by an isopeptide bond between its C-terminal α-carboxyl group and the lysine residues of target proteins (Humbard *et al.*, 2010a) (Fig. 8.5). Furthermore, an E1-like enzyme termed ubiquitin-like activating enzyme of Archaea (UbaA) is required for sampylation (Miranda *et al.*, 2011). Additional proteins that may provide specificity to the sampylation pathway (e.g. E2 and E3 homologues) have yet to be identified. RING (really interesting new gene) domain E3 ubiquitin ligases (Borden and Freemont, 1996; Budhidarmo *et al.*, 2012) harbour a zinc finger motif that is common to many small hypothetical proteins of Archaea. While zinc fingers can mediate a wide range of protein activities (e.g. DNA binding, protein–protein interaction), the function(s) of small proteins with these motifs is only recently studied in Archaea (Tarasov *et al.*, 2008). Thus, in archaeal sampylation, it is unclear whether the E1-like UbaA catalyses the first step of a divergent ubiquitin-like E1–E2–E3 cascade or directly conjugates SAMPs to their protein targets.

In analogy to ubiquitylation, sampylation is thought to be regulated and reversible. In eukaryotes, deubiquitylases (DUBs) are responsible for the removal of ubiquitin from protein conjugates and have an important role in cell regulation (Burrows and Johnston, 2012). DUBs are of two major types: cysteine proteases and JAMM [Jab1/Mov34/Mpr1 Pad1 N-terminal+ (MPN+)] domain metalloproteases (Komander, 2010). Of these, the JAMM domain homologues are conserved in Archaea and, thus, predicted to mediate desampylation (Humbard *et al.*, 2010a; Maupin-Furlow, 2012). Archaeal JAMM domain protein function is yet to be demonstrated. However, the crystal structure of an *Archaeoglobus fulgidus* JAMM domain protein (termed AfJAMM) has been used to model the active sites of the eukaryotic DUB Rpn11/Poh1 subunit of 26S proteasomes and deneddylase Csn5/Jab1 subunit of the COP9 signalosome (CSN) (Tran *et al.*, 2003; Ambroggio *et al.*, 2004).

In analogy to ubiquitin, sampylation is thought to alter protein structure, enzyme activity and the types of protein partners that would associate including proteasomes, transcription factors and others (Maupin-Furlow, 2012). However, the physiological reason for the formation of ubiquitin-like protein conjugates in Archaea (and hyperthermophilic bacteria) remains to be determined. Interestingly, SAMPs not only form isopeptide bonds with protein targets, but are also required for sulfur transfer to biomolecules including molybdenum cofactor (MoCo) and tRNA (similar to Urm1 of eukaryotes and TtuB of the bacterium *Thermus thermophilus*) (Miranda *et al.*, 2011; Wang *et al.*, 2011; Shigi, 2012). Consistent with this physiological association, SAMP encoding genes are linked to with molybdenum cofactor (MoCo) biosynthesis and tRNA modification (Makarova and Koonin, 2010). Sampylation differs from the system of protein conjugation recently predicted for the archaeon *Candidatus Caldiarchaeum subterraneum* (based on metagenomics) (Nunoura *et al.*, 2011) and some bacteria (based on analogy) (Burroughs *et al.*, 2011). This latter system is rare in Archaea (restricted to *Candidatus Caldiarchaeum subterraneum*) and appears to incorporate not only ubiquitin activating E1 homologues but also ubiquitin-conjugating E2 and ubiquitin ligase E3 homologues (Nunoura *et al.*, 2011).

Sampylation may be linked to proteasome function. The levels of SAMP1-protein conjugates are higher in proteasome mutants than wild-type cells providing indirect evidence for a connection between these two systems (Humbard *et al.*, 2010a). However, *ubaA* (encoding the single ubiquitin-activating E1 homologue of *Hfx. volcanii*) is dispensable while proteasomal CP genes are essential (Miranda *et al.*, 2011). Thus, not all proteasome functions are linked to sampylation.

N-end rule pathway

The N-end rule is a pathway in which the N-terminal residue of a protein determines its half-life (Varshavsky, 2011; Tasaki *et al.*, 2012). Not only can the type of N-terminal residue influence protein stability but also its modification by acetylation, proteolytic cleavage, oxidation, deamidation, leucylation, arginylation, and phenylalanylation. N-end rule pathways function in bacteria and eukaryotes and are presumed to occur in Archaea (Varshavsky, 2011; Tasaki *et al.*, 2012).

In bacteria such as *E. coli*, the N-end rule pathway is streamlined (Erbse *et al.*, 2006; Varshavsky, 2011; Tasaki *et al.*, 2012). A ClpS adaptor protein recognizes proteins with destabilizing N-termini and targets them for degradation by ClpAP. N-terminal residues that are destabilizing include residues that are bulky hydrophobic (Leu, Phe, Trp, Tyr) or susceptible to leucylation/phenylalanylation (Asp, Glu, Arg, Lys, Met).

The N-end rule pathway is less direct in eukaryotes than in bacteria (Varshavsky, 2011; Tasaki *et al.*, 2012). Components of the ubiquitylation system (i.e. E3 ubiquitin ligases) are used in eukaryotes to recognize proteins with destabilizing N-termini and modify these proteins with covalent poly-ubiquitin chains. The poly-ubiquitylated proteins are then recognized by 26S proteasomes for destruction. Destabilizing N-termini include bulky hydrophobic residues (Ile, Leu, Phe, Trp, Tyr), basic residues (Arg, His, Lys), residues susceptible to oxidation (Cys) or deamidation (Asn, Gln), and residues that undergo arginylation (Glu, Asp, oxidized Cys). Recent evidence also reveals an 'Ac/N-end' rule pathway operates in yeast, in which an E3 ubiquitin ligase (Doa10) ubiquitylates substrates carrying exposed N-terminal residues that are Nα-acetylated (Met, Ala, Val, Ser, Thr, Cys) and, thus, targets them for degradation by 26 proteasomes (Hwang *et al.*, 2010). While yet to be demonstrated, proteins with 'hidden' acetylated N-termini (e.g. due to rapid folding, interaction with chaperones, or assembly into appropriate multisubunit complexes) are thought to have degradation signals that are inaccessible for ubiquitinylation by E3 ubiquitin ligases. Thus, the protein would be stabilized. Contrastingly, delayed or defective protein folding would keep acetylated N-termini exposed and available for ubiquitylation and ultimate destruction by proteasomes.

An N-end rule pathway is thought to exist in Archaea (but yet to be demonstrated) (Varshavsky, 2011). Glimpses into an archaeal Ac/N-end rule pathway are provided by study of α-type subunits of 20S proteasomes (Humbard *et al.*, 2009; Varshavsky, 2011). These α-type proteins are acetylated at their N-terminal Met in diverse Archaea (e.g. species of *Haloferax, Halobacterium, Natronomonas, Sulfolobus* and *Methanothermobacter*) (Humbard *et al.*, 2006; Falb *et al.*, 2006; Mackay *et al.*, 2007; Enoki *et al.*, 2011). Using *Hfx. volcanii* as a model system, modification of the N-terminal penultimate residue of α1 is found to alter the Nα-acetylation profiles of α1 and the concentration of α1 protein in the cell (Humbard *et al.*, 2009). In particular, α1 variants that were only partially Nα-acetylated were at dramatically higher levels than the fully Nα-acetylated α1, and the preferred form of α1 that was assembled into 20S proteasomes was Nα-acetylated. These findings can now be explained in light of the Ac/N-end rule pathway developed for yeast (Hwang *et al.*, 2010; Varshavsky, 2011). Based on this model, α1 proteins with exposed Nα-acetylated N-termini would be less stable than unmodified forms of α1 unless assembled with proteasomal partners. Association of α1 with proteasomal partners may obstruct recognition of the Nα-acetylated domain of α1 by an Ac/N-end rule pathway and prevent its proteolytic destruction.

Phosphorylation

Phosphorylation is a covalent form of post-translational modification that occurs at His, Asp, Ser, Thr or Tyr residues of proteins. Halophilic Archaea were demonstrated to phosphorylate all of these different types of residues (Rudolph *et al.*, 1995; Kirkland *et al.*, 2008; Aivaliotis *et al.*, 2009). The cellular advantages of using phosphorylation are that the process is rapid and reversible and can generate conformational changes in protein structure that mediate an array of biological responses (Johnson and Barford, 1993). A number of reviews and genomic surveys are available that highlight the phosphorylation of archaeal proteins and the enzymes (protein kinases/phosphatases) likely to mediate and/or regulate this form of posttranslational modification (Leonard *et al.*, 1998; Kennelly and Potts, 1999; Kennelly, 2003; Eichler and Adams, 2005; Tyagi *et al.*, 2010).

Phosphorylation often regulates proteolysis. A classic example is in *E. coli*, where the response regulator RssB is a σ^S recognition factor, whose affinity for RpoS is enhanced by phosphorylation of its receiver domain (Hengge-Aronis, 2002). RssB~P delivers the RpoS cargo to the ClpXP protease for degradation. Likewise in eukaryotes, phosphorylation can serve as a molecular signature (or 'phosphodegron') that leads to recruitment of E3 ubiquitin ligases followed by ubiquitylation and cleavage/hydrolysis of the phosphoprotein by 26S proteasomes (Chen *et al.*, 1995; Lin *et al.*, 2006; Zou *et al.*, 2011; Omnus *et al.*, 2011).

Phosphorylation may be linked to proteasome function in Archaea. In support of this possibility, an *Hfx. volcanii panA* mutant (deficient in synthesis of the proteasomal AAA+ PAN-A) displays a striking increase in the number of proteins that are phosphorylated compared to its parent strain (Kirkland *et al.*, 2008). A number of phosphosites were identified in these strains by enrichment of phosphopeptides using immobilized metal affinity chromatography (IMAC) and metal oxide affinity chromatography (MOAC) (in parallel and sequentially) followed by tandem mass spectrometry (MS/MS) (Kirkland *et al.*, 2008). Included in this analysis were phosphosites that mapped to Cdc6–1/Orc1–1 Ser321 and pyruvate kinase Thr533 that were unique to the proteasomal *panA* mutant (vs. parent). Proteins corresponding to these phosphopeptides are speculated to be substrates of the PAN-A AAA+ of proteasomes that depend on their phosphorylation status for destruction.

Conclusions

We now recognize that proteolysis and protein conjugation mechanisms are important in regulating a broad array of cellular activities ranging from cellular homeostasis and protein quality to metabolism, transcription and cell cycle control. Consistent with this insight, proteases and protein conjugation systems are present in Archaea that appear important in regulating cell function. Most notable are the intramembrane proteases, energy-dependent proteases (Lon and proteasomes), and the recently discovered archaeal form of ubiquitylation, termed sampylation. Sampylation is likely to regulate cells by targeting proteins for proteasome-mediated proteolysis, modulating enzyme activity, altering association with protein partners, and/or facilitating other activities of the cell.

References

Adam, Z., Adamska, I., Nakabayashi, K., Ostersetzer, O., Haussuhl, K., Manuell, A., Zheng B., Vallon, O., Rodermel, S.R., Shinozaki, K., *et al.* (2001). Chloroplast and mitochondrial proteases in *Arabidopsis*. A proposed nomenclature. Plant Physiol. *125*, 1912–1918.

Aivaliotis, M., Macek, B., Gnad, F., Reichelt, P., Mann, M., and Oesterhelt, D. (2009). Ser/Thr/Tyr protein phosphorylation in the archaeon *Halobacterium salinarum* – a representative of the third domain of life. PLoS One *4*, e4777.

Akopian, T.N., Kisselev, A.F., and Goldberg, A.L. (1997). Processive degradation of proteins and other catalytic properties of the proteasome from *Thermoplasma acidophilum*. J. Biol. Chem. *272*, 1791–1798.

Alba, B.M., Leeds, J.A., Onufryk, C., Lu, C.Z., and Gross, C.A. (2002). DegS and YaeL participate sequentially in the cleavage of RseA to activate the s^E-dependent extracytoplasmic stress response. Genes Dev. *16*, 2156–2168.

Albers, S.V., Szabo, Z., and Driessen, A.J. (2003). Archaeal homolog of bacterial type IV prepilin signal peptidases with broad substrate specificity. J. Bacteriol. *185*, 3918–3925.

Alexandrescu, A.T., Gittis, A.G., Abeygunawardana, C., and Shortle, D. (1995). NMR structure of a stable "OB-fold" sub-domain isolated from staphylococcal nuclease. J. Mol. Biol. *250*, 134–143.

Alexandrescu, A.T., Jaravine, V.A., Dames, S.A., and Lamour, F.P. (1999). NMR hydrogen exchange of the OB-fold protein LysN as a function of denaturant: the most conserved elements of structure are the most stable to unfolding. J. Mol. Biol. *289*, 1041–1054.

Ambroggio, X.I., Rees, D.C., and Deshaies, R.J. (2004). JAMM: a metalloprotease-like zinc site in the proteasome and signalosome. PLoS Biol. *2*, E2.

An, F.Y., Sulavik, M.C., and Clewell, D.B. (1999). Identification and characterization of a determinant (*eep*) on the *Enterococcus faecalis* chromosome that is involved in production of the peptide sex pheromone cAD1. J. Bacteriol. *181*, 5915–5921.

Bar-Nun, S., and Glickman, M.H. (2012). Proteasomal AAA-ATPases: structure and function. Biochim. Biophys. Acta *1823*, 67–82.

Bardy, S.L., and Jarrell, K.F. (2002). FlaK of the archaeon *Methanococcus maripaludis* possesses preflagellin peptidase activity. FEMS Microbiol. Lett. *208*, 53–59.

Bardy, S.L., and Jarrell, K.F. (2003). Cleavage of preflagellins by an aspartic acid signal peptidase is essential for flagellation in the archaeon *Methanococcus voltae*. Mol. Microbiol. *50*, 1339–1347.

Barthelme, D., and Sauer, R.T. (2012). Identification of the Cdc48•20S proteasome as an ancient AAA+ proteolytic machine. Science 337, 843–846.

Barthelme, D., and Sauer, R.T. (2013). Bipartite determinants mediate an evolutionarily conserved interaction between Cdc48 and the 20S peptidase. Proc. Natl. Acad. Sci. U.S.A. *110*, 3327–3332.

Bertaggia, E., Coletto, L., and Sandri, M. (2012). Posttranslational modifications control FoxO3 activity during denervation. Am. J. Physiol. Cell Physiol. *302*, C587–C596.

Besche, H., and Zwickl, P. (2004). The *Thermoplasma acidophilum* Lon protease has a Ser-Lys dyad active site. Eur. J. Biochem. *271*, 4361–4365.

Besche, H., Tamura, N., Tamura, T., and Zwickl, P. (2004). Mutational analysis of conserved AAA+ residues in the archaeal Lon protease from *Thermoplasma acidophilum*. FEBS Lett. *574*, 161–166.

Borden, K.L., and Freemont, P.S. (1996). The RING finger domain: a recent example of a sequence-structure family. Curr. Opin. Struct. Biol. *6*, 395–401.

Botos, I., Melnikov, E.E., Cherry, S., Kozlov, S., Makhovskaya, O.V., Tropea, J.E., Gustchina, A., Rotanova, T.V., and Wlodawer, A. (2005). Atomic-resolution crystal structure of the proteolytic domain of *Archaeoglobus fulgidus lon* reveals the conformational variability in the active sites of lon proteases. J. Mol. Biol. *351*, 144–157.

Botting, C.H., Talbot, P., Paytubi, S., and White, M.F. (2010). Extensive lysine methylation in hyperthermophilic crenarchaea: potential implications for protein stability and recombinant enzymes. Archaea *2010*, 106341.

Brown, M.S., Ye, J., Rawson, R.B., and Goldstein, J.L. (2000). Regulated intramembrane proteolysis: a control mechanism conserved from bacteria to humans. Cell *100*, 391–398.

Brown, R.S. (2005). Zinc finger proteins: getting a grip on RNA. Curr. Opin. Struct. Biol. *15*, 94–98.

Budhidarmo, R., Nakatani, Y., and Day, C.L. (2012). RINGs hold the key to ubiquitin transfer. Trends Biochem. Sci. *37*, 58–65.

Burroughs, A.M., Balaji, S., Iyer, L.M., and Aravind, L. (2007). Small but versatile, the extraordinary functional and structural diversity of the β-grasp fold. Biol. Direct *2*, 18.

Burroughs, A.M., Iyer, L.M., and Aravind, L. (2011). Functional diversification of the RING finger and other binuclear treble clef domains in prokaryotes and the early evolution of the ubiquitin system. Mol. Biosyst. *7*, 2261–2277.

Burrows, J.F., and Johnston, J.A. (2012). Regulation of cellular responses by deubiquitinating enzymes, an update. Front. Biosci. *17*, 1184–1200.

Calo, D., Kaminski, L., and Eichler, J. (2010). Protein glycosylation in Archaea: sweet and extreme. Glycobiology *20*, 1065–1076.

Cardozo, C., and Michaud, C. (2002). Proteasome-mediated degradation of tau proteins occurs independently of the chymotrypsin-like activity by a nonprocessive pathway. Arch. Biochem. Biophys. *408*, 103–110.

Cha, S.-S., An, Y.J., Lee, C.R., Lee, H.S., Kim, Y.-G., Kim, S.J., Kwon, K.K., De Donatis, G.M., Lee, J.-H., Maurizi, M.R., *et al.* (2010). Crystal structure of Lon protease: molecular architecture of gated entry to a sequestered degradation chamber. EMBO J. *29*, 3520–3530.

Chamieh, H., Guetta, D., and Franzetti, B. (2008). The two PAN ATPases from *Halobacterium* display N-terminal heterogeneity and form labile complexes with the 20S proteasome. Biochem. J. *411*, 387–397.

Chamieh, H., Marty, V., Guetta, D., Perollier, A., and Franzetti, B. (2012). Stress regulation of the PAN-proteasome system in the extreme halophilic archaeon *Halobacterium*. Extremophiles *16*, 215–225.

Chen, J.C., Hottes, A.K., McAdams, H.H., McGrath, P.T., Viollier, P.H., and Shapiro, L. (2006). Cytokinesis signals truncation of the PodJ polarity factor by a cell cycle-regulated protease. EMBO J. *25*, 377–386.

Chen, Z., Hagler, J., Palombella, V.J., Melandri, F., Scherer, D., Ballard, D., and Maniatis, T. (1995). Signal-induced site-specific phosphorylation targets IkBa to the ubiquitin-proteasome pathway. Genes Dev. *9*, 1586–1597.

Chin, D.T., Goff, S.A., Webster, T., Smith, T., and Goldberg, A.L. (1988). Sequence of the *lon* gene in *Escherichia coli*. A heat-shock gene which encodes the ATP-dependent protease La. J. Biol. Chem. *263*, 11718–11728.

Ciechanover, A. (2010). Intracellular protein degradation: from a vague idea through the lysosome and the ubiquitin-proteasome system and onto human diseases and drug targeting. Medicina (Buenos Aires) *70*, 105–119.

Clewell, D.B. (2011). Tales of conjugation and sex pheromones: a plasmid and enterococcal odyssey. Mob. Genet. Elements *1*, 38–54.

Denny, J.B. (2004). Growth-associated protein of 43 kDa (GAP-43) is cleaved nonprocessively by the 20S proteasome. Eur. J. Biochem. *271*, 2480–2493.

Djuranovic, S., Hartmann, M.D., Habeck, M., Ursinus, A., Zwickl, P., Martin, J., Lupas, A.N., and Zeth, K. (2009). Structure and activity of the N-terminal substrate recognition domains in proteasomal ATPases. Mol. Cell *34*, 580–590.

Dougan, D.A., Mogk, A., Zeth, K., Turgay, K., and Bukau, B. (2002a). AAA+ proteins and substrate recognition, it all depends on their partner in crime. FEBS Lett. *529*, 6–10.

Dougan, D.A., Reid, B.G., Horwich, A.L., and Bukau, B. (2002b). ClpS, a substrate modulator of the ClpAP machine. Mol. Cell *9*, 673–683.

Eichler, J., and Adams, M.W. (2005). Posttranslational protein modification in Archaea. Microbiol. Mol. Biol. Rev. *69*, 393–425.

Enoki, M., Shinzato, N., Sato, H., Nakamura, K., and Kamagata, Y. (2011). Comparative proteomic analysis of *Methanothermobacter themautotrophicus* ΔH in pure culture and in co-culture with a butyrate-oxidizing bacterium. PLoS One *6*, e24309.

Erbse, A., Schmidt, R., Bornemann, T., Schneider-Mergener, J., Mogk, A., Zahn, R., Dougan, D.A., and Bukau, B. (2006). ClpS is an essential component of the N-end rule pathway in *Escherichia coli*. Nature *439*, 753–756.

Erez, E., Fass, D., and Bibi, E. (2009). How intramembrane proteases bury hydrolytic reactions in the membrane. Nature *459*, 371–378.

Falb, M., Aivaliotis, M., Garcia-Rizo, C., Bisle, B., Tebbe, A., Klein, C., Konstantinidis, K., Siedler, F., Pfeiffer, F., and Oesterhelt, D. (2006). Archaeal N-terminal protein maturation commonly involves N-terminal acetylation, a large-scale proteomics survey. J. Mol. Biol. *362*, 915–924.

Feng, L., Yan, H., Wu, Z., Yan, N., Wang, Z., Jeffrey, P.D., and Shi, Y. (2007). Structure of a site-2 protease family intramembrane metalloprotease. Science *318*, 1608–1612.

Ferrington, D.A., Sun, H., Murray, K.K., Costa, J., Williams, T.D., Bigelow, D.J., and Squier, T.C. (2001). Selective degradation of oxidized calmodulin by the 20 S proteasome. J. Biol. Chem. *276*, 937–943.

Finley, D., Tanaka, K., Mann, C., Feldmann, H., Hochstrasser, M., Vierstra, R., Johnston, S., Hampton, R., Haber, J., McCusker, J., *et al.* (1998). Unified nomenclature for subunits of the *Saccharomyces cerevisiae* proteasome regulatory particle. Trends Biochem. Sci. 23, 244–245.

Flynn, J.M., Levchenko, I., Seidel, M., Wickner, S.H., Sauer, R.T., and Baker, T.A. (2001). Overlapping recognition determinants within the ssrA degradation tag allow modulation of proteolysis. Proc. Natl. Acad. Sci. U.S.A. *98*, 10584–10589.

Flynn, J.M., Neher, S.B., Kim, Y.I., Sauer, R.T., and Baker, T.A. (2003). Proteomic discovery of cellular substrates of the ClpXP protease reveals five classes of ClpX-recognition signals. Mol. Cell *11*, 671–683.

Franzetti, B., Schoehn, G., Garcia, D., Ruigrok, R.W., and Zaccai, G. (2002). Characterization of the proteasome from the extremely halophilic archaeon *Haloarcula marismortui*. Archaea *1*, 53–61.

Fu, G.K., and Markovitz, D.M. (1998). The human LON protease binds to mitochondrial promoters in a single-stranded, site-specific, strand-specific manner. Biochemistry *37*, 1905–1909.

Fu, G.K., Smith, M.J., and Markovitz, D.M. (1997). Bacterial protease Lon is a site-specific DNA-binding protein. J. Biol. Chem. *272*, 534–538.

Fukui, T., Eguchi, T., Atomi, H., and Imanaka, T. (2002). A membrane-bound archaeal Lon protease displays ATP-independent proteolytic activity towards unfolded proteins and ATP-dependent activity for folded proteins. J. Bacteriol. *184*, 3689–3698.

Gamsjaeger, R., Liew, C.K., Loughlin, F.E., Crossley, M., and Mackay, J.P. (2007). Sticky fingers: zinc-fingers as protein-recognition motifs. Trends Biochem. Sci. *32*, 63–70.

Gerega, A., Rockel, B., Peters, J., Tamura, T., Baumeister, W., and Zwickl, P. (2005). VAT, the thermoplasma homolog of mammalian p97/VCP, is an N domain-regulated protein unfoldase. J. Biol. Chem. *280*, 42856–42862.

Gil, M.A., Sherwood, K.E., and Maupin-Furlow, J.A. (2007). Transcriptional linkage of *Haloferax volcanii* proteasomal genes with non-proteasomal gene neighbours including RNase P, MOSC domain and SAM-methyltransferase homologues. Microbiology UK *153*, 3009–3022.

Golde, T.E., Wolfe, M.S., and Greenbaum, D.C. (2009). Signal peptide peptidases: a family of intramembrane-cleaving proteases that cleave type 2 transmembrane proteins. Semin. Cell. Dev. Biol. *20*, 225–230.

Gonzalez, M., Frank, E.G., Levine, A.S., and Woodgate, R. (1998). Lon-mediated proteolysis of the *Escherichia coli* UmuD mutagenesis protein: *in vitro* degradation and identification of residues required for proteolysis. Genes Dev. *12*, 3889–3899.

Gottesman, S. (1999). Regulation by proteolysis: developmental switches. Curr. Opin. Microbiol. *2*, 142–147.

Gottesman, S., and Maurizi, M.R. (1992). Regulation by proteolysis: energy-dependent proteases and their targets. Microbiol. Rev. *56*, 592–621.

Gottesman, S., Roche, E., Zhou, Y., and Sauer, R.T. (1998). The ClpXP and ClpAP proteases degrade proteins with carboxy-terminal peptide tails added by the SsrA-tagging system. Genes Dev. *12*, 1338–1347.

Groll, M., Brandstetter, H., Bartunik, H., Bourenkow, G., and Huber, R. (2003). Investigations on the maturation and regulation of archaebacterial proteasomes. J. Mol. Biol. *327*, 75–83.

Gur, E., Biran, D., and Ron, E.Z. (2011). Regulated proteolysis in Gram-negative bacteria – how and when? Nat. Rev. Microbiol. *9*, 839–848.

Hall, T.M. (2005). Multiple modes of RNA recognition by zinc finger proteins. Curr. Opin. Struct. Biol. *15*, 367–373.

Hengge-Aronis, R. (2002). Signal transduction and regulatory mechanisms involved in control of the s^S (RpoS) subunit of RNA polymerase. Microbiol. Mol. Biol. Rev. *66*, 373–395.

Hinnerwisch, J., Fenton, W.A., Furtak, K.J., Farr, G.W., and Horwich, A.L. (2005). Loops in the central channel of ClpA chaperone mediate protein binding, unfolding, and translocation. Cell *121*, 1029–1041.

Hochstrasser, M. (2009). Origin and function of ubiquitin-like proteins. Nature *458*, 422–429.

Hsieh, F.-C., Chen, C.-T., Weng, Y.-T., Peng, S.-S., Chen, Y.-C., Huang, L.-Y., Hu, H.-T., Wu, Y.-L., Lin, N.-C., and Wu, W.-F. (2011). Stepwise activity of ClpY (HslU) mutants in the processive degradation of *Escherichia coli* ClpYQ (HslUV) protease substrates. J. Bacteriol. *193*, 5465–5476.

Hu, J., Xue, Y., Lee, S., and Ha, Y. (2011). The crystal structure of GXGD membrane protease FlaK. Nature *475*, 528–531.

Humbard, M.A., Stevens, S.M. Jr., and Maupin-Furlow, J.A. (2006). Posttranslational modification of the 20S proteasomal proteins of the archaeon *Haloferax volcanii*. J. Bacteriol. *188*, 7521–7530.

Humbard, M.A., Zhou, G., and Maupin-Furlow, J.A. (2009). The N-terminal penultimate residue of 20S proteasome a1 influences its N^a-acetylation and protein levels as well as growth rate and stress responses of *Haloferax volcanii*. J. Bacteriol. *191*, 3794–3803.

Humbard, M.A., Miranda, H.V., Lim, J.M., Krause, D.J., Pritz, J.R., Zhou, G., Chen. S., Wells, L., and Maupin-Furlow, J.A. (2010a). Ubiquitin-like small archaeal modifier proteins (SAMPs) in *Haloferax volcanii*. Nature *463*, 54–60.

Humbard, M.A., Reuter, C.J., Zuobi-Hasona, K., Zhou, G., and Maupin-Furlow, J.A. (2010b). Phosphorylation and methylation of proteasomal proteins of the haloarcheon *Haloferax volcanii*. Archaea *2010*, 481725.

Hwang, C.S., Shemorry, A., and Varshavsky, A. (2010). N-terminal acetylation of cellular proteins creates specific degradation signals. Science *327*, 973–977.

Inobe, T., Kraut, D.A., and Matouschek, A. (2008). How to pick a protein and pull at it. Nat. Struct. Mol. Biol. *15*, 1135–1136.

Ishii, Y., Sonezaki, S., Iwasaki, Y., Miyata, Y., Akita, K., Kato, Y., and Amano, F. (2000). Regulatory role of C-terminal residues of SulA in its degradation by Lon protease in *Escherichia coli*. J. Biochem. *127*, 837–844.

Jennings, L.D., Lun, D.S., Medard, M., and Licht, S. (2008). ClpP hydrolyzes a protein substrate processively in the absence of the ClpA ATPase: mechanistic studies of ATP-independent proteolysis. Biochemistry *47*, 11536–11546.

Jeong, Y.J., Jeong, B.C., and Song, H.K. (2011). Crystal structure of ubiquitin-like small archaeal modifier protein 1 (SAMP1) from *Haloferax volcanii*. Biochem. Biophys. Res. Commun. *405*, 112–117.

Johnson, L.N., and Barford, D. (1993). The effects of phosphorylation on the structure and function of proteins. Annu. Rev. Biophys. Biomol. Struct. *22*, 199–232.

Kaczowka, S.J., and Maupin-Furlow, J.A. (2003). Subunit topology of two 20S proteasomes from *Haloferax volcanii*. J. Bacteriol. *185*, 165–174.

Karadzic, I.M., Maupin-Furlow, J., Humbard, M., Prunetti, L., Singh, P., and Goodlett, D.R. (2012). Chemical cross-linking, mass spectrometry, and *in silico* modeling of proteasomal 20S core particles of the haloarchaeon *Haloferax volcanii*. Proteomics *12*, 1806–1814.

Kennelly, P.J. (2003). Archaeal protein kinases and protein phosphatases: insights from genomics and biochemistry. Biochem. J. *370*, 373–389.

Kennelly, P.J., and Potts, M. (1999). Life among the primitives: protein *O*-phosphatases in prokaryotes. Front. Biosci. *4*, D372–D385.

Kim, A.C., Oliver, D.C., and Paetzel, M. (2008). Crystal structure of a bacterial signal peptide peptidase. J. Mol. Biol. *376*, 352–366.

Kinch, L.N., Ginalski, K., and Grishin, N.V. (2006). Site-2 protease regulated intramembrane proteolysis: sequence homologs suggest an ancient signaling cascade. Protein Sci. *15*, 84–93.

Kirkland, P.A., and Maupin-Furlow, J.A. (2009). Stabilization of an archaeal DNA-sliding clamp protein, PCNA, by proteasome-activating nucleotidase gene knockout in *Haloferax volcanii*. FEMS Microbiol. Lett. *294*, 32–36.

Kirkland, P.A., Reuter, C.J., and Maupin-Furlow, J.A. (2007). Effect of proteasome inhibitor clasto-lactacystin-β-lactone on the proteome of the haloarchaeon *Haloferax volcanii*. Microbiology *53*, 2271–2280.

Kirkland, P.A., Gil, M.A., Karadzic, I.M., and Maupin-Furlow, J.A. (2008). Genetic and proteomic analyses of a proteasome-activating nucleotidase A mutant of the haloarchaeon *Haloferax volcanii*. J. Bacteriol. *190*, 193–205.

Kirstein, J., Moliere, N., Dougan, D.A., and Turgay, K. (2009). Adapting the machine, adaptor proteins for Hsp100/Clp and AAA+ proteases. Nat. Rev. Microbiol. *7*, 589–599.

Kisselev, A.F., Akopian, T.N., and Goldberg, A.L. (1998). Range of sizes of peptide products generated during degradation of different proteins by archaeal proteasomes. J. Biol. Chem. *273*, 1982–1989.

Kisselev, A.F., Akopian, T.N., Woo, K.M., and Goldberg, A.L. (1999). The sizes of peptides generated from protein by mammalian 26 and 20 S proteasomes. Implications for understanding the degradative mechanism and antigen presentation. J. Biol. Chem. *274*, 3363–3371.

Klug, A. (1999). Zinc finger peptides for the regulation of gene expression. J. Mol. Biol. *293*, 215–218.

Komander, D. (2010). Mechanism, specificity and structure of the deubiquitinases. Subcell. Biochem. *54*, 69–87.

Koonin, E.V., Makarova, K.S., Rogozin, I.B., Davidovic, L., Letellier, M.C., and Pellegrini, L. (2003). The rhomboids: a nearly ubiquitous family of intramembrane serine proteases that probably evolved by multiple ancient horizontal gene transfers. Genome Biol. *4*, R19.

Kowit, J.D., and Goldberg, A.L. (1977). Intermediate steps in the degradation of a specific abnormal protein in *Escherichia coli*. J. Biol. Chem. *252*, 8350–8357.

Kusmierczyk, A.R., Kunjappu, M.J., Kim, R.Y., and Hochstrasser, M. (2011). A conserved 20S proteasome assembly factor requires a C-terminal HbYX motif for proteasomal precursor binding. Nat. Struct. Mol. Biol. *18*, 622–629.

Large, A.T., Kovacs, E., and Lund, P.A. (2002). Properties of the chaperonin complex from the halophilic archaeon *Haloferax volcanii*. FEBS Lett. *532*, 309–312.

Leonard, C.J., Aravind, L., and Koonin, E.V. (1998). Novel families of putative protein kinases in bacteria and archaea: evolution of the "eukaryotic" protein kinase superfamily. Genome Res. *8*, 1038–1047.

Levchenko, I., Seidel, M., Sauer, R.T., and Baker, T.A. (2000). A specificity-enhancing factor for the ClpXP degradation machine. Science *289*, 2354–2356.

Lewis, A.P., and Thomas, P.J. (1999). A novel clan of zinc metallopeptidases with possible intramembrane cleavage properties. Protein Sci. *8*, 439–442.

Licht, S., and Lee, I. (2008). Resolving individual steps in the operation of ATP-dependent proteolytic molecular machines: from conformational changes to substrate translocation and processivity. Biochemistry *47*, 3595–3605.

Lin, D.I., Barbash, O., Kumar, K.G., Weber, J.D., Harper, J.W., Klein-Szanto, A.J., Rustgi, A., Fuchs, S.Y., and Diehl. J.A. (2006). Phosphorylation-dependent ubiquitination of cyclin D1 by the $SCF^{FBX4-\alpha B\ crystallin}$ complex. Mol. Cell *24*, 355–366.

Liu, T., Lu, B., Lee, I., Ondrovicova, G., Kutejova, E., and Suzuki, C.K. (2004). DNA and RNA binding by the mitochondrial lon protease is regulated by nucleotide and protein substrate. J. Biol. Chem. *279*, 13902–13910.

Lowe, J., Stock, D., Jap, B., Zwickl, P., Baumeister, W., and Huber, R. (1995). Crystal structure of the 20S proteasome from the archaeon *T. acidophilum* at 3.4 A resolution. Science *268*, 533–539.

Lu, B., Liu, T., Crosby, J.A., Thomas-Wohlever, J., Lee, I., and Suzuki, C.K. (2003). The ATP-dependent Lon protease of *Mus musculus* is a DNA-binding protein that is functionally conserved between yeast and mammals. Gene *306*, 45–55.

Lupas, A., Flanagan, J.M., Tamura, T., and Baumeister, W. (1997). Self-compartmentalizing proteases. Trends Biochem. Sci. *22*, 399–404.

McAdams, H.H., and Shapiro, L. (2009). System-level design of bacterial cell cycle control. FEBS Lett. *583*, 3984–3991.

Mackay, D.T., Botting, C.H., Taylor, G.L., and White, M.F. (2007). An acetylase with relaxed specificity catalyses protein N-terminal acetylation in *Sulfolobus solfataricus*. Mol. Microbiol. *64*, 1540–1548.

Makarova, K.S., and Koonin, E.V. (2010). Archaeal ubiquitin-like proteins: functional versatility and putative ancestral involvement in tRNA modification revealed by comparative genomic analysis. Archaea *2010*, 61014405.

Martin, A., Baker, T.A., and Sauer, R.T. (2008a). Diverse pore loops of the AAA+ ClpX machine mediate unassisted and adaptor-dependent recognition of ssrA-tagged substrates. Mol. Cell *29*, 441–450.

Martin, A., Baker, T.A., and Sauer, R.T. (2008b). Pore loops of the AAA+ ClpX machine grip substrates to drive translocation and unfolding. Nat. Struct. Mol. Biol. *15*, 1147–1151.

Matthews, J.M., and Sunde, M. (2002). Zinc fingers – folds for many occasions. IUBMB Life *54*, 351–355.

Maupin-Furlow, J. (2012). Proteasomes and protein conjugation across domains of life. Nat. Rev. Microbiol. *10*, 100–111.

Maupin-Furlow, J.A., Gil, M.A., Karadzic, I.M., Kirkland, P.A., and Reuter, C.J. (2004). Proteasomes: perspectives from the Archaea. Front. Biosci. *9*, 1743–1758.

Maupin-Furlow, J.A., Gil, M.A., Humbard, M.A., Kirkland, P.A., Li, W., Reuter, C.J., and Wright, A.J. (2005). Archaeal proteasomes and other regulatory proteases. Curr. Opin. Microbiol. *8*, 720–728.

Medalia, N., Beer, A., Zwickl, P., Mihalache, O., Beck, M., Medalia, O., and Navon, A. (2009). Architecture and molecular mechanism of PAN, the archaeal proteasome regulatory ATPase. J. Biol. Chem. *284*, 22952–22960.

Meyer, H., Bug, M., and Bremer, S. (2012). Emerging functions of the VCP/p97 AAA-ATPase in the ubiquitin system. Nat. Cell Biol. *14*, 117–123.

Miranda, H.V., Nembhard, N., Su, D., Hepowit, N., Krause, D.J., Pritz, J.R., Phillips, C., Söll, D., and Maupin-Furlow J.A. (2011). E1- and ubiquitin-like proteins provide a direct link between protein conjugation and sulfur transfer in archaea. Proc. Natl. Acad. Sci. U.S.A. *108*, 4417–4422.

Ng, S.Y., Chaban, B., and Jarrell, K.F. (2006). Archaeal flagella, bacterial flagella and type IV pili: a comparison of genes and posttranslational modifications. J. Mol. Microbiol. Biotechnol. *11*, 167–191.

Ng, S.Y., Chaban, B., VanDyke, D.J., and Jarrell, K.F. (2007). Archaeal signal peptidases. Microbiology *153*, 305–314.

Ng, S.Y., VanDyke, D.J., Chaban, B., Wu, J., Nosaka, Y., Aizawa, S., and Jarrell, K.F. (2009). Different minimal signal peptide lengths recognized by the archaeal prepilin-like peptidases FlaK and PibD. J. Bacteriol. *191*, 6732–6740.

Nunoura, T., Takaki, Y., Kakuta, J., Nishi, S., Sugahara, J., Kazama, H., Chee, G.-J., Hattori, M., Kanai, A., Atomi, H., *et al.* (2011). Insights into the evolution of Archaea and eukaryotic protein modifier systems revealed by the genome of a novel archaeal group. Nucleic Acids Res. *39*, 3204–3223.

O'Brien, K.M., Schufreider, A.K., McGill, M.A., O'Brien, K.M., Reitter, J.N., and Mills, K.V. (2010). Mechanism of protein splicing of the *Pyrococcus abyssi lon* protease intein. Biochem. Biophys. Res. Commun. *403*, 457–461.

Ogura, T., and Wilkinson, A.J. (2001). AAA+ superfamily ATPases: common structure – diverse function. Genes Cells *6*, 575–597.

Omnus, D.J., Pfirrmann, T., Andreasson, C., and Ljungdahl, P.O. (2011). A phosphodegron controls nutrient-induced proteasomal activation of the signaling protease Ssy5. Mol. Biol. Cell *22*, 2754–2765.

Park, S.-C., Jia, B., Yang, J.-K., Van, D.L., Shao, Y.G., Han, S.W., Jeon, Y.-J., Chung, C.H., and Cheong, G.-W. (2006). Oligomeric structure of the ATP-dependent protease La (Lon) of *Escherichia coli*. Mol. Cells *21*, 129–134.

Pietrokovski, S. (2001). Intein spread and extinction in evolution. Trends Genet. *17*, 465–472.

Pohlschröder, M., Ghosh, A., Tripepi, M., and Albers, S.V. (2011). Archaeal type IV pilus-like structures – evolutionarily conserved prokaryotic surface organelles. Curr. Opin. Microbiol. *14*, 357–363.

Pradhan, S., Chin, H.G., Esteve, P.O., and Jacobsen, S.E. (2009). SET7/9 mediated methylation of non-histone proteins in mammalian cells. Epigenetics *4*, 383–387.

Ranjan, N., Damberger, F.F., Sutter, M., Allain, F.H., and Weber-Ban, E. (2010). Solution structure and activation mechanism of ubiquitin-like small archaeal modifier proteins. J. Mol. Biol. *405*, 1040–1055.

Ravid, T., and Hochstrasser, M. (2008). Diversity of degradation signals in the ubiquitin-proteasome system. Nat. Rev. Mol. Cell Biol. *9*, 679–690.

Rawson, R.B., Zelenski, N.G., Nijhawan, D., Ye, J., Sakai, J., Hasan, M.T. Chang, T.Y., Brown, M.S., and Goldstein, J.L. (1997). Complementation cloning of S2P, a gene encoding a putative metalloprotease required for intramembrane cleavage of SREBPs. Mol. Cell *1*, 47–57.

Reuter, C.J., and Maupin-Furlow, J.A. (2004). Analysis of proteasome-dependent proteolysis in *Haloferax volcanii* cells, using short-lived green fluorescent proteins. Appl. Environ. Microbiol. *70*, 7530–7538.

Reuter, C.J., Kaczowka, S.J., and Maupin-Furlow, J.A. (2004). Differential regulation of the PanA and PanB proteasome-activating nucleotidase and 20S proteasomal proteins of the haloarchaeon *Haloferax volcanii*. J. Bacteriol. *186*, 7763–7772.

Reuter, C.J., Uthandi, S., Puentes, J.A., and Maupin-Furlow, J.A. (2010). Hydrophobic carboxy-terminal residues dramatically reduce protein levels in the haloarchaeon *Haloferax volcanii*. Microbiology UK *156*, 248–255.

Rigas, S., Daras, G., Tsitsekian, D., and Hatzopoulos, P. (2012). The multifaceted role of Lon proteolysis in seedling establishment and maintenance of plant organelle function: living from protein destruction. Physiol. Plant. *145*, 215–223.

Roman-Hernandez, G., Hou, J.Y., Grant, R.A., Sauer, R.T., and Baker, T.A. (2011). The ClpS adaptor mediates staged delivery of N-end rule substrates to the AAA+ ClpAP protease. Mol. Cell *43*, 217–228.

Rotanova, T.V., Melnikov, E.E., Khalatova, A.G., Makhovskaya, O.V., Botos, I., Wlodawer, A., and Gostchina, A. (2004). Classification of ATP-dependent proteases Lon and comparison of the active sites of their proteolytic domains. Eur. J. Biochem. *271*, 4865–4871.

Rotanova, T.V., Botos, I., Melnikov, E.E., Rasulova, F., Gustchina, A., Maurizi, M.R., and Wlodawer, A. (2006). Slicing a protease: structural features of the ATP-dependent Lon proteases gleaned from investigations of isolated domains. Protein Sci. *15*, 1815–1828.

Rudner, D.Z., Fawcett, P., and Losick, R. (1999). A family of membrane-embedded metalloproteases involved in regulated proteolysis of membrane-associated transcription factors. Proc. Natl. Acad. Sci. U.S.A. *96*, 14765–14770.

Rudolph, J., Tolliday, N., Schmitt, C., Schuster, S.C., and Oesterhelt, D. (1995). Phosphorylation in Halobacterial signal transduction. EMBO J. *14*, 4249–4257.

Sakoh, M., Ito, K., and Akiyama, Y. (2005). Proteolytic activity of HtpX, a membrane-bound and stress-controlled protease from *Escherichia coli*. J. Biol. Chem. *280*, 33305–33310.

Sakr, S., Cirinesi, A.M., Ullers, R.S., Schwager, F., Georgopoulos, C., and Genevaux, P. (2010). Lon protease quality control of presecretory proteins in *Escherichia coli* and its dependence on the SecB and DnaJ (Hsp40) chaperones. J. Biol. Chem. *285*, 23506–23514.

Sastre, D.E., Paggi, R.A., and De Castro, R.E. (2011). The Lon protease from the haloalkaliphilic archaeon *Natrialba magadii* is transcriptionally linked to a cluster of putative membrane proteases and displays DNA-binding activity. Microbiol. Res. *166*, 304–313.

Serrano, M., Hovel, S., Moran, C.P. Jr., Henriques, A.O., and Volker, U. (2001). Forespore-specific transcription of the *lonB* gene during sporulation in *Bacillus subtilis*. J. Bacteriol. *183*, 2995–3003.

Shigi, N. (2012). Post-translational modification of cellular proteins by a ubiquitin-like protein in bacteria. J. Biol. Chem. *287*, 17558–17577.

Smith, D.M., Kafri, G., Cheng, Y., Ng, D., Walz, T., and Goldberg, A.L. (2005). ATP binding to PAN or the 26S ATPases causes association with the 20S proteasome, gate opening, and translocation of unfolded proteins. Mol. Cell *20*, 687–698.

Smith, D.M., Fraga, H., Reis, C., Kafri, G., and Goldberg, A.L. (2011). ATP binds to proteasomal ATPases in pairs with distinct functional effects, implying an ordered reaction cycle. Cell *144*, 526–538.

Sonezaki, S., Ishii, Y., Okita, K., Sugino, T., Kondo, A., and Kato, Y. (1995). Overproduction and purification of SulA fusion protein in *Escherichia coli* and its degradation by Lon protease *in vitro*. Appl. Microbiol. Biotechnol. *43*, 304–309.

Soppa, J. (2010). Protein acetylation in archaea, bacteria, and eukaryotes. Archaea *2010*, 820681.

Stadtmueller, B.M., and Hill, C.P. (2011). Proteasome activators. Mol. Cell *41*, 8–19.

Stahlberg, H., Kutejova, E., Suda, K., Wolpensinger, B., Lustig, A., Schatz, G., Engel, A., and Suzuki, C.K. (1999). Mitochondrial Lon of *Saccharomyces cerevisiae* is a ring-shaped protease with seven flexible subunits. Proc. Natl. Acad. Sci. U.S.A. *96*, 6787–6790.

Streif, S., Oesterhelt, D., and Marwan, W. (2010). A predictive computational model of the kinetic mechanism of stimulus-induced transducer methylation and feedback regulation through CheY in archaeal phototaxis and chemotaxis. BMC Syst. Biol. *4*, 27.

Suzuki, C.K., Rep, M., van Dijl, J.M., Suda, K., Grivell, L.A., and Schatz, G. (1997). ATP-dependent proteases that also chaperone protein biogenesis. Trends Biochem. Sci. *22*, 118–123.

Swamy, K.H., and Goldberg, A.L. (1982). Subcellular distribution of various proteases in *Escherichia coli*. J. Bacteriol. *149*, 1027–1033.

Szabo, Z., Albers, S.V., and Driessen, A.J. (2006). Active-site residues in the type IV prepilin peptidase homologue PibD from the archaeon *Sulfolobus solfataricus*. J. Bacteriol. *188*, 1437–1443.

Tarasov, V.Y., Besir, H., Schwaiger, R., Klee, K., Furtwangler, K., Pfeiffer, F., and Oesterhelt, D. (2008). A small protein from the bop-brp intergenic region of *Halobacterium salinarum* contains a zinc finger motif and regulates *bop* and *crtB1* transcription. Mol. Microbiol. *67*, 772–780.
Tasaki, T., Sriram, S.M., Park, K.S., and Kwon, Y.T. (2012). The N-end rule pathway. Annu. Rev. Biochem. *81*, 261–289.
Thompson, M.W., Singh, S.K., and Maurizi, M.R. (1994). Processive degradation of proteins by the ATP-dependent Clp protease from *Escherichia coli*. Requirement for the multiple array of active sites in ClpP but not ATP hydrolysis. J. Biol. Chem. *269*, 18209–18215.
Torres-Arancivia, C., Ross, C.M., Chavez, J., Assur, Z., Dolios, G., Mancia, F., and Ubarretxena-Belandia, I. (2010). Identification of an archaeal presenilin-like intramembrane protease. PLoS One *5*, e13072.
Tran, H.J., Allen, M.D., Löwe, J., and Bycroft, M. (2003). Structure of the Jab1/MPN domain and its implications for proteasome function. Biochemistry *42*, 11460–11465.
Tripepi, M., Imam, S., and Pohlschröder, M. (2010). *Haloferax volcanii* flagella are required for motility but are not involved in PibD-dependent surface adhesion. J. Bacteriol. *192*, 3093–3102.
Tucker, P.A., and Sallai, L. (2007). The AAA+ superfamily – a myriad of motions. Curr. Opin. Struct. Biol. *17*, 641–652.
Tyagi, N., Anamika, K., and Srinivasan, N. (2010). A framework for classification of prokaryotic protein kinases. PLoS One *5*, e10608.
Urban, S., and Shi, Y. (2008). Core principles of intramembrane proteolysis: comparison of rhomboid and site-2 family proteases. Curr. Opin. Struct. Biol. *18*, 432–441.
Varshavsky, A. (2011). The N-end rule pathway and regulation by proteolysis. Protein Sci. *20*, 1298–1345.
Venkatesh, S., Lee, J., Singh, K., Lee, I., and Suzuki, C.K. (2012). Multitasking in the mitochondrion by the ATP-dependent Lon protease. Biochim. Biophys. Acta *1823*, 56–66.
Voges, D., Zwickl, P., and Baumeister, W. (1999). The 26S proteasome: a molecular machine designed for controlled proteolysis. Annu. Rev Biochem *68*, 1015–1068.
Wang, F., Liu, M., Qiu, R., and Ji, C. (2011). The dual role of ubiquitin-like protein Urm1 as a protein modifier and sulfur carrier. Protein Cell *2*, 612–619.
Wang, J., Song, J.J., Seong, I.S., Franklin, M.C., Kamtekar, S., Eom, S.H., and Chung, C.H. (2001). Nucleotide-dependent conformational changes in a protease-associated ATPase HslU. Structure *9*, 1107–1116.
Wang, R., Chait, B.T., Wolf, I., Kohanski, R.A., and Cardozo, C. (1999). Lysozyme degradation by the bovine multicatalytic proteinase complex (proteasome): evidence for a nonprocessive mode of degradation. Biochemistry *38*, 14573–14581.
Wilson, H.L., Aldrich, H.C., and Maupin-Furlow, J. (1999). Halophilic 20S proteasomes of the archaeon *Haloferax volcanii*: purification, characterization, and gene sequence analysis. J. Bacteriol. *181*, 5814–5824.
Wilson, H.L., Ou, M.S., Aldrich, H.C., and Maupin-Furlow, J. (2000). Biochemical and physical properties of the *Methanococcus jannaschii* 20S proteasome and PAN, a homolog of the ATPase (Rpt) subunits of the eucaryal 26S proteasome. J. Bacteriol. *182*, 1680–1692.
Wolfe, M.S. (2009). Intramembrane-cleaving proteases. J. Biol. Chem. *284*, 13969–13973.
Wolfe, M.S. (2010). Structure, mechanism and inhibition of γ-secretase and presenilin-like proteases. Biol. Chem. *391*, 839–847.
Ye, J., Rawson, R.B., Komuro, R., Chen, X., Dave, U.P., Prywes, R., Brown, M.S., and Goldstein, J.L. (2000). ER stress induces cleavage of membrane-bound ATF6 by the same proteases that process SREBPs. Mol. Cell *6*, 1355–1364.
Yu, Y.T., and Kroos, L. (2000). Evidence that SpoIVFB is a novel type of membrane metalloprotease governing intercompartmental communication during *Bacillus subtilis* sporulation. J. Bacteriol. *182*, 3305–3309.
Yu, Y., Smith, D.M., Kim, H.M., Rodriguez, V., Goldberg, A.L., and Cheng, Y. (2010). Interactions of PAN's C-termini with archaeal 20S proteasome and implications for the eukaryotic proteasome–ATPase interactions. EMBO J. *29*, 692–702.
Zhang, F., Hu, M., Tian, G., Zhang, P., Finley, D., Jeffrey, P.D., and Shi, Y. (2009a). Structural insights into the regulatory particle of the proteasome from *Methanocaldococcus jannaschii*. Mol. Cell *34*, 473–484.
Zhang, F., Wu, Z., Zhang, P., Tian, G., Finley, D., and Shi, Y. (2009b). Mechanism of substrate unfolding and translocation by the regulatory particle of the proteasome from *Methanocaldococcus jannaschii*. Mol. Cell *34*, 485–496.

Zhou, G., Kowalczyk, D., Humbard, M.A., Rohatgi, S., and Maupin-Furlow, J.A. (2008). Proteasomal components required for cell growth and stress responses in the haloarchaeon *Haloferax volcanii*. J. Bacteriol. *190*, 8096–8105.

Zolghadr, B., Klingl, A., Rachel, R., Driessen, A.J., and Albers, S.V. (2011). The bindosome is a structural component of the *Sulfolobus solfataricus* cell envelope. Extremophiles *15*, 235–244.

Zou, C., Butler, P.L., Coon, T.A., Smith, R.M., Hammen, G., Zhao, Y., Chen, B.B., and Mallampalli, R.J. (2011). LPS impairs phospholipid synthesis by triggering β-transducin repeat-containing protein (β-TrCP)-mediated polyubiquitination and degradation of the surfactant enzyme acyl-CoA:lysophosphatidylcholine acyltransferase I (LPCAT1). J. Biol. Chem. *286*, 2719–2727.

Zwickl, P., Ng, D., Woo, K.M., Klenk, H.P., and Goldberg, A.L. (1999). An archaebacterial ATPase, homologous to ATPases in the eukaryotic 26 S proteasome, activates protein breakdown by 20 S proteasomes. J. Biol. Chem. *274*, 26008–26014.

Index

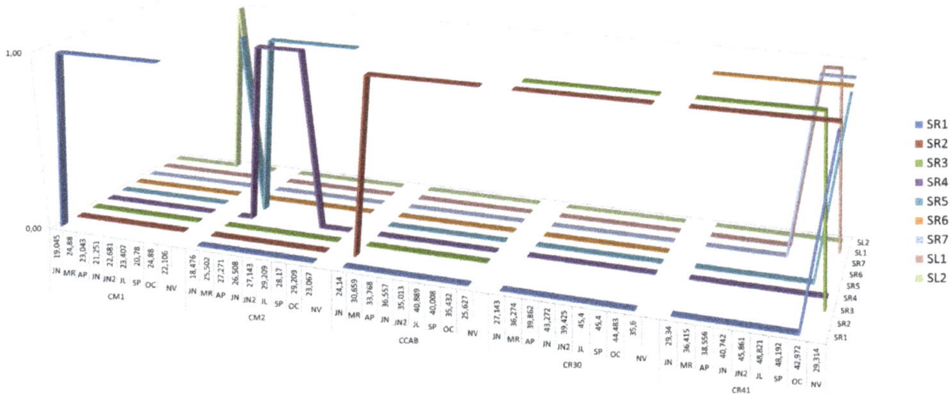

Figure 2.1 Presence (1 in the vertical axis) or absence (0) of several Bacteroidetes phylotypes in five different ponds (CM1, CM2, CCAB, CR30, and CR41) of Santa Pola solar salterns at nine time points during year 2009. For every pond and sampling date, the corresponding salinity values (numbers above months labelling) expressed in% are indicated. Phylotypes related to *Salinibacter ruber* and to *Salisaeta longa* are labelled, respectively, as SR and SL.

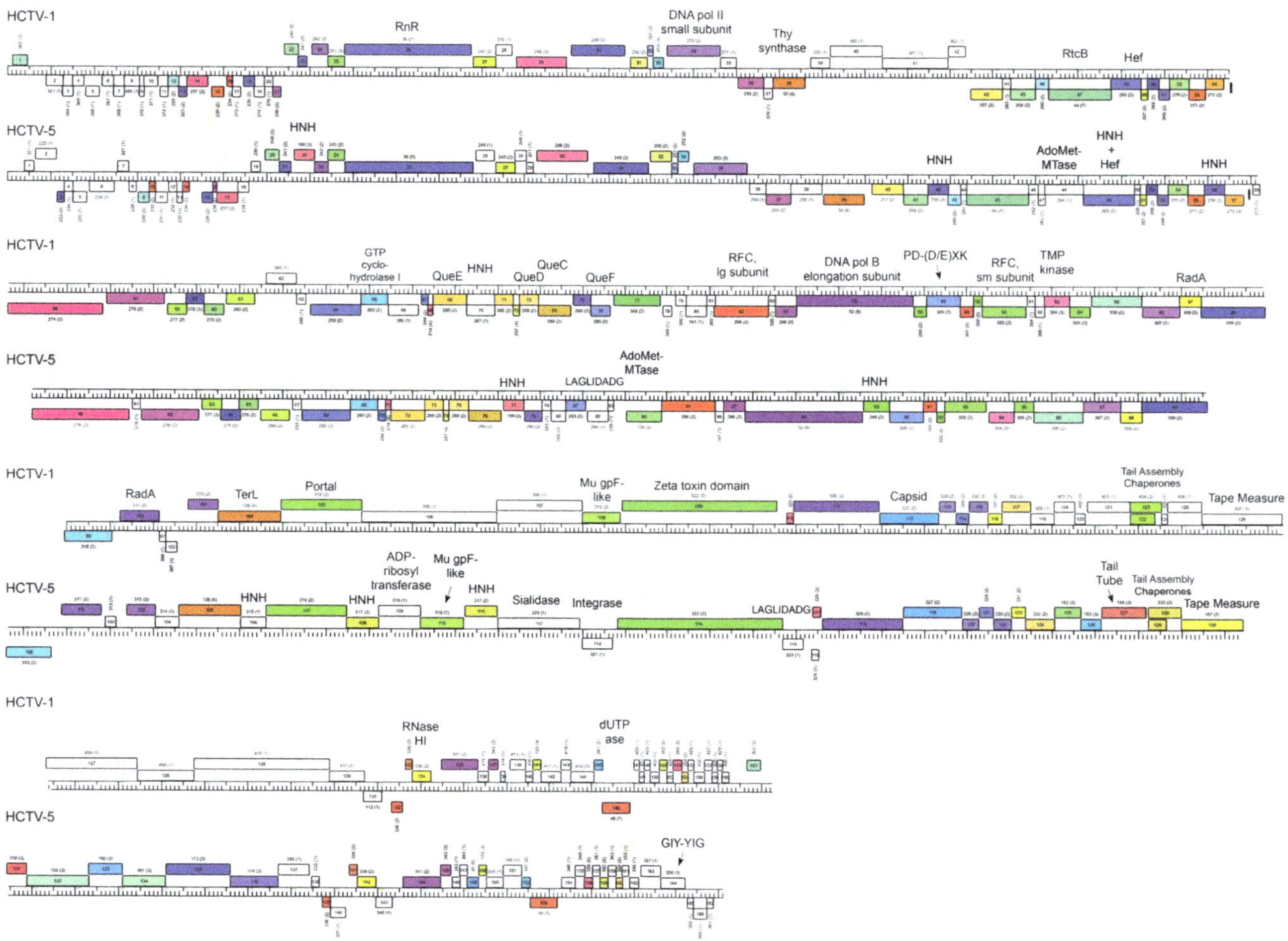
HCTV-1
RnR
DNA pol II small subunit
Thy synthase
RtcB
Hef
HCTV-5
HNH
HNH
AdoMet-MTase
HNH + Hef
HNH
HCTV-1
GTP cyclo-hydrolase I
QueE
HNH
QueC QueD
QueF
RFC, lg subunit
DNA pol B elongation subunit
PD-(D/E)XK
RFC, sm subunit
TMP kinase
RadA
HCTV-5
HNH
LAGLIDADG
AdoMet-MTase
HNH
HCTV-1
RadA
TerL
Portal
Mu gpF-like
Zeta toxin domain
Capsid
Tail Assembly Chaperones
Tape Measure
HCTV-5
HNH
HNH
ADP-ribosyl transferase
Mu gpF-like
HNH
Sialidase
Integrase
LAGLIDADG
Tail Tube
Tail Assembly Chaperones
Tape Measure
HCTV-1
RNase HI
dUTP ase
HCTV-5
GIY-YIG

Figure 4.2 (opposite page) Genomic organization of the haloarchaeal tailed viruses HCTV-1 and HCTV-5. Nucleotide sequence similarity is depicted by shading between the genomes according to colour spectrum determined in Phamerator (Cresawn *et al.*, 2011) with purple representing the highest identity. Protein families (pfams) as determined in Phamerator are numbered above each ORF. RnR, ribonucleotide reductase; Thy, thymidylate; AdoMet-Mtase, S-adenosylmethionine-dependent methyltransferase; RFC, replication factor; TMP, thymidine monophosphate; TerL, terminase large subunit. Figure reproduced from Senčilo *et al.* (2013).

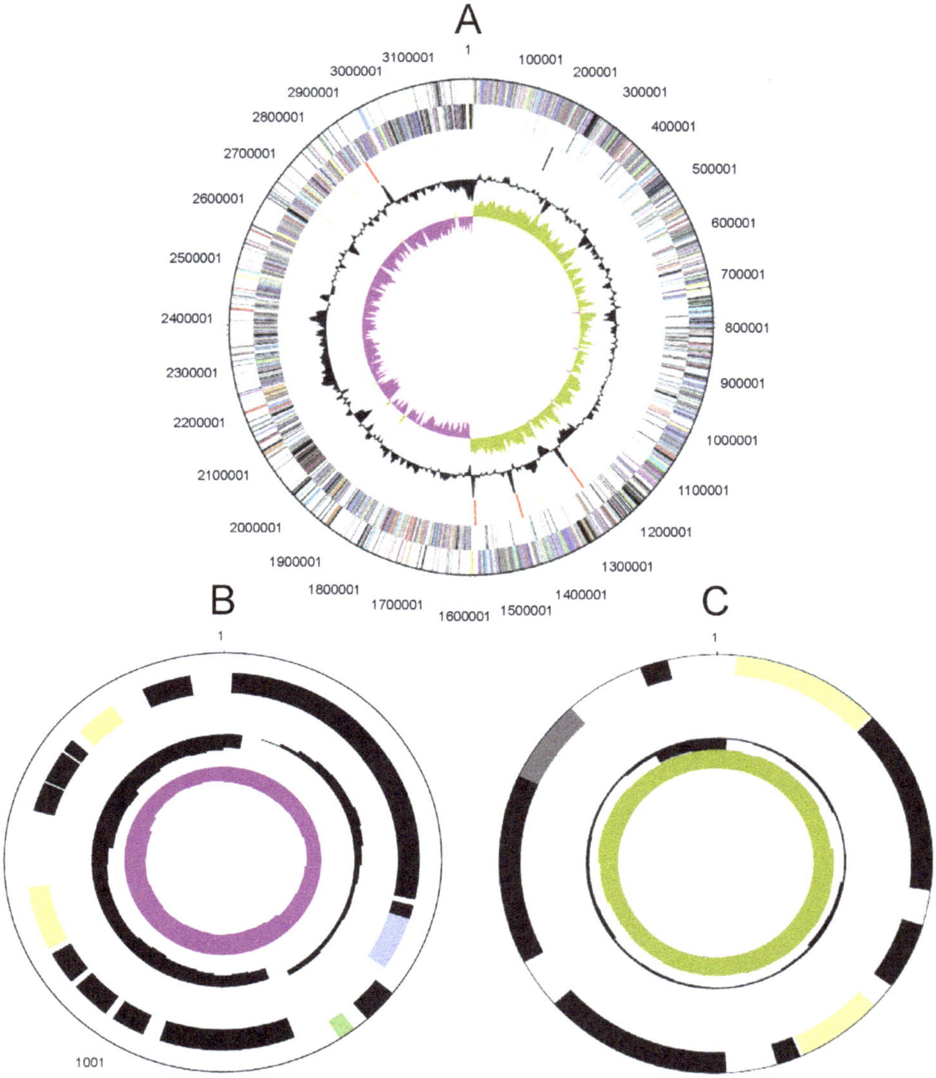

Figure 5.1 Circular representation of the *N. thermophilus* chromosome (A), plasmid pNTHE01 (B) and plasmid pNTHE02 (C). The distribution of coding sequences is depicted by the coloured boxes according to functional categories. The first two circles from the outside represent the predicted coding sequences on the plus and minus strands, respectively. The two innermost circles represent the per cent G+C and G+C skew, respectively. Colours represent the following: dark grey, hypothetical proteins; light grey, conserved hypothetical and unknown function; brown, general function prediction; red, replication and repair; green, energy metabolism; blue, carbon and carbohydrate metabolism; cyan, lipid metabolism; magenta, transcription; yellow, translation; orange, amino acid metabolism; pink, metabolism of cofactors and vitamins; light red, purine and pyrimidine metabolism; lavender, signal transduction; sky blue, cellular processes. The third circle represents red for rRNA and green for tRNA genes.

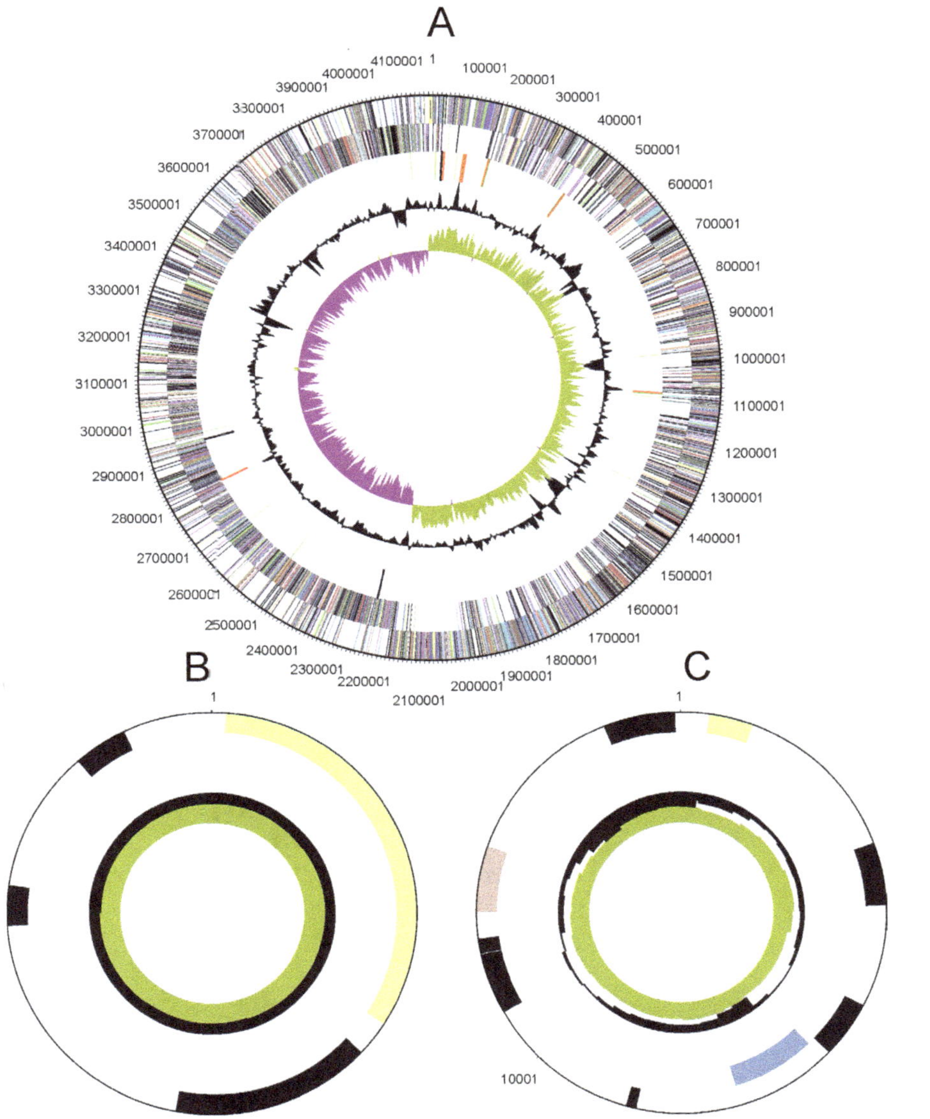

Figure 5.3 Circular representation of the *H. halophilus* chromosome (A), plasmid PL3 (B) and plasmid PL16 (C). The distribution of coding sequences is depicted by the coloured boxes according to functional categories. The first two circles from the outside represent the predicted coding sequences on the plus and minus strands, respectively. The two innermost circles represent the per cent G+C and G+C skew, respectively. Colours represent the following: dark grey, hypothetical proteins; light grey, conserved hypothetical and unknown function; brown, general function prediction; red, replication and repair; green, energy metabolism; blue, carbon and carbohydrate metabolism; cyan, lipid metabolism; magenta, transcription; yellow, translation; orange, amino acid metabolism; pink, metabolism of cofactors and vitamins; light red, purine and pyrimidine metabolism; lavender, signal transduction; sky blue, cellular processes. The third circle represents red for rRNA and green for tRNA genes.

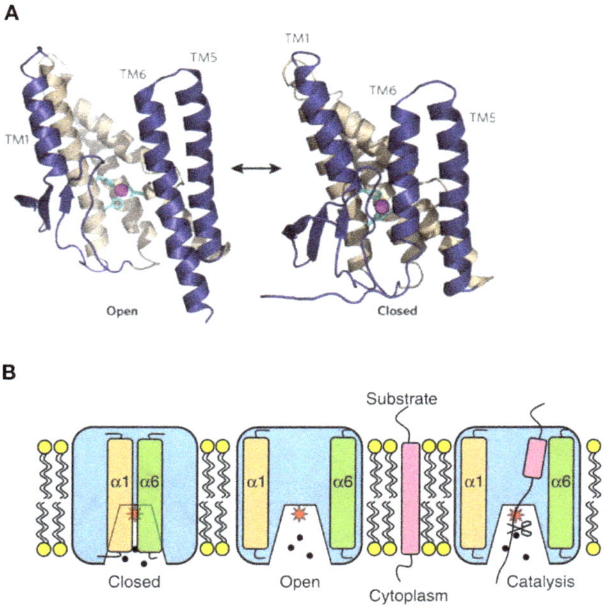

Figure 8.2 Intramembrane metalloprotease (mjS2P). (A) Crystal structures suggest mjS2P adopts an open and closed conformation that is used to gate substrate access to the active site. Transmembrane helices mobile between these two conformations (TM1 and TM5-TM6 in dark blue), catalytic zinc ion (purple ball), and side chains of residues coordinating zinc (cyan sticks) are indicated. (B) Cartoon of S2P mechanism based on crystal structure with water molecules (black dots) thought to access the catalytic active site (red asterisk) through a channel from the cytosol. [Figure modified from (Feng *et al.*, 2007; Erez *et al.*, 2009) with permission].

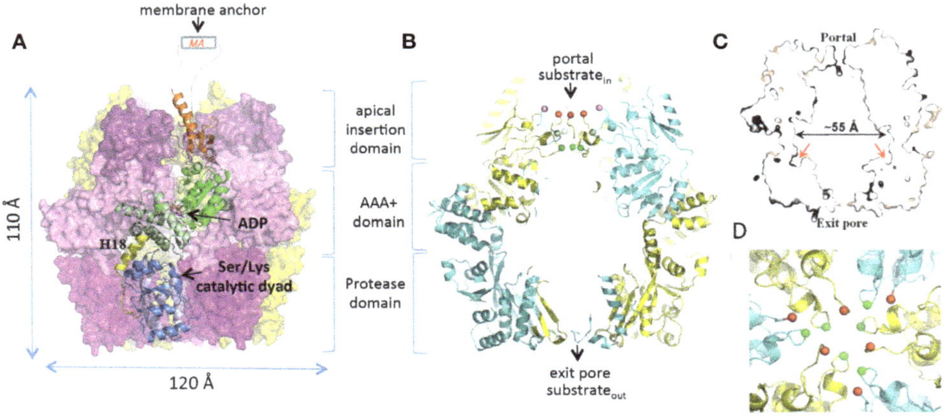

Figure 8.3 Archaeal B-Lon protease. (A) Crystal structure of *Thermococcus onnurineus* B-Lon (*Ton*LonB) hexamer with five subunits in surface representation and one monomer as a ribbon diagram. Insert 1 (which is attached to the deleted membrane anchor) is depicted as an orange ribbon. ADP binding site is indicated as stick figure and lavender balls indicate positions where Ser-Lys catalytic dyad for protein degradation likely residues. (B) Ribbon diagram of *Ton*LonB with a vertical section through the centre. Substrate entry portal with residues likely gating the opening indicated (Cα atoms of F216 and M275 in red and green balls, respectively). Substrate exit pore is also indicated. (C) Surface rendering of *Ton*LonB vertical slab with red arrows indicating substrate binding grooves. (D) Top view of TonLonB residues gating the entry portal indicated as above. [Figure modified from (Cha *et al*., 2010) with permission].

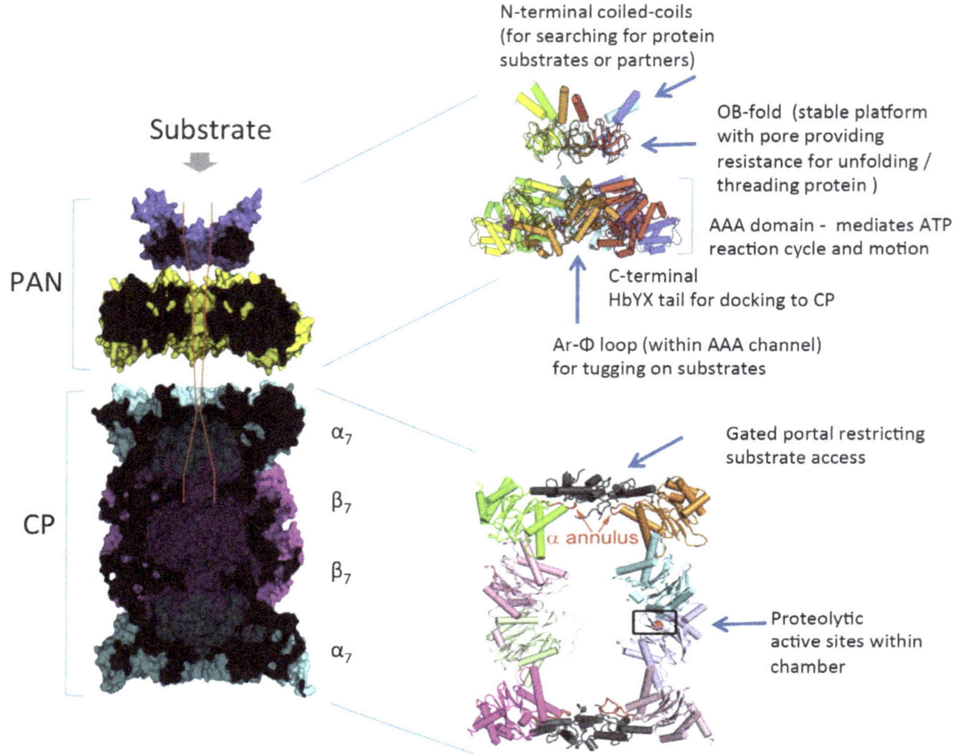

Figure 8.4 Archaeal proteasomes. Proteasomes are composed of a catalytic core particle (CP) that catalyses the hydrolysis of peptide bonds. CPs are barrel-like structures composed of four stacked heptameric rings of α- and β-type subunits (indicated by α_7 and β_7, respectively). CPs have a central channel with narrow openings on each end that are gated by α subunit N-termini. The proteolytic active sites formed by the N-terminal Thr residues of β-type subunits (indicated in red) line the central cavity of the CP. CPs associate with a variety of regulators including the archaeal PAN. PAN is a hexameric ATPase that docks (via its C-terminal HbYX motif) to the ends of CPs. PAN has coiled-coil, OB-fold, Ar-Φ loop and AAA motor components that coordinate to unfold proteins and facilitate their passage into the central chamber of the CP for destruction. [Figure modified from (Zhang *et al.*, 2009a; Stadtmueller and Hill, 2011) with permission].

www.ingramcontent.com/pod-product-compliance
Lightning Source LLC
LaVergne TN
LVHW011852200926
848484LV00002B/440